Dynamic
Stratigraphy

Dynamic Stratigraphy

An Introduction
to Sedimentation and Stratigraphy

SECOND EDITION

Robley K. Matthews

Department of Geological Sciences
Brown University

Prentice-Hall, Inc., Englewood Cliffs, New Jersey 07632

Library of Congress Cataloging in Publication Data

Matthews, Robley K. (date)
   Dynamic stratigraphy.

   Includes bibliographies and index.
   1. Geology, Stratigraphic. 2. Sedimentation and
deposition. I. Title.
QE651.M354 1984   551.7   83-9672
ISBN   0-13-222109-8

Editorial/production supervision
   and interior design: Paula Martinac
Cover design: George Cornell
Manufacturing buyer: John B. Hall

Cover photographs: Modern reef tract of Belize (*left*) and a Pleistocene coral reef terrace on Barbados (*right*). The well-documented tectonic framework of Belize allows formulation of a quantitative model for coral reef sedimentation. Application of this model to the coral reef terraces of Barbados is fundamental to the development of the Quaternary sea-level dynamics model. This model, in turn, appears applicable throughout much of the Phanerozoic.

Printed in the United States of America

10 9 8 7 6 5 4 3 2 1

ISBN 0-13-222109-8

PRENTICE-HALL INTERNATIONAL, INC., *London*
PRENTICE-HALL OF AUSTRALIA PTY. LIMITED, *Sydney*
EDITORA PRENTICE-HALL DO BRASIL, LTDA., *Rio de Janeiro*
PRENTICE-HALL CANADA INC., *Toronto*
PRENTICE-HALL OF INDIA PRIVATE LIMITED, *New Delhi*
PRENTICE-HALL OF JAPAN, INC., *Tokyo*
PRENTICE-HALL OF SOUTHEAST ASIA PTE. LTD., *Singapore*
WHITEHALL BOOKS LIMITED, *Wellington, New Zealand*

To all those scientists
who stop to ask,
"What's the data?"
and
to Nancy, Betty Jo, Gretchen, Sandra, and Charles

Contents

V Dynamics in the Context of Time *373*

Preface
to the Second Edition

In many ways, the thing I like best about the first edition of this book is the preface. I think I caught the science of stratigraphy at an interesting time. My goals in approaching the stratigraphic record and in teaching students remain the same. The second edition concentrates on the big picture and continues to develop the promise that *"much can be known* if stratigraphers approach the record properly."

Dynamic Stratigraphy takes a "geophysical" approach to the subject of stratigraphy. Whereas the geologist tends to work from small detail toward larger and larger synthesis, the geophysicist tends to work from grand simple models that become more complicated only as required by new data. Hopefully, this book provides those simple models for stratigraphy upon which geologists can build as their profession becomes more complicated. This is far preferable to the "unlearning process" through which many of the older generation have had to pass as new knowledge rendered common wisdom untenable.

The major additions to the second edition are new sections concerning "geologic context of sediment accumulation" and "dynamics in the context of time." Plate tectonics was a fairly recent breakthrough at the time of the first edition; hopefully, the growing maturity of this exciting field is captured in the new Part III. Likewise, whereas the first edition concluded with discussion of "cyclicity" emphasizing largely the Pleistocene and the late Paleozoic, the second edition broadens this concept to "dynamics" in general and provides an overview of the entire Phanerozoic, taking into account plate reconstructions, seismic stratigraphy, and the isotopic and geologic evidence for the history of continental ice volume.

Acknowledgments. I continue to express my great appreciation to a number of colleagues at Brown University. To the former list, I would add D. W. Forsyth, J. W. Head, J. F. Hermance, E. M. Parmentier, and especially W. L. Prell.

An army of students has contributed to the content of the second edition through library research and data compilation. I thank George Hogeman, Phil King, Betty Jo Matthews, Gretchen Matthews, Sandra Matthews, Peter Staugaard, Paul Wagner, and especially Ken Ransom and Don Swann. Manuscript typing and editorial assistance were cheerfully provided by Christine Sherlock.

Preliminary drafts of the manuscript were reviewed by the following professors, to whom I express sincere thanks: K. R. Aalto, Humboldt State University; Pamela Hallock Muller, University of Texas of the Permian Basin; Harvey M. Sachs, Princeton University; Randolph P. Steinen, University of Connecticut; and Walter H. Wheeler, University of North Carolina at Chapel Hill. Nevertheless, I accept full responsibility for any errors or misconceptions the book may contain.

R. K. Matthews

Preface
to the First Edition

There are commonly three stages to the development of a field of science. First, the identification and cataloging of the fundamental building blocks that are to be the subject of this field. Second, the identification and cataloging of the major configurations that exist among the fundamental materials. Finally, the consideration of the kinetics by which these configurations were achieved. Upon reaching the kinetics level, all aspects of the science can be considered simultaneously for the first time.

The field of global tectonics has recently passed through one of these development cycles. Continental crust, oceanic crust, mid-ocean ridges, deep-sea trenches, and velocity structure of the upper mantle were the building blocks. Global seismicity and magnetic patterns along mid-ocean ridges revealed the major configuration of lithosphere plates. Dating by radiometric methods, magnetic stratigraphy, and biostratigraphy revealed kinetics of plate motions, both present and past. Thus, the years since the mid-sixties have given us a new, dynamic framework within which to consider the stratigraphic record.

Stratigraphy itself has been moving systematically toward its own stage three in a somewhat slower fashion. We have cataloged the sediment types, the formations, and the biostratigraphic time framework. We have seen that map configuration of Recent sedimentary facies can often be related to facies configurations in Ancient lithologic units. We are beginning to think of lithologic sequences in terms of sedimentation rates, rates of eustatic sea-level fluctuation, and rates of tectonic deformation.

If we are to fully appreciate the stratigraphic record, all aspects of the science must be considered within the dynamic context. This book attempts to pull together the essentials of (a) dynamics of Recent sedimentation, (b) dynamics of tectonism on the present earth's surface, and (c) dynamics of Quaternary eustatic sea-level fluctuations. With these basic input parameters, we begin to devise models which generate various stratigraphic relationships from the interplay of the various rates involved. Examples from the stratigraphic record tend to confirm the potential utility of such models.

This book is intended as an introduction suitable for all earth science majors, not just for those anticipating a career in stratigraphic geology. While it is important that future stratigraphers get a start in the right directions, it is equally important that future geophysicists, meteorologists, etc. acquire confidence that stratigraphers really can make accurate statements about the history of the earth. The book is designed to instill such a confidence; *not* that the student knows the stratigraphic record, but rather that *much can be known* if stratigraphers approach the record properly.

This book was developed for a one-semester course intended for sophomores and juniors. The format of Sections III and IV [Parts IV and V in the second edition] is easily expandable by addition of more complicated Ancient examples. In this manner, the book could serve as a focal point for a two-semester undergraduate course or Sections III and IV could serve as the starting point for a graduate-level course.

Finally, some important topics in stratigraphy are omitted. Lacustrine deposits, deep-water evaporites, flysch-molasse, glacial sediments, and desert sediments simply do not fit the format of Section III. It is suggested that these topics be deferred to a more advanced stratigraphy course.

Acknowledgments.   To a large extent, this book is the product of Brown University; it would not have come out the same if I had written it anywhere else. I am particularly grateful for the interaction with W. M. Chapple, B. J. Giletti, J. Imbrie, L. F. Laporte, and T. A. Mutch.

Preliminary drafts of the manuscript, or portions of it, were reviewed by L. V. Benson, W. M. Chapple, R. S. Harrison, S. I. Husseini, G. deV. Klein, K. J. Mesolella, R. C. Murray, T. A. Mutch, N. D. Smith, and R. P. Steinen, to whom I express sincere thanks. Nevertheless, I accept full responsibility for any errors or misconceptions the book may contain.

The manuscript was typed primarily by Margaret T. Cummings. J. A. Creaser and Paul R. Jones contributed significantly to the preparation and organization of illustrative materials.

Finally, I acknowledge with thanks the sufferings of Nancy, Betty, Gretchen, Sandra, and Charles throughout the writing process.

R. K. Matthews

Introduction

—————————————————————— I

Chapter 1 examines the fundamental motivations for human interest in sedimentation and stratigraphy. Chapter 2 attempts to demonstrate that the stratigraphic record can be approached on a rational basis. We shall note, however, that the situation becomes rather complicated. These complications must be brought under our intellectual control if we are to explore the interesting questions posed in Chapter 1. Parts II and III are intended to provide some necessary background. Parts IV and V go on to investigate Recent sedimentary environments and to apply this knowledge, along with sound stratigraphic principles, to specific sequences of Ancient rocks.

The Importance
of Sedimentary Rocks

1

Human interest in sedimentary rocks can be conveniently divided into two parts: a curiosity about the history of the earth and a desire to utilize the natural resources contained within the sediments.

The History of the Earth

The written record of human experience with the earth is extremely limited. In the New World, it is measured in hundreds of years; in the eastern Mediterranean, thousands of years. Yet evidences in the stratigraphic record, which we are about to discuss, lead us to recognize that human history on earth goes back as far as a few million years. Indeed, complex forms of life have existed on the face of the earth for hundreds of millions of years. Thus, if we should for any reason wish to consider the history of the earth, we must look beyond the written record of human experience with the planet, for this written record covers only a trivial amount of time.

The Stratigraphic Record, the Archives of Earth History. Sediments have been deposited in various environments throughout the history of the earth. These sediments, today preserved as sequences of sedimentary rocks, provide a spectacular wealth of information concerning earth history. Consider Figure 1.1, for example. Limestones of Mesozoic age overlie an angular unconformity. Alternating layers of sandstone and shale of Paleozoic age underlie the unconformity. If we know how to recognize sedimentary environments in the stratigraphic record, we have a glimpse here of the various environments that have occupied this spot on the face of the earth at varying times in its history. Furthermore, the rocks contain fossils. Thus, we have also a record of the kinds of life that existed in these various environments

3

Figure 1.1 Photograph of an outcrop, Sahara desert, Libya, North Africa. Paleozoic sandstones and shales are unconformably overlain by Mesozoic carbonate rocks. The area is today a desert. Thus, the rocks provide a record of changing conditions on the face of the earth.

at the various times in earth history. Finally, the existence of the angular unconformity records structural deformation of this portion of the earth's crust at some time in the past.

Figure 1.1 gives us one small glimpse at the history of one small spot on the face of the earth. How does that history fit together with the history of other areas? How do the life-forms preserved in these rocks fit together with life-forms preserved in other rocks? These are the broad questions that first led human beings to intellectual consideration of the stratigraphic record and then to the accumulation of a data base sufficiently large that they could begin to ask important questions concerning the history of earth.

Recognition of Large-Scale Earth Processes. Life-forms occupying the various environments on the face of the earth have evolved with time. The orientation of the earth's magnetic field has from time to time become reversed. Both of these interesting facts are the subject of considerable scholarly activity, and both subjects have a voluminous literature. We shall discuss them in more detail later. For the time being, let us accept them as fact and note how stratigraphy can be used to document a large-scale earth process.

Ocean-floor spreading is one of the most gigantic and all-encompassing earth processes yet documented. For the present, let us consider the process only as an example of how stratigraphic data can aid in our understanding of the dynamics of the earth. Figure 1.2 schematically depicts two types of stratigraphic data that allow documentation of the process of ocean-floor spreading.

The late Cenozoic magnetic history of the earth was first worked out in volcanic stratigraphic sequences on mid-ocean islands. Then it was recognized that magnetic patterns in the sea floor could be explained by continuous generation of new oceanic crust beneath mid-ocean ridges and rises. As new crust cooled, it took on its magnetic properties within the field existing at the time of cooling. Thus, as the magnetic field reversed, each new generation of oceanic crust took on magnetic properties that were distinct from the crust preceding it.

A further test of the concept of ocean-floor spreading is provided by the biostratigraphic estimation of the age of the sediment overlying oceanic crust. As indicated in Figure 1.2, it can be shown that oceanic crust closer to the mid-ocean ridge is overlain by the younger sediment, whereas oceanic crust farther removed from the ridge is overlain by older sediments.

In this manner, sea-floor spreading rates of centimeters to tens of centimeters per year are estimated. Such data, for example, lead us to understand earthquakes

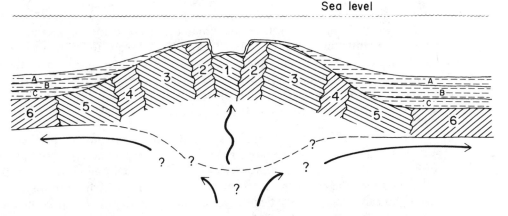

Figure 1.2 Schematic cross section depicting stratigraphic evidence concerning the generation of new oceanic crust along mid-ocean ridges. The crust itself, labeled (1–6), acquired its magnetic properties at the time of formation. Thus, the various segments of crust possess unique magnetic properties indicative of the time when that particular segment of crust was formed. Magnetic surveys, such as indicated in Figure 8.1, may therefore be interpreted in the light of the earth's magnetic history and the sea-floor spreading hypothesis. Biostratigraphic determination of the relative age of deep-ocean sediments overlying the crust provides an independent check on the sea-floor spreading hypothesis. In this schematic diagram, the age of sediments overlying the basalt crust becomes younger toward the center of the ridge.

in the context of lithosphere dynamics on a global scale. Earthquakes are not "freaks of nature" or "acts of God"; they are simply the inevitable and recurring consequences of a very large-scale earth process. Clearly, an understanding of such large-scale earth processes is important and relevant to human beings.

Dynamics of "The Environment." To many people "the environment" seems both important and relevant, but what can they do about it? Once again, we must note that human experience on earth is limited to an extremely short time. Indeed, we face numerous environmental hazards, the likes of which humankind has never experienced. Consider for example the Antarctic Ice Cap.

The volume of ice perched above sea level on the Antarctic continent is sufficient to raise the sea level by some 50 meters if it should melt or slide into the sea. Throughout the world, many large cities are built at or near sea level. A 50-meter sea-level rise would be an environmental problem of the first magnitude! Is such a rise likely to happen? This is essentially a very large-scale engineering question and the data are not conclusive. Has such a rise ever happened in the past? Now this is a stratigraphic question that we might be able to answer! The deep-sea sediments adjacent to the Antarctic continent would contain the record of such an event, had it occurred. Piston cores and rotary drill cores allow us access to the stratigraphic record contained in these and other deep-sea sediments. Work continues on this fascinating question.

Utility to Humankind

Natural processes, acting over large periods of geologic time, have concentrated many of the raw materials utilized by human beings. Consider for example common table salt. Sodium chloride is a major component of seawater. It would be a simple matter to design a processing plant to evaporate seawater and recover the sodium chloride for human use. However, nature has already constructed such "facilities" numerous times in the past. The stratigraphic record contains enough available rock salt to serve our needs for some time to come.

Fossil Fuels. The industrialized Western world runs on fossil fuels: coal, oil, and natural gas. Petroleum exploration is particularly demanding upon the sedimentologist and the stratigrapher. Modern drilling techniques allow us to probe 5 to 8 kilometers into the earth in search of concentrations of hydrocarbons. Earth processes and earth history must be unraveled and shown to have occurred in a sequence that experience tells us is favorable for petroleum accumulation (see Figure 1.3). In the past, petroleum exploration activities have provided major employment for sedimentologists and stratigraphers. These activities will undoubtedly remain important for some years to come.

Other Uses for Holes in Rocks. Exploitation of groundwater resources also requires considerable knowledge of sedimentology and stratigraphy. The well must encounter rock with good porosity and permeability if water is to be extracted in

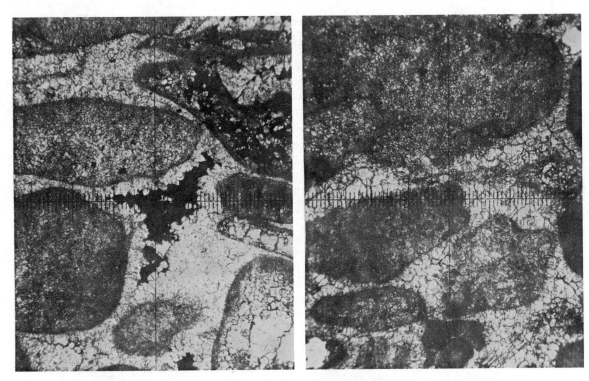

Figure 1.3 Photomicrographs of a core of carbonate rocks in an oil field. Both sediments were deposited as relatively clean, washed skeletal sands. The rock on the left has its pore space filled with petroleum, whereas the rock on the right has its pore space completely filled with calcite cement. The former rock type will produce oil; the latter will not. The petroleum geologist must attempt to understand the time and the mechanism of emplacement of both petroleum and calcite cement in order to predict where to find oil.

large volumes. Likewise, aquifer recharge and potential pollution of acquifers involve questions requiring considerable stratigraphic knowledge.

Although we have long been interested in taking things out of holes in rocks, we are becoming more and more interested in putting things into holes in rocks; that is, in subsurface waste disposal. Where will the waste go once it is pumped down into the earth? Once again, the answer will require considerable sedimentologic and stratigraphic knowledge. If we do not want these wastes turning up in municipal water supplies, we had best pay close attention to what we are doing from the beginning.

Earth Processes, Finite Resources, and the Future of Humankind. Sedimentologists and stratigraphers think on a different time scale from most people. This ability is perhaps the greatest gift that such scientists have to offer humankind. The problem of finite resources provides the most clear-cut example of the divergence between popular opinion and the thought processes of the stratigrapher.

Economists talk of an equilibrium between supply and demand. The Western world will consume copious quantities of petroleum products, for example, if the price is right. In general, the price is right, because past petroleum exploration activities discovered abundant quantities of petroleum at relatively small cost. To the economist's mind, there is an equilibrium here. To the stratigrapher's mind, this idea is absurd. True equilibrium can never exist, simply because new petroleum concentrations form much more slowly than the rate at which existing supplies are being exploited.

Thus, even as the geologist explores for new reserves of oil and gas, the geologist is also the person in the best position to point out to the world's nations that eventually there will be no supply at any price. How and when we phase out fossil fuels is a fundamental problem facing us today.

Sedimentologists and stratigraphers can potentially offer similar insight into innumerable problems concerning earth processes. A perfect short-term engineering solution to an environmental problem may run afoul of some long-term earth process. The long-term earth process is likely to remain beyond the range of everyday engineering thought. It will likely fall to the sedimentologist and stratigrapher to keep society advised concerning the hazards of long-term disequilibrium between human beings and the natural processes of their environment.

Selected References

Cook, Earl. 1975. Man, energy, society. W. H. Freeman and Co., San Francisco. 478 p.
Socially oriented consideration of energy resources.

Denton, G. H., and S. C. Porter. 1970. Neoglaciation. Sci. American 222 (6): 100–111.
Stratigraphic data record the history of glacial advance and retreat beyond the time scale of human observation, yet within a time range that clearly has implication concerning future climates.

Halbouty, M. T. 1967. Our profession's challenge and responsibility. Bull. Amer. Assoc. Petrol. Geol. 51: 124–125.
Then-president of the American Association of Petroleum Geologists gives some views concerning the role of geologists in a changing world.

———— (ed.). 1971. Geology of giant petroleum fields. Amer. Assoc. Petrol. Geol. Mem. 14. 575 p.

———— (ed.). 1980. Giant oil and gas fields of the decade 1968–1978. Amer. Assoc. Petrol. Geol. Mem. 30. 596 p.
Sequel to Halbouty (1971). It is interesting to see how a decade of dramatic price increases changes the definition of a "giant field."

Kukla, G. J., R. K. Matthews, and J. M. Mitchell (eds.). 1972. The end of the present interglacial. Quaternary Res. 2: 261–269.
Whole issue devoted to the application of stratigraphic data to the problem of global climatic forecasting.

NATIONAL ACADEMY OF SCIENCES-NATIONAL RESEARCH COUNCIL (NAS–NRC). 1969. Resources and man. W. H. Freeman and Co., San Francisco. 259 p.
  The impending finiteness of resources. Energy resources chapter especially interesting as a petroleum geologist's view of the exploration picture prior to the "energy crisis" of the 1970's.

SKINNER, BRIAN J. 1969. Earth resources. Prentice-Hall, Inc., Englewood Cliffs, N.J. 150 p.
  Introductory level; easy reading; overview.

The Present
as the Key to the Past

2

The major tenet of this book is that late Cenozoic processes and history provide a model that will help us to understand the details of earth history recorded by the stratigraphic record. Sediments are being deposited today in a wide variety of environments. We must study them in detail and learn to infer the depositional environment of ancient sediments by analogy with Recent sediments. With recognition of depositional environment comes a whole series of insights concerning the processes that must have been active and the sequence of events required to produce a certain stratigraphic sequence.

As often happens with such statements, "the present as the key to the past" sounds deceptively simple. In fact, this approach requires sufficient sophistication that we shall have to cover a large amount of background material (Parts II and III) before we can come to grips with the major subject matter of the book (Parts IV and V). In the meantime, the following discussion is offered as a simplified overview of how the environmental approach can make the stratigraphic record come alive in our minds.

A Rational Approach
to Sandstone, Shale, and Limestone

Figure 2.1 depicts a stratigraphic sequence that commonly occurs, with varying dimensions, throughout the stratigraphic record. Figure 2.2 portrays lithologic correlation of three measured sections containing this sequence of lithologies. In early stratigraphic work, there was a tendency to equate lithology with time. With this simple view of the stratigraphic record, a geologist of the old school might have looked at Figures 2.1 and 2.2 and written a scenario of earth history that would read as follows: "A time" of deformation and peneplanation was followed by "a time'' of marine sandstone deposition. (The quotation marks are added for special significance in later discussions.) The sandy nature of this basal marine unit indicates that nearby mountains stood high at ''this time.'' As the mountains became worn down, ''the time'' of sandstone deposition gave way to ''a time'' of marine shale deposition. As the sources of clastic sediment supply became completely peneplained, there came ''a time'' of marine carbonate deposition. Subsequent to the deposition of the marine limestone, there was ''a time'' of tectonic rejuvenation of the source area, leading once again to deposition of marine shale and finally to deposition of marine sandstone.

The preceding outline of earth history is probably a well-reasoned *ad hoc* explanation of the data. This is the so-called ''layer-cake'' approach to stratigraphy. Things may have happened just that way. We cannot argue conclusively against it on the basis of the limited amount of data presented. On the other hand, this explanation does not fit our study of Recent sediments. Thus, we are led to ask if the data contained in Figures 2.1 and 2.2 could be equally well explained in terms of processes and products with which we are familiar from studying Recent sediments.

Figure 2.3 presents an exceedingly simplified and generalized schematic cross section of a situation common in Recent sedimentation. Scale of the model and depth relationship among the sediment types may vary widely. On the one hand, Figure 2.3 may generally describe the transition from intertidal deltaic sediments to globigerina ooze accumulating in oceanic depths. On the other hand, in some Recent environments, the transition from clastics to carbonates may involve little or no

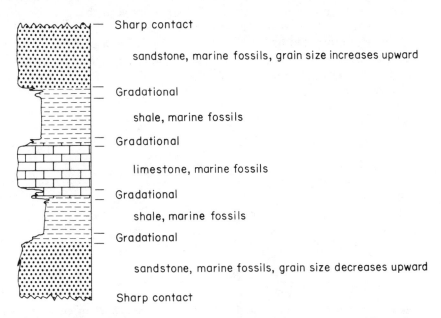

Figure 2.1 Common sequential relationships among sandstone, shale, and limestone.

Datum: Top of sandstone

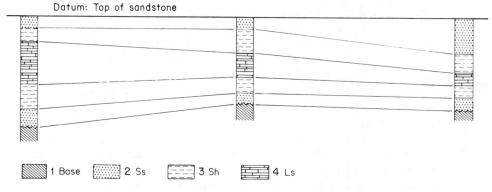

1 Base   2 Ss   3 Sh   4 Ls

Figure 2.2 Lithologic correlation of three measured sections, displaying variations on the general sedimentary sequence depicted in Figure 2.1. Pattern (1) represents basement rock; (2), sandstone; (3), shale; and (4), limestone. By the *ad hoc* model discussed in the text, each lithology may be taken to represent "a time" in earth history: "a time" of sandstone sedimentation, "a time" of shale sedimentation, and so on.

change in water depth. For the moment, let us accept the following discussion as a reasonable and moderate generalization useful only to convey an initial feeling of security. As we move into Part IV, we shall develop more specific models based on specific examples of Recent sedimentation.

Marine sands are commonly high-energy nearshore deposits: deltas, beaches, and the like. Marine shales usually occur in deeper waters seaward of the high-energy nearshore sand deposits. Still farther seaward, beyond the influence of clay input from the land area, clear-water carbonate sedimentation occurs. Here, in the absence of a high influx of terrigenous clay, carbonate-secreting organisms such as foraminifers, molluscs, corals, bryozoa, and calcareous algae produce biogenic accumulations of calcium carbonate that are quite similar to limestones of the stratigraphic record.

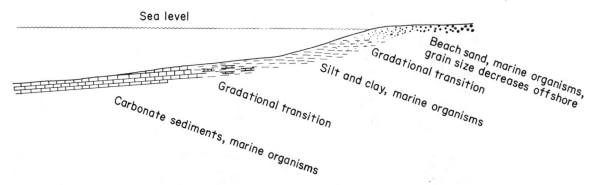

Figure 2.3 Hypothetical cross section depicting a common relationship among sand, clay, and carbonate sediments in the Recent epoch.

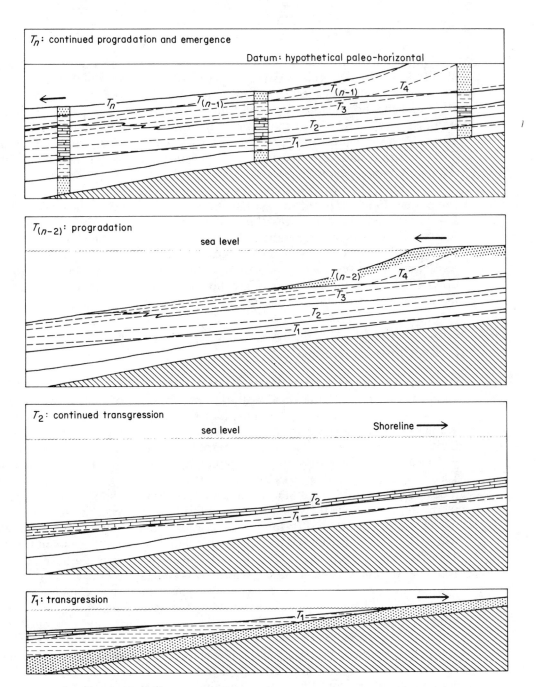

T_n: continued progradation and emergence

Datum: hypothetical paleo-horizontal

$T_{(n-2)}$: progradation

sea level

T_2: continued transgression

sea level   Shoreline ⟶

T_1: transgression

Figure 2.4 Dynamic model for the generation of the stratigraphic sections depicted in Figure 2.2. Transgression and regression superimposed upon the general model of Figure 2.3 generates an alternate hypothesis concerning the earth history recorded by the sections. The sequence of events begins with transgression at time T_1. Note that time lines T_1 through T_{n-1} cut across sediment types. Throughout the deposition of the entire sequence, sand, shale, and limestone are all being deposited somewhere within the model.

To the modern sedimentologist, therefore, Figure 2.2 might suggest transgression followed by regression. As the sea level rises, the site of high-energy sand accumulation might be expected to move landward. Likewise, the environments of shale and limestone deposition would migrate landward. Thus, marine sandstone becomes overlain by marine shale, which in turn becomes overlain by marine limestone. With subsequent regression, shale comes to overlie limestone, and the sandstone in turn comes to overlie shale. Such a reinterpretation of the data in Figures 2.1 and 2.2 are given in Figure 2.4.

Note that the two interpretations of each set of data are quite different. For the present time, suffice it to say that the interpretation given in Figures 2.3 and 2.4 "makes sedimentologic sense" in that all sediment types exist at the same time, whereas the previous interpretation of "a time" of sand, "a time" of clay sedimentation, and so forth seems rather foreign to the sedimentologist familiar with the Recent epoch. In subsequent chapters, we shall discuss additional sedimentological criteria that might indicate more clearly how we should interpret the earth's history as recorded by stratigraphic sequences.

Sedimentation Rates
and the Stratigraphic Record

Although the preceding examination of a hypothetical stratigraphic example encourages our trying to understand stratigraphy in terms of Recent sedimentation, comparison of Recent sedimentation rates with the thickness of the total stratigraphic record complicates the problem. Consider, for example, sedimentation rates in Recent calcium carbonate environments. Reasonable rates for Recent shallow-water calcium carbonate vertical accumulation rates are from 0.1 to 1 meter per 1000 years. Similar deposits occur in Mississippian through Permian strata over much of the central United States. Mississippian through the Permian periods represent approximately 10^8 years. Thus, application of Recent sedimentation rates would suggest that some 10^4 to 10^5 meters of sediment should have accumulated within that time. In reality, these sediments seldom exceed 10^3 meters in thickness. Thus, we have a problem. During the late Paleozoic, there could have been 10 to 100 times as much sediment accumulation as is actually recorded.

Another way of fitting Recent sedimentation data to stratigraphic record is by considering the lengths of time represented by the Recent and by the classical stratigraphic units of Ancient deposits. Recent shallow-marine sedimentation began some 5000 years ago as the post-Wisconsin transgression brought sea level up to approximately where it now stands. From our knowledge of sedimentation dynamics within 5000 years, we must attempt to build models that will apply to the stratigraphic record. In contrast, biostratigraphic zonation of Ancient rock sequences usually provides us with working units of geologic time that are on the order of 1 to 10 million years long. If the processes involved in Recent sedimentation are indeed responsible for the sedimentation of stratigraphic units representing 1 to 10 million years, then we must suspect (1) that our Recent sedimentation model has barely begun to run its course, or (2) that our Recent sedimentation model has been repeated over and over again within a single biostratigraphic interval, or (3) that the record is missing for large portions of many biostratigraphic intervals, or (4) some combination of (1), (2), and (3).

Thus, understanding the stratigraphic record in terms of Recent sedimentation will not be as simple as we might have originally anticipated. To begin with, we must study the Recent sediments as we see them today. Next, we must seek to understand the dynamics of Recent sedimentation over the short time interval for which it has been operating. Then we must construct a dynamic model that will extend Recent sediment models to a time scale appropriate to the stratigraphic record. Finally, we must apply these models to the stratigraphic record in an iterative fashion; that is, crude models leading to an improved understanding of the stratigraphic record, which in turn leads to an improved model, which in turn leads to still a better understanding of the stratigraphic record, and so on. These four activities are treated in Parts IV and V. But first we must organize the materials and dynamics with which we shall be dealing. This organization is the subject of Parts II and III.

Selected References

IMBRIE, J., and N. NEWELL (eds.). 1964. Approaches to paleoecology. John Wiley & Sons, New York. 432 p.
Collection of topical papers. Fairly advanced level.

LAPORTE, L. F. 1968. Ancient environments. Prentice-Hall, Inc., Englewood Cliffs, N. J. 115 p.
Introductory treatment of analogies between Recent sediments and Ancient sedimentary rocks.

Sediments, Time,
and the Stratigraphic Record

II

In the following chapters, we shall begin to think carefully of sediments, sedimentary rocks, and sequences of sedimentary rocks. What are their properties and how should we go about studying them? What are some reasonable conclusions that we can draw from various observations? What are the pitfalls?

At this stage, our approach must be somewhat traditional. Many generations of stratigraphers and sedimentologists have gone before us. They have done some things well, some things poorly, and a great many things that are somewhere in between. Yet, if each new generation is to build upon the structure left it by previous generations, then each new generation must acquaint itself with what has gone before. Thus, in Part II, we gather some basic facts and ideas upon which to build.

The Properties of Sediments and Sedimentary Rocks

3

The sedimentary cycle begins with the mechanical and chemical weathering of preexisting rocks. Mechanical processes produce boulders, sand, and even clay-sized particles. Chemical processes make new minerals, predominantly clays, out of old minerals, take soluble salts into aqueous solution, and leave behind a chemically inert residue that is predominantly sand-sized quartz. The solid products of mechanical and chemical weathering become the terrigenous clastic sediments of the stratigraphic record. Soluble salts may ultimately become the chemical rocks, such as carbonates and evaporites, or the cement within the terrigenous clastic sediments.

The following discussion concerns what happens after the weathering process. Sediments continually accumulate in a variety of sedimentary environments. Which of their properties can be used to decipher the earth history recorded in the accumulated layers?

The Basic Elements of Classification of Sedimentary Rocks

Rocks with certain attributes recur over and over again in the stratigraphic record. We shall consider the following classifications to be equally applicable to sediments and sedimentary rocks. A sediment is just a rock that has not been cemented; a sedimentary rock is just a sediment that has been cemented—whichever way you choose to look at it.

Grains, Matrix, and Cement. In the broadest sense, sediments consist of these three components in varying proportions. By grains, we generally mean particles of sand size or larger. Matrix refers to detrital sediment of silt or clay size. Cement is the mineral matter that is a chemical precipitate within the interstices of a sediment composed of larger particles. Where grains are abundant, the sediment is referred to as a *conglomerate* (large grains) or a *sandstone* (sand-sized grains). Where large particles are not present, the terrigenous clastic rock will be a *siltstone* or a *shale*. Finally, if terrigenous sediment forms an insignificant portion of the rock, we speak of chemical sediments; *limestones, dolomites,* and *evaporites* are the most common.

Further classification of sediment types is strongly dependent upon our study methods. In particular, conglomerates, sandstones, and limestones can be studied quite effectively in hand specimens and in thin sections with a relatively low-power optical microscope. These sediments are traditionally classified on the basis of hand-specimen and thin-section observations. On the other hand, shales, dolomites, and evaporites commonly require X-ray diffraction mineralogy data and chemical data to be meaningfully classified. These studies are traditionally beyond the scope of the stratigrapher-sedimentologist, who instead concentrates on studying outcrops, hand specimens, and thin sections. Thus, our discussions concerning the classification of sediments will center around the classification of sandstones and limestones.

Composition of Sandstones. The composition of sand grains provides important information concerning the provenance and predepositional history of the sediments. Rock fragments and mineralogical suites commonly allow the distinction of granitic, metamorphic, volcanic, and sedimentary source areas. Recognition of these general provenances is one factor in the classification of sandstones.

Compositional maturity of the sand is another major concern underlying most classifications of sandstones. Rigorous weathering and transportation tend to reduce many igneous and metamorphic minerals to clays. However, quartz, zircon, tourmaline, and rutile generally resist chemical alteration and thus remain as the ultrastable sand-sized component even after the sediment has undergone weathering and transporting.

Textural Maturity of Sandstones. The concept of the maturity of a sandstone may also be applied to the overall texture of the sediment. For example, prolonged transportation of sediment should ultimately result in separation of the various-sized fractions. Clay minerals, for example, should become separated from sand-sized particles, because clay is transported in true suspension (*suspended load*), whereas sand tends to be bounced or dragged along the bottom (*bed load*). If a sand contains a relatively large amount of clay, it is considered texturally immature. Conversely, clean sands are considered to be texturally mature. In this fashion, the presence or absence of fine-grained matrix is regarded as a major element in most classifications of sandstones.

Textural maturity may be further refined by consideration of the rounding of sand grains. Angular sand particles are the result of mechanical and chemical weathering. Either rigorous or prolonged transportation tends to wear down the sharp edges and produces rounded sand grains.

In general, compositional maturity and textural maturity go together. We can expect, however, complex deviation from this generality. Compositional maturity and textural maturity, as observed in any hand specimen of sedimentary rock, are the end products of chemical and mechanical processes acting in the source area, during transportation, and during and following the ultimate deposition in the sedimentary environment.

Classification of Sandstones. Guided by a general concern for the provenance of the sediment and for its general state of maturity, geologists have formulated numerous classifications of sandstones. Four of these schemes are indicated in Figures 3.1 through 3.4. Although memorization of these classifications is not necessarily desirable, you will find the various rock names frequently used throughout the literature. Unfortunately, no single classification suffices, for the simple reason that older literature may use any of these classifications or even some other classification.

All of these classifications are essentially arbitrary. Each geologist has spent a certain amount of time working on the petrology of sandstones and has found it useful to make certain distinctions. Modern statistical techniques may provide a

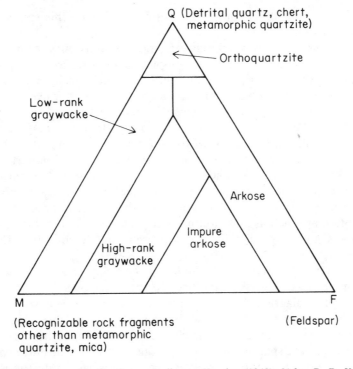

Figure 3.1 Sandstone classification according to Krynine (1948). [After P. D. Krynine, "The Megascopic Study and Field Classification of Sedimentary Rocks," *Jour. Geol.,* **56,** 130–165 (1948).]

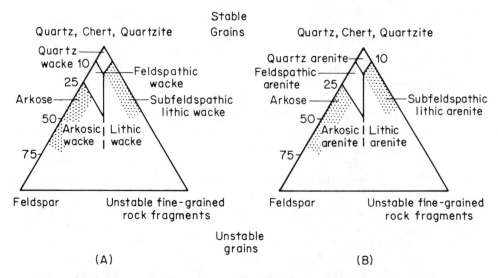

Figure 3.2 Sandstone classification according to Gilbert (1954). Rocks that are named wacke contain significant mud matrix; rocks named arenite contain essentially clean sands.

more rigorous basis for classification. We discuss this subject later under natural classification.

Figure 3.1 presents the classification of sandstones according to Krynine (1948). Triangular diagrams are quite popular among those workers who classify sandstones. Although tripartite division is not necessarily a fundamental property of sandstones, three components are all we can handle conveniently in a single graphic illustration. In the Krynine classification, quartz and chert constitute the supermature subdivision of the classification. F (feldspar) may be viewed as shorthand for a granitic source area. M (rock fragments and mica) may generally be read as a volcanic or low-rank metamorphic source area. Orthoquartzite, arkose, and graywacke are the major sandstone types recognized in this classification. Note that texture per se does not enter into this classification. One orthoquartzite might contain as much as 20% clay minerals. Alternatively, another orthoquartzite might contain as much as 20% well-rounded feldspar sand grains. These two orthoquartzites would be genetically quite different, but the classification would place them together.

Figure 3.2 presents the Gilbert (1954) classification of sandstone. This classification uses two words to describe the sandstone. The first word, based on the triangular diagram, is an adjective describing the general composition of the sand grains. The second word describes the presence or absence of fine-grained detrital matrix. Once again quartz and chert denote the supermature end of the classification spectrum; feldspar denotes granitic source area; and unstable rock fragments denote volcanic or metamorphic source area. In comparison to the Krynine classification (Figure 3.1), note especially the appearance of a number of special words in the impure regions around the quartz-chert corner of the diagram. A quartz arenite of

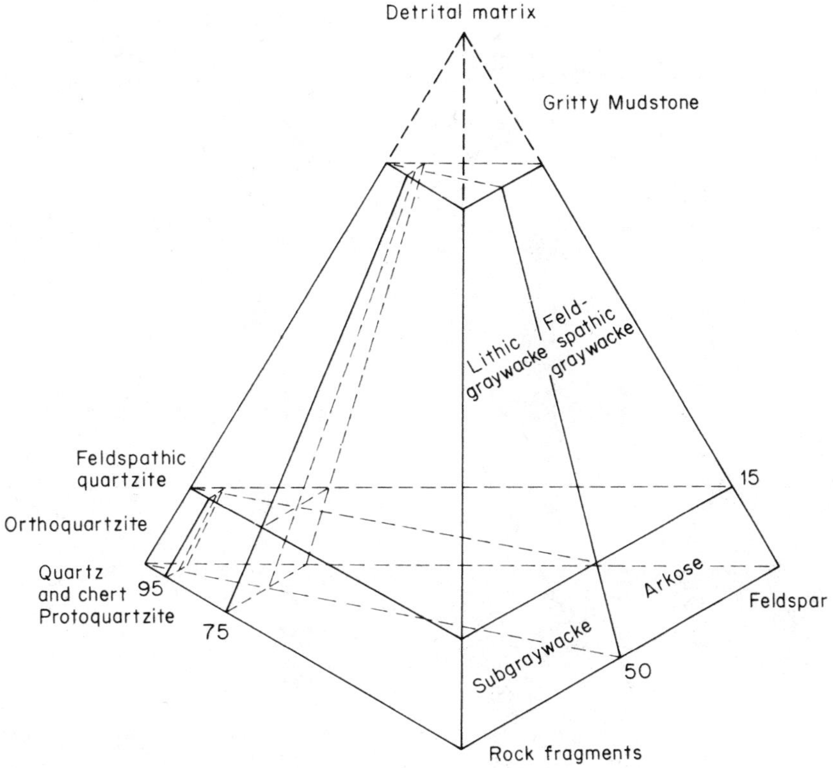

Cement or matrix		Detrital matrix exceeds 15% Chemical cement absent	Detrital matrix less than 15%. Voids empty or filled with Chemical cement			
Sand or detrital fraction	Feldspar exceeds rock fragments	**Graywackes**: Feldspathic graywacke	**Arkosic** Arkose	**Sandstones** Subarkose or feldspathic Ss	**Orthoquartzites**	Chert <5%
	Rock fragments exceeds felds	Lithic graywacke	**Lithic** Subgraywacke	**Sandstones** Protoquartzites		Chert >5%
	Quartz content	Variable; generally <75%	< 75%	> 75% < 95%		> 95%

Figure 3.3 Sandstone classification according to Pettijohn (1954, 1957).

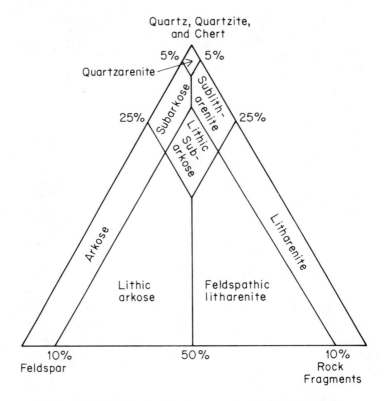

Figure 3.4 Sandstone classification according to McBride (1963).

the Gilbert classification must contain greater than 90% quartz, whereas the ortho-quartzite of the Krynine classification need contain only 80% quartz. Krynine's graywacke may be either a lithic wacke or a lithic arenite, depending on the presence or absence of a fine-grained detrital matrix.

Figure 3.3 summarizes the classification of sandstones after Pettijohn (1954, 1957). By and large, the input concepts of this classification are similar to those of Gilbert (1954). Only the names and style of presentation have changed. The use of the word "orthoquartzite" is now restricted to rocks containing greater than 95% quartz.

Figure 3.4 presents a sandstone classification of McBride (1963). Once again, the major elements of the classification are those of Gilbert (1954), yet we find still more names added to the classification on the basis of arbitrary subdivision within the triangular diagram.

Classification of Fine-Grained Clastic Rocks. Because of small particle size, these rocks are difficult to classify on the basis of petrographic examination, a tech-nique that is very useful for classifying sandstones and carbonates. A useful field classification of fine-grained clastic rocks, however, can be based upon presence or absence of fissile bedding, a crude estimation of grain size, and color on outcrop. We shall follow the general classification of Blatt (1982).

If the fine-grained clastic rock exhibits fissile bedding, it is referred to as a shale; without fissile bedding, it is a ''-stone.'' These two general names are used with precedent grain-size modifiers. If silt is abundant, the rock is a silt-shale or a silt stone. If the rock contains only clay, it is a clay-shale or a clay-stone. If both silt and clay-size particles are present, the rock is a mud-shale or mud-stone. Grain size is conveniently determined on outcrop by use of hand lense and teeth. If the rock is predominantly silt-size material, this can be determined with hand lense. If silt-size particles are not visible, try chewing a small fragment. If the sediment is gritty, silt is present. If smooth, only clay minerals are present.

The color of many sedimentary rocks is nondiagnostic because high initial permeability allows through-going solutions to impart color changes at any time in the history of the sediment. In contrast, many fine-grained clastic rocks have sufficiently low initial permeability so that color is retained as an attribute of the primary sediment. The color of shales, mud-stones and clay-stones is dictated by the oxidation state of iron minerals and by relative abundance of fine-grained particulate organic matter. Color may range from black to greys and greens to yellows, browns, and reds, depending on the oxidation state of iron minerals.

One of the most spectacular lithologies within this classification scheme is the fissile black clay-shale. This rock type is quite common among the deposits from Paleozoic epeiric seas which once covered North America. The black shale environment is set up when prevailing winds cause influx of low-oxygen sea water, derived from within the oxygen minimum of a nearby deep ocean. So long as bottom-water is replaced by low-oxygen water from the nearby ocean, benthic organisms cannot thrive and only clay minerals and particulate organic matter come to rest on the floor of the shallow sea. In contrast, if there is some shift in paleoceanography by which the circulation of this shallow sea taps oxygenated water in the nearby ocean, these same paleogeographic situations can suddenly support abundant benthic life, resulting in the formation of interbedded limestones among the black shales.

Classification of Limestones. The classification of limestones runs somewhat parallel to the sandstone classification of Gilbert. The classifications of Folk (1962) and of Dunham (1962) are the most generally accepted ones. Both schemes name the sediment on the basis of grain types and the amount of mud matrix.

The grain types of carbonate rocks are commonly the chemical or biochemical products of the sedimentary environment. Whereas terrigenous clastic grains were derived from preexisting rock and transported to depositional environment, the grains within carbonate sediments are commonly precipitated from the water by chemical or biochemical processes active in the depositional environment. Folk (1962) recognizes four basic categories of grain types. These are bioclastic debris, oolites, intraclasts, and pellets. Bioclastic debris is the skeletal remains of carbonate-secreting organisms that live in or near the depositional environment. Oolites are sand-sized grains constructed by concentric laminations of calcium carbonate deposited under agitated water conditions. Intraclasts are small chips of semi-indurated calcium carbonate sediment commonly formed by alternate wetting and desiccation in intratidal and supratidal carbonate environments. Pellets are the semi-indurated to indurated excreta of benthonic organisms.

In the Folk classification, abbreviations identifying the various grain types constitute the prefix of a shorthand word that describes the rock type. If the rock contains no mud, it is a sparite (taken from "spar," which is the cement for the sediment). If the sediment contains considerable mud matrix, it is a micrite ("micrite" being a contraction for microcrystalline calcite). Thus, a pelsparite is a clean sand composed of pellets, and a oopelmicrite is a muddy sand containing pellets and oolites. If we wish to emphasize that the sand grains are larger than 2 millimeters, the rock may be named a rudite. Pelsparrudites and oopelmicrudites, therefore, contain larger grains than pelsparites and oopelmicrites. Adjectives may also be added to name rocks like mollusc-bearing oopelmicrudite, and so on.

The basic elements of the Folk classification of carbonate rocks is given in Figure 3.5. As with the sandstone classification discussed earlier, the boundaries among the various sediment names are placed at arbitrary percentages.

		> 10% Allochems Allochemical Rocks		< 10% Allochems Microcrystalline Rocks		
		Sparry Calcite Cement > Microcrystalline Ooze Matrix	Microcrystalline Ooze Matrix > Sparry Calcite Cement	1–10% Allochems	< 1% Allochems	Undisturbed Bioherms Rocks
		Sparry Allochemical rocks	Microcrystalline Allochemical Rocks			
	> 25% Intraclasts	Intrasparrudite Intrasparite	Intramicrudite Intramicrite	Intraclasts: Intraclast-bearing Micrite	Micrite	
	> 25% Oolites	Oosparrudite Oosparite	Oomicrudite Oomicrite	Oolites Oolite-bearing Micrite		Biolithite
< 25% Intraclasts < 25% Oolites / Volume Ratio of Fossils to Pellets: > 3:1		Biosparrudite Biosparite	Biomicrudite Biomicrite	Fossils Fossiliferous		
3:1–1:3		Biopelsparite	Biopelmicrite	Pellets: Pelletiferous Micrite		
< 1:3		Pelsparite	Pelmicrite			

Vertical axis labels: Volumetric Allochem Composition; Most Abundant Allochem

Figure 3.5 A simplified version of the classification of limestones according to Folk (1962).

Dunham's classification hinges around the concept of "grain support" versus "mud support" of the depositional fabric. Note that this emphasis on the ability of the grains to support the depositional fabric allows wide variation in such parameters as the percentage of mud. Dunham contends that the percentage of mud is not the fact that should demand our attention. This approach carries with it some very important implications. For example, if a rock is truly mud-supported, then the mud and the grains must have been deposited at the same time and in an environment that was quiet enough to allow mud to accumulate. If, on the other hand, the sediment is grain-supported, then we may entertain the suggestion that the grains were deposited under high-energy conditions and that the mud filtered down in among the grains at some later time. Thus, our interpretation of the conditions of sedimentation may vary considerably, depending on whether we think the depositional fabric to be grain-supported or mud-supported. The classification of Dunham (1962) is given in Figure 3.6. The names mudstone, wackestone, packstone, and grainstone are widely used.

Natural Classification. With the exception of the Dunham limestone classification, we have noted considerable usage of arbitrary boundaries. If the only purpose of a classification is communication among people who have memorized it, then there is nothing wrong with arbitrary boundaries. If, on the other hand, we think these various names convey important sedimentological distinctions, we may be making our studies unnecessarily complicated. Imbrie and Purdy (1962) pointed this fact out in their classification of modern Bahamian carbonate sediments.

Consider a group of samples for which there is quantitative data concerning the abundance of *A, B,* and *C.* A brief glance at the triangle diagrams in Figure

Dunham Classification of Carbonate Rocks According to Depositional Texture				
Depositional Texture Recognizable				*Depositional Texture Not Recognizable*
Original components not bound together during deposition			*Original components bound together during deposition*	
Contains mud (particles of clay and fine silt size)	Lacks mud and is grain-supported		Shown by intergrown skeletal matter, lamination contrary to gravity, or sediment-floored cavities that are roofed over by organic or questionably organic matter and are too large to be interstices.	*Crystalline Carbonate*
Mud-supported	*Grain-supported*			(Subdivide according to classifications designed to bear on physical texture or diagenesis.)
Less than 10% grains	More than 10% grains			
Mudstone	*Wackestone*	*Packstone*	*Grainstone*	*Boundstone*	

Figure 3.6 Classification of carbonate rocks according to Dunham (1962).

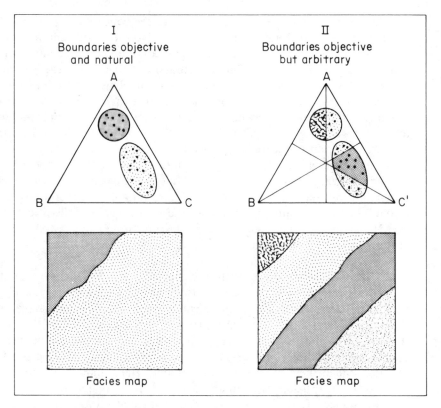

Figure 3.7 Hypothetical example indicating the utility of natural classification as opposed to arbitrary classification. Where two groups of data exist in nature, to rely on an arbitrary classification to subdivide the data would be unnecessarily complicated. (After Imbrie and Purdy, 1962, with modification.)

3.7 confirms that the samples fall into two naturally occurring populations. Observe, however, that arbitrary classification of these sediments might require that we name four lithologies. Clearly, we are complicating the problem when we employ an arbitrary classification in a situation like this one.

Instead, there are available multivariate statistical techniques, which allow the recognition of naturally occurring clusters in *n*-dimensional space. *N*-dimensional space is simply a mathematical abstraction beyond the three-dimensional array portrayed in Figure 3.7 (See Imbrie, 1964, for further discussion.)

The Concept of Sedimentary Facies. The word *facies* is commonly used to denote the product of an environment. Thus, a highly deformed terrain containing many rock types might be mapped as a single *metamorphic facies* if we wish to convey that all of the rocks within that terrain have been subjected to the same temperature-pressure conditions. Similarly, the concept of *sedimentary facies* may group many different lithologies under the concept that all of these lithologies represent various subenvironments of an overall sedimentary environment which can be given a convenient name. For example, a conglomeration of oyster reefs, fossiliferous clays, supratidal algal stromatolites, and peats might all be conveniently lumped together as "lagoonal facies," inasmuch as all of these rock types are quite typically deposited in the lagoonal environment.

In general, it will be convenient to think of the major environments reviewed in Part IV as producing characteristic sedimentary facies: the braided-stream facies, the meandering-stream facies, the deltaic facies, and so on. However, the concept of sedimentary facies can be twisted to suit one's needs. For example, in petroleum exploration, one may wish to define facies in a fashion more restricted to porous and permeable rock units as opposed to the surrounding impermeable rocks. Such a practical orientation toward facies definition is perfectly appropriate so long as one bears in mind the overriding definition that a sedimentary facies is the product of a sedimentary environment.

Size Distribution

Naming the sediment according to its composition is just the first small step toward learning what the sediment can tell us about its history. Size and size distribution within sediments is also of significance for studying the paleogeography and depositional environments. Standard terminology concerning description of mean grain size in sediments is given in Table 3.1. Size data are usually reported in millimeters or in phi units, where

$$\phi = -\log_2 X, \qquad (3.1)$$

X being the size measure in millimeters.

The standard for description and comparison of size-distribution data is the normal distribution function:

$$f(z) = \frac{\exp(-z^2/2)}{\sqrt{2\pi}}. \qquad (3.2)$$

In this equation, $z = (x - \mu)/\sigma$, where μ is the population mean and σ is the population standard deviation. Where size data are in phi units, size distribution is expected to be normally distributed; where the data are in millimeters, the data are expected to be lognormally distributed.

As indicated in Figure 3.8, the function in Eq. (3.2) plots as a straight line on a cumulative probability graph. The mean is the 50th percentile, and the standard deviation is the difference in size between any spread of 34 percentiles. In common usage, deviation is reported as $(\phi_{84} - \phi_{16})/2$. The mean is used to name the size of the sediment as indicated in Table 3.1; the standard deviation, or sorting, is used as a first estimate of textural maturity. Sediments that have a standard deviation of .35 ϕ are considered well sorted; .50 ϕ, moderately well sorted; .70 ϕ, moderately sorted; 1.0 ϕ, poorly sorted; and 2.0 ϕ, very poorly sorted.

Skewness is a measure of the departure of a size distribution from normality

Table 3.1 Particle-Size Nomenclature

	Name	Millimeters (mm)	Size in Microns (μ)	Phi φ	U.S. Standard Sieve Mesh #
GRAVEL	Boulder				
		256.0	—	−8.0	—
	Cobble				
		64.0	—	−6.0	—
	Pebble				
		4.0	—	−2.0	5
	Granule				
		2.0	—	−1.0	10
SAND	Very coarse sand				
		1.0	—	0.0	18
	Coarse sand				
		0.5	500.0	1.0	35
	Medium sand				
		—	250.0	2.0	60
	Fine sand				
		—	125.0	3.0	120
	Very fine sand				
		—	62.5	4.0	230
MUD	Coarse silt				
		—	31.0	5.0	—
	Medium silt Fine silt Very fine silt				
		—	3.9	8.0	—
	Clay				

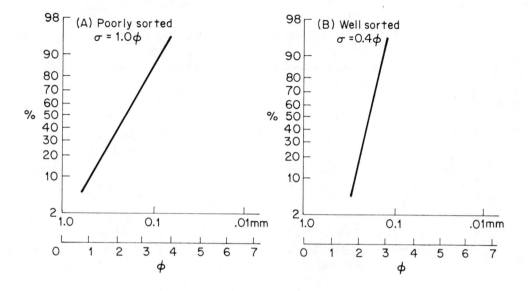

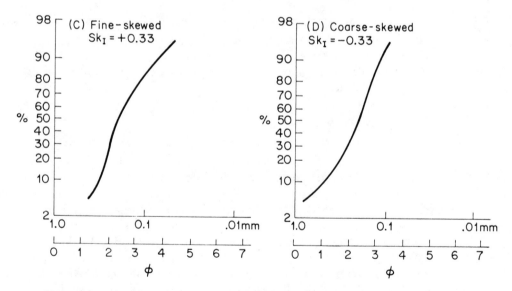

Figure 3.8 Various examples of size distribution plotted with a cumulative probability function as the Y-axis. Rather than attempt to construct the traditional bell-shaped frequency distribution curve, we find it convenient to plot size data by using a cumulative probability function as the Y-axis. On such a plot, a normal distribution of the form in Eq. (3.2) is a straight line. Thus, variations from normality are easily detectable upon visual examination.

or lognormality. If a sediment sample contains excess fine-grained material, it is said to be (+) skewed; if it contains excess coarse-grained material, it is (−) skewed. In theory, skewness should compare the symmetry of size distribution about the mean for each pair of percentiles of the curve. In practice, Folk (1980) suggests:

$$\text{Sk}_I = \frac{\phi_{16} + \phi_{84} - 2\,\phi_{50}}{2\,(\phi_{84} - \phi_{16})} + \frac{\phi_5 + \phi_{95} - 2\,\phi_{50}}{2\,(\phi_{95} - \phi_5)}, \tag{3.3}$$

which shows inclusive graphic skewness as an adequate measure for most purposes. The comparison of a large number of size-distribution curves suggest that Sk_I of $+.5$ is strongly fine-skewed; $+.2$, fine-skewed; $+.1$ to $-.1$, more or less symmetrical; $-.2$ coarse-skewed; and $-.5$ strongly coarse-skewed. Figure 3.8 plots some curves having these attributes.

Size distribution data are often gathered as part of the overall description of sediments. However, size distribution data are treated as an especially important data set with regard to identification of sediments associated with glaciation. Tillite (the ancient rock equivalent of glacial till) is characterized by extremely large standard deviation of particle size. While it is possible for hand specimens of tillite to resemble hand specimens of mud-flow material, paleogeographic context or shape and orientation characteristics of pebbles will usually allow clear distinction between these two rock types.

Another example of size variation resulting from the action of ice is the occurrence of ice-rafted detritus in otherwise fine-grained sediments. At the extreme of this phenomenon are cobble-size dropstones in shales. On a less spectacular scale, "ice-rafted detritus" may show up as coarser grained sand in an otherwise fine-grained sediment.

Stratification

Sedimentary rocks are commonly arranged in layers. Indeed, the very name of this subject, *stratigraphy,* comes from this attribute.

In the field description of sedimentary sequences, it is customary to make note of the thickness of stratification and its degree of development. McKee and Weir (1953) provide a simple and convenient discussion of terminology. Strata less than 1 centimeter thick are referred to as *laminae.* Strata thicker than 1 centimeter are *beds,* ranging from very thin to very thick: 1 to 5 centimeters is very thin; 5 to 60 centimeters, thin; 60 to 120 centimeters, thick; and greater than 120 centimeters, very thick. To these words are added such modifiers as "well-developed," "irregular," and so on, as appropriate.

If a single bed exhibits decreasing average grain size from bottom to top, it may be referred to as a *graded* bed. Similarly, if grain size increases upward in a bed, it is referred to as a *reverse-graded* bed. Beds involving other evidences of stratification are said to contain primary structures and require further discussion.

Primary Structures

Sediments deposited by moving air or water commonly show a depositional fabric indicative of the conditions under which sedimentation occurred. Furthermore, activities of burrowing organisms may impart recognizable characteristics to the sediment even though the organism is not preserved. These and similar features are referred to as *primary sedimentary structures;* that is, rock fabrics indicating the conditions of deposition.

Primary sedimentary structures are especially valuable to the sedimentologist-stratigrapher because important interpretations can often be made by visually examining the outcrop or core. Our discussions of sedimentary environments will rely heavily on primary structures as a quick and easy way to distinguish several important environments.

Laminar and Turbulent Flow. When a liquid or a gas moves in such a fashion that the particles remain in parallel flow lines, the flow is said to be laminar. When the net forward movement of the liquid or gas is accomplished by irregular, non-parallel motion of the particles, the flow is said to be turbulent. All of the features that we shall discuss are the result of turbulent flow. Turbulent flow is further divided into tranquil (sometimes called streaming, lower, or subcritical) and rapid (shooting, upper, or supercritical) flow regime.

Flume Studies. A flume provides a convenient method for investigating bed form and primary structures as a function of increasing flow velocity with flow depth held constant. Figure 3.9 depicts the results of a series of these experiments carried out by Simons *et al.* (1961). With low current velocity, small ripples were formed. As velocity increases, the height of the ripples increases. With still greater velocity, the shape of the large ripples begins to flatten down. With even greater velocity, a planar bed form is achieved and no sediment is deposited. This picture is the transition from tranquil to rapid flow regime. Finally, with still greater stream velocity, antidunes are formed. Sediment accumulation on antidunes occurs on the updip side of the bed form; the antidune may actually migrate upstream with continuing sedimentation.

Simons *et al.* (1961) discuss only the evolution of bed form with increasing velocity. Let us now translate these discussions into equivalent primary structures.

Consider the small changes in flow velocity associated with the passage of water over a tranquil-ripple bed form. Inasmuch as the depth of the flow is finite and the top of the flow is more or less a smooth surface, then the velocity of flow must increase as the water passes from the trough up onto the crest of the ripple. Similarly, the velocity must decrease as the water passes from the crest of the ripple on toward the next trough. Figure 3.10 depicts these relationships. With increasing velocity, the bottom sediment will be mobilized by the flow. With decreasing velocity, some of this sediment load will be dropped back to the stream bed. Con-

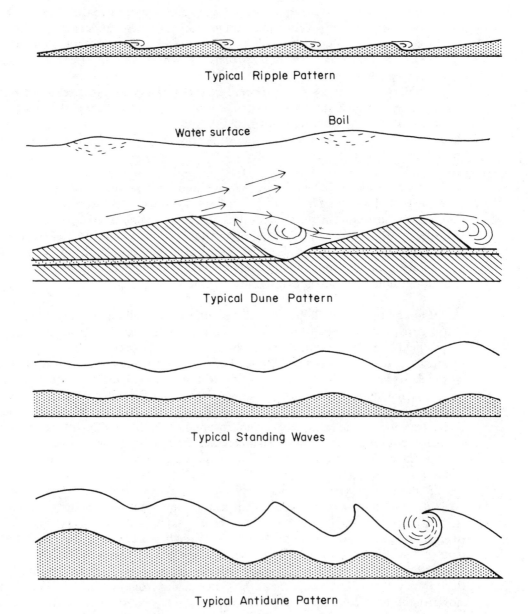

Water surface

Typical Ripple Pattern

Water surface   Boil

Typical Dune Pattern

Typical Standing Waves

Typical Antidune Pattern

Figure 3.9 Schematic summary of flume studies relating bed form in medium-sized sand to flow regime. The upper two diagrams represent tranquil flow, and the lower diagram represents rapid turbulent flow. (After Simons *et al.,* 1961.)

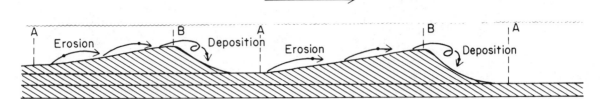

Figure 3.10 Illustration of erosion and deposition associated with ripple migration and the resultant development of planar cross-stratification. As the tranquil flow passes from point A to point B, velocity will be increasing; as the flow passes from B to A, velocity is decreasing. Increasing velocity results in net erosion of the sediment, whereas decreasing velocity results in net deposition.

sequently, material from the upstream side of a ripple will be transferred (by *erosion*) to the downstream side of the same ripple (causing *sedimentation*). The net result is the formation of high-angle cross-stratification (*lower flow regime*).

With increasing velocity the sediment is transported somewhat further beyond the crest of the ripple. This process results in lowering the angle of cross-stratification to the point where the bed form becomes planar as critical flow (*upper flow regime*) is approached.

Most of the primary sedimentary structures of interest to us form in the lower flow regime. Supercritical flow is sometimes encountered in turbidity currents (where high velocities are attained by density currents) and in beach and alluvial fan deposits (where upper flow regime results from extremely shallow water depth).

Cross-Stratification. Flume studies are carried out in long straight channels in the laboratory under very specific conditions. In sharp contrast, natural channels are seldom straight, seldom of constant depth, and seldom operate under specified conditions. Thus, we must supplement knowledge gained from flume studies with empirical observations on cross strata as they are actually observed under field conditions.

McKee and Weir (1953) suggest a convenient nomenclature and classification for cross-stratification in sandstones (see Figure 3.11). Cross strata that owe their origin to the migration of a single ripple form are referred to as a *set*. Similar sets of cross strata are referred to as a *coset*. Where cosets of differing morphology are usually arranged sequentially in the stratigraphic section, the sequence of cosets may be referred to as a *composite set*.

Individual sets of cross strata are classified as simple, planar, or trough. Simple cross strata do not have erosional boundaries. Planar cross strata have erosional set boundaries that are more or less flat planes. Trough cross strata have lower set boundaries that are curved surfaces of erosion. This scheme is further modified by the designation of small, medium, and large sets. Small-scale sets of cross strata are less than 30 centimeters thick. Medium-scale sets are 30 centimeters to 6 meters thick. Large-scale sets are thicker than 6 meters.

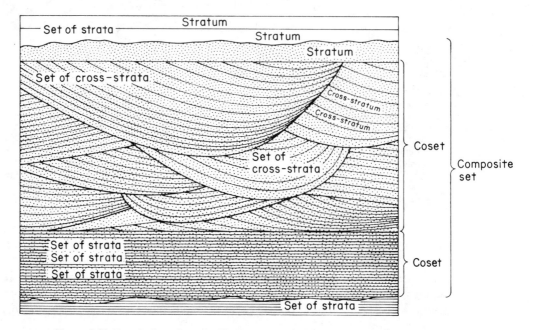

Figure 3.11 Terminology for stratified and cross-stratified units as proposed by McKee and Weir (1953).

Figures 3.12, 3.13, and 3.14 provide examples of naturally occurring cross-stratification associated with alluvial and marginal-marine sedimentation.

The azimuth of dip of cross strata can provide important information concerning directions of sediment transport. The data are typically gathered by making numerous measurements of directional properties of cross strata using a Brunton compass. These data are then plotted by class intervals of compass bearing from 0 to 360 degrees. The resulting construct, a "rose diagram," provides a graphic presentation of average transport direction. Unidirectional transport perpendicular to paleotopography characterizes gravity transport, such as in stream environments or deep-water turbidity current environments. Unidirectional transport parallel to regional paleotopography characterizes marine environments. Bimodal transport directions perpendicular to paleotopography characterize intertidal environments.

Vertical Sequences of Cross Strata. Cross strata can be studied set by set; each set records the hydrodynamic conditions under which it was sedimented. On the other hand, certain vertical sequences of cross strata commonly repeat themselves in sedimentary rocks. By recognizing the broad general significance of these vertical sequences, we can rapidly identify in the field major features of the paleogeography in which sedimentation occurred. Geologists have developed two models that particularly relate vertical sequences of primary structures to flow regime in natural

Figure 3.12 Cross-stratification shown in dip section through sand deposits of a modern meandering stream. Cosets of medium-scale high-angle, planar cross strata are overlain by cosets of low-angle simple cross strata. Blade width on the large knife is approximately 4 centimeters. (Photo by L. V. Benson.)

environments. These are the point-bar model of fluvial sedimentation and the turbidite model for deep-water sedimentation from turbidity currents.

The point-bar model and related interpretations concerning flow regime are given in Figure 3.15. The deep portion of the channel carries the most water and is floored with the coarsest-grained sediment within the system. High current velocity combined with relatively deep water commonly places this environment in transitional to lower flow regime. Closer to the point bar, stream velocity is slightly lower, but water depth is much less. In this position, decreased water depth more than compensates for slightly lower stream velocities, and the combination usually places this environment in transitional to upper flow regime. Simple low-angle cross-stratification, such as depicted in Figures 3.12 and 3.13, are typical of this condition. On top of the bar, water depth is extremely shallow and current velocity is minimal. Such a combination places this environment well down into lower flow regime. Small-scale, high-angle cross-stratification is typical of this environment. (See the

Figure 3.13 Strike section revealing cross-stratification in sand deposits of a modern meandering stream. Scale at the bottom of the photograph is 15 centimeters total length. Simple low-angle cross strata, such as seen in the upper portions of Figure 3.12, are overlain by medium-scale trough cross strata, which are in turn overlain by cosets of small scale simple cross strata. In the upper portion of the photograph, note that burrows and plant roots have reworked the sediment to the extent that primary cross-stratification has been almost obliterated. The burrows, root structures, and general obliteration of the cross-stratified fabric are themselves the sedimentological record of subaerial exposure of this river-sand deposit. (Photo by L. V. Benson.)

Figure 3.14 An example of simple low-angle cross-stratification in beach sands of the high intertidal environment. Extremely shallow water and the relative rapidity with which swash comes on to and off the beach places this environment in the transitional to upper flow regime. (Photo by R. P. Steinen.)

Point-bar model for stream sedimentation

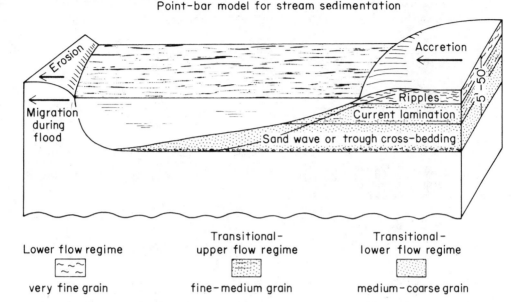

Figure 3.15 Schematic representation of the vertical distribution of sedimentary structures and grain size on a point bar. [After G. S. Visher, "Fluvial Processes as Interpreted from Ancient and Recent Fluvial Deposits," *Primary Structures and Hydrodynamic Interpretation,* ed. G. V. Middleton. Soc. Econ. Paleontologists and Mineralogists Spec. Pub. **12,** 116–132 (1965).]

middle portions of Figure 3.13.) Stream sedimentation is discussed in more detail in Chapter 11.

The hydrodynamic interpretation of a typical turbidite bed is presented in Figure 3.16. Turbidites are the sedimentary record of turbidity currents. Presumably, sediments occupying the outer portions of the continental shelf are (1) gravitationally unstable and (2) very poorly consolidated. After an initial perturbation of this unstable condition, by an earthquake perhaps, large masses of unconsolidated sediment and their contained fluid may flow downslope into adjacent deep water. Because this turbid suspension of sediment is more dense than the surrounding water, the turbidity current gains speed as it flows down the slope. As the turbidity current reaches the bottom of the slope, its velocity begins to decrease and sedimentation is initiated. Initial sedimentation may occur within the upper flow regime and generally involves very coarse material. Upper flow regime then gives way to transitional and lower flow regimes accompanied by decrease in grain size. Finally, normal processes of deep-water clay sedimentation again dominate the area, and the upper fine-grained portion of the typical sequence is deposited. The erosion of the upper portions of the sequence precedes the deposition of the next *A* layer.

Tool Marks. The cohesive properties of muddy sediments provide the opportunity for the formation of current-associated markings on bedding planes. Clay

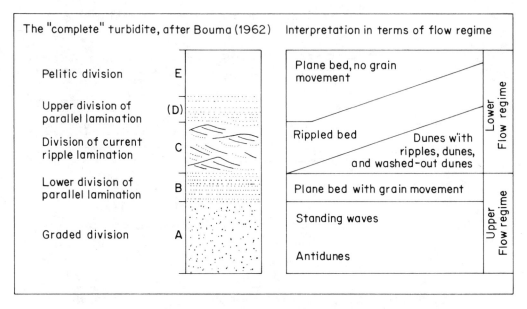

Figure 3.16 Schematic representation of the "complete turbidite" according to Bouma, with hydrodynamic interpretation according to Walker (1967). [After R. G. Walker, "Turbidite Sedimentary Structures and their Relationship to Proximal and Distal Depositional Environments," *Jour. Sed. Petrology* **37**, 25–43 (1967).]

mineral particles are difficult to resuspend once they have been deposited on the sediment-water interface. Thus, relatively strong currents may flow across sedimented mud without inducing large-scale erosion of the sediment-water interface (Figure 3.17). Such currents may drag or roll larger objects, like plant debris or perhaps even pebbles, across a mud bottom and plow a small furrow into the mud. These furrows, or grooves, are often preserved by the rapid deposition of sand-sized material over the mud surface. In the field, these features are most readily observed as groove casts on the bottom side of sandstone beds; the shales below are usually too friable to produce slabs large enough to observe the actual grooves. Groove casts can provide valuable information concerning paleo-current directions in marine sequences.

Flute casts are a somewhat similar phenomena. In this case, the mud at the sediment-water interface is locally disturbed by vortex currents associated with gravity flow down submarine slopes.

The rasping action of glacial ice commonly gouges distinct furrows into the underlying sediment or rock. These are referred to as "glacial striations." Further, glacial striation on pebbles may serve as important confirmation of tillite origin for poorly sorted clastic sediments. Likewise, glacial striations on dropstones in clays may also serve as confirmation for ice-rafted origin of the dropstones.

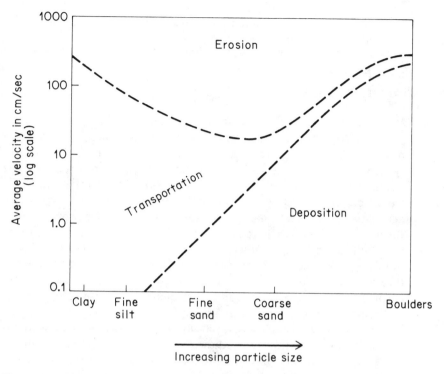

Figure 3.17 Diagram relating erosion, transportation, and deposition of various-sized particles to flow velocity of water. For coarse sand and larger particles, a relatively small increase in velocity will cause the sediment-water interface to change from a surface of net deposition to a surface of net erosion. In sharp contrast, particles of clay-size can be transported by very slow-moving currents; yet once the clay minerals are sedimented, very large currents are required to erode them. (Adapted from Hjulstrom, 1939.)

Bioturbation and Desiccation Features. Although moving water and wind impart the most dramatic primary structures to the sedimentary record, other processes can leave important imprints on the fabric of the rock. The burrowing activity of marine benthonic organisms has long been recognized as producing burrow-mottled fabrics that are in themselves diagnostic of specific depositional environments, even though the organisms that produced the burrows are not preserved (see Rhoads, 1967, for example). Similarly, the roots of plants move the sediment around as they grow. This activity will both disrupt any preexisting primary fabrics (see Figure 3.13) and create structures and fabrics diagnostic of the vegetated cover itself. Again, these primary structures may be preserved, although the plants that produce them are totally lost from the geologic record.

Desiccation features, such as mud cracks, are commonly found quite well preserved, especially in intertidal and lower supratidal sediments. In these environments, desiccation features are an extremely valuable environmental datum, for they record those sedimentary environments that were alternately wet during high tide and dry during low tide.

Sequential Relationships in Sedimentary Rocks

We have noted earlier that a single bed, hand specimen, or thin section of rock can yield considerable information concerning the origin of the sediment and the conditions of deposition. Next, we must integrate this information for all of the rock types falling within the scope of the particular area or problem. The basic unit of stratigraphic synthesis is the *measured section* through the lithologic sequence. We would prefer to study each time-rock unit in a map view. We would like to walk across the old depositional surfaces and study the facies changes in much the same fashion as we study facies changes in Recent sediments today. Unfortunately, time-rock units are most typically overlain by other time-rock units, thus precluding random access to the old depositional surfaces. Consequently, we are compelled to direct our attention to specific vertical sequences made available to us by fault escarpments, stream erosion, road cuts, quarry operations, and drill holes.

Thus, the object of most stratigraphic work is to synthesize numerous measured sections into maps and cross sections that portray the paleogeography and earth history recorded within the sediments. It is extremely important that we learn to make maximum use of observations concerning sequential relationships within measured sections of strata.

Unconformities. Where it can be demonstrated that the sedimentary sequences above a surface are not in sedimentological or temporal continuity with the rocks below the surface, the surface is said to be an *unconformity*. The recognition of unconformities is of prime importance in stratigraphic work. These surfaces must bound the major rock units and time-rock units within the area under consideration.

Within a measured section, unconformities may have visible lithologic properties associated with long periods of subaerial exposure (see Figure 3.18). Erosion surfaces and fossilized soil zones may be discernible on the material beneath the unconformity. Conglomerates, gravels, or other terrestrial lag deposits may begin the sequence immediately above the unconformity. Where strata above and below the unconformity can be shown to have angular relationships, owing to tectonic activity that occurred during the time represented by the unconformity, the surface is referred to as an angular unconformity. Where angularity cannot be demonstrated, the surface is commonly referred to as a disconformity. Clearly, a disconformity in one set of outcrops may be demonstrated to be an angular unconformity elsewhere. The terms are not genetic, simply descriptive (see Figure 3.19).

Recognition of off-lap and on-lap relationships associated with major unconformities is often a primary key to unraveling stratigraphic complexities.

Major regression and transgressions that produce the intervening unconformity surface are usually gradual and pulsating events. The last records of marine dep-

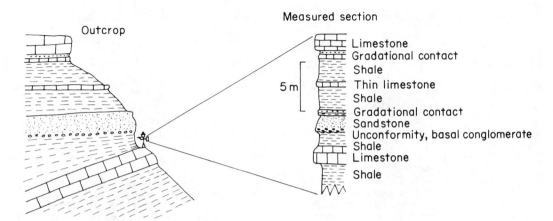

Figure 3.18 Translation of outcrop observations into a description of a measured section. The geologist is measuring the section with a hand level. Holding the level to his eye, he locates the next convenient spot to reach in his traverse up the hill. Then he describes the strike, dip, and lithologic properties of the rocks that lie between his feet and the next spot up the hill. By the time he climbs to the top of the hill, he has the thickness and lithologic description of the sequence. In this schematic representation, the unconformity is easy to recognize. There is an angular relationship between the beds above and below the unconformity, and there is a pronounced basal conglomerate immediately above the unconformity.

osition below an unconformity, for example, may be much older on the higher portions of the land area and much younger on its seaward flanks (see Figure 3.20). Conversely, the first appearance of marine sediments over the unconformity surface may be much younger over the former high areas than in the basin sediments flanking the high areas.

Thus, it is a rational and widely accepted stratigraphic practice to solve difficult

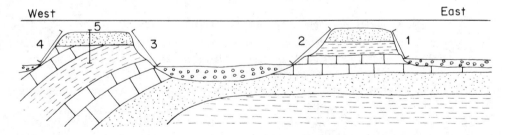

Figure 3.19 Schematic representation concerning the recognition of a major unconformity that is not readily apparent in individual measured sections. When the geologist begins to work this area, the unconformity between the upper sandstone and the shale below is not obvious in measured sections (1), (2), and (3). As the geologist measures section (4), he notes the occurrence of limestone where previous sections would have led him to anticipate shale. Measured section (5) confirms the existence of a major unconformity at the base of the upper sandstone.

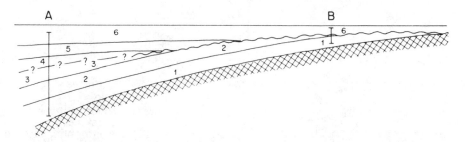

Figure 3.20 Off-lap and on-lap relationships associated with a major unconformity. Six biostratigraphic zones are well defined in apparently continuous sedimentation in measured section A. Only zones (1) and (6) occur in measured section B, separated by an easily recognizable unconformity surface. The apparent continuity of sedimentation within section A suggests that we are dealing with off-lap and on-lap relations in the area between sections A and B. As we move from section A toward section B, more and more of the time represented by sedimentation in section A will be represented by missing section.

correlation problems by pushing units up or down into major regional unconformity surfaces.

Gradational Contacts.   Just as the recognition of an unconformity surface demands that we think of everything above as separate from everything below, the recognition of gradational contacts between contrasting lithologies demands that we think of these units as genetically related.

Gradational contacts can take several forms. Sometimes a shale unit, for example, becomes more sandy toward the top as it gradationally passes upward into sandstone. In other situations, the shale unit may begin to show an increasing number of small sandstone beds alternating with otherwise normal shale. Both situations herald the nearby proximity of sandstone deposition. In both cases, the continuing passage of time brings sandstone deposition over a spot that was previously dominated by shale deposition.

Bedding Planes.   Unconformity surfaces clearly indicate that a portion of geologic time is not represented. On the other hand, gradational contacts demonstrate that sedimentation is more or less continuous throughout the transition from one depositional stage to another. Between these two conditions, there is undoubtedly a vast no-man's-land: situations in which we are not really sure whether we are dealing with continuous deposition or not.

In this connection, bedding planes pose a particularly intriguing problem. Many bedding planes can be attributed to events that make good sedimentological sense. Storm deposits, progradation, channel migration, and ashfalls, for example, produce beds that can be easily understood in terms of observations concerning Recent sedimentation. Contrasting lithology from one bed to the next tells of deposition under differing conditions, similar to environmental fluctuations that we observe today.

On the other hand, some bedding planes separate sediments that appear to be virtually identical. Such bedding planes are particularly troublesome in subtidal normal marine environments. When we observe similar environments today, we find that bioturbation by the marine benthos thoroughly homogenizes the sediment, so that all of the Recent accumulation forms essentially a single bed. Therefore, how can presumably similar sediments in the stratigraphic record exhibit bedding on the scale of centimeters to tens of centimeters?

One possibility is that each bed records a period of deposition and that each bedding plane records a period of nondeposition during which the underlying bed became sufficiently lithified so as to resist the activities of burrowing organisms during the next depositional phase. This possibility leads us to consider a fundamental philosophical point concerning the continuity or discontinuity of the stratigraphic record.

Traditionally, stratigraphers have presumed that measured sections record continuous deposition unless an unconformity can be demonstrated. The fact that some bedding planes were extremely difficult to explain was just one of the messy problems that they learned to ignore. However, a geologist favoring strict interpretation of the evidence available from Recent sedimentation may contend that the very existence of these bedding planes is in itself good evidence of the lack of continuity of sedimentation. Indeed, if we take Recent sediments to represent the formation of a bed, this bed represents 5000 years of geologic time, whereas the bedding plane separating the Recent from Pleistocene deposits commonly represents the passage of more than 100,000 years of geologic time. (See Bloom, 1972, for example.)

For the present, let us simply recognize that these two points of view exist. One stratigrapher may express the opinion that thousands of meters of shallow-marine sediments were deposited more or less continuously and without intermittent subaerial exposure. Another stratigrapher may look at the same sequence of rocks and propose subaerial exposure surfaces every few meters. The latter will cite data from the study of Recent sediments, and the former will say that we cannot make that strict an interpretation of Recent sediment data. The accommodation of these two points of view is the subject of Parts IV and V of this book.

Electrical Properties of Sedimentary Rocks

Much effort goes into the study of sedimentary rocks that are below the surface of the earth. This is the region where the petroleum geologist seeks oil and where the engineering geologist tries to determine whether or not a proposed dam will actually hold water. Such studies are accomplished by drilling holes into the rocks. Although it is possible to core the rock and bring a piece back up to the surface for study, it is far more economical to drill the hole with a bit that pulverizes the rock as it drills. This pulverization leaves us, however, with very little to study firsthand.

We therefore rely heavily upon measurement of the electrical properties of the sedimentary rocks after the hole has been drilled. Electrodes are lowered down the borehole on a wire line. As the electrodes are raised slowly up through the hole, a continuous record is made of the electrical properties of the various lithologies encountered.

Resistivity and *spontaneous potential* are the two electrical properties most commonly measured. Taken individually, each set of data is inconclusive; but taken together, these two measurements provide a good indication of some important lithologic distinctions. Shale or porous rock filled with salt water will be a good conductor. Such intervals in the borehole will have low resistivity. On the other hand, porous rocks that are filled with petroleum will have high resistivity, as will impermeable rocks that are tightly cemented with quartz or calcite.

With proper salinity relationships between drilling mud and formation water, spontaneous potential provides a direct indication of the permeability of the rock. If the rock is permeable, a spontaneous potential will be generated; if the rock is impermeable, there will be little or no spontaneous potential. Figure 3.21 illustrates the various possible combinations of resistivity and spontaneous potential data that allow distinction among important lithologic attributes of the sediments encountered in a borehole.

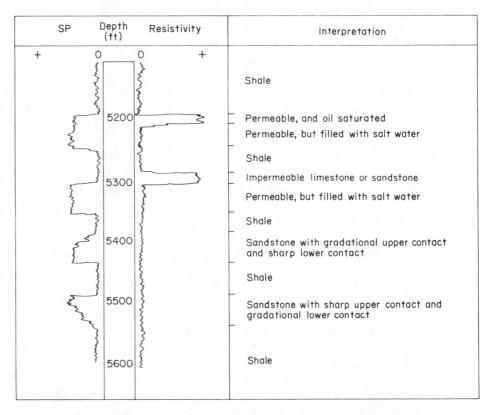

Figure 3.21 Lithologic properties suggested by various combinations of spontaneous potential and resistivity in petroleum exploration boreholes.

Note further that the contacts between sandstones and shales can be investigated in considerable detail from electric logs. The abruptness of upper and lower contacts of sands may allow us to distinguish marginal-marine sands as opposed to alluvial or turbidite sands. Given the availability of an occasional core to check the interpretation, we may be able to map these various sand types over large areas with considerable confidence.

Some Basic Graphic Constructs of Stratigraphy

Geologists use maps and cross sections to organize their data, to solve problems, and to communicate ideas to other geologists. Production of useful maps and cross sections commonly involved reduction of verbal thoughts to quantitative distinctions that can be put on a map. When a pattern begins to emerge among these numbers on the maps and cross sections, an understanding of regional sedimentation and stratigraphy emerges with it. A few of the graphic constructs in common usage among stratigraphers are discussed briefly below.

Stratigraphic Cross Sections. Whereas the structural geologist commonly constructs a cross section as a literal slice through the earth, the stratigrapher often finds it advantageous to remove the structural complications from consideration and construct a stratigraphic cross section. Two types of stratigraphic cross sections are in common usage. Horizontal distance serves as the X-axis in both constructs. In a *lithostratigraphic cross section,* actual thickness of the sediments in the various sections under consideration serves as the Y-axis. Lithologic sections are hung from an easily recognizable stratigraphic level (the *datum* of the cross section), and variations within each section are plotted as distance above or below that datum. Common datums for hanging lithostratigraphic sections would include unconformities, ash beds, or other prominent lithologic or biologic markers. A properly constructed lithostratigraphic cross section can often be viewed as a paleogeologic reconstruction depicting geology soon after deposition but before structural deformation.

A *time-stratigraphic cross section* uses time or time-stratigraphic units as the Y-axis. All well-defined time lines among the various sections under consideration will appear as horizontal lines. Other geologic information is placed on the cross section by linear interpolation of sediment thickness between time lines. Such a data presentation is especially appropriate if there is large topographic relief within the depositional environment. Further, this presentation is extremely compatible with development of paleogeographic maps at various time slices.

Facies Maps and Cross Sections. The concept of facies (discussed above) can be viewed as the first and least quantitative step toward quantification of the geologist's verbal description of the rocks. Instead of trying to comment on each and every small layer of rock, the geologist groups several rock types under the concept of "the sedimentary record of a depositional environment." He or she gives the environment a convenient name, like lagoonal, bank-margin complex, or outer shelf, and begins to prepare maps and cross sections depicting facies relationships. *Paleogeographic maps* go beyond the concept of facies to depict the actual environment in geographic view for some time in the past. Whereas a facies map or cross section would leave the discussion in terms of the sedimentary record of environment, the paleogeographic map and cross section would attempt to portray the environment as it was at the time of deposition.

Isopachous Contour Maps. The next step in the quantification of geological description is commonly to produce numbers that can then become the basis for isopachous contour maps. "Isopachous" simply means equal thickness. A straightforward application of the concept would be to map the thickness of a lithologic unit. Alternatively, an isopachous map can be prepared with regard to some property of a stratigraphic unit. For example, a net sand map might be constructed from electric log data by adding up all portions of a section that have a spontaneous potential above a certain value. By putting these values on a map and contouring them, it may be possible to delineate the trend of former shoreline features, for example. Note further that combinations of various isopachous maps can provide important clues to geologic interpretation. For example, it would be important to know where the largest net sand accumulation occurs within the context of the total thickness of a biostratigraphic unit.

Seismic Stratigraphy. The oil exploration business has found it useful to produce several stratigraphic constructs from seismic data. Figure 3.22 presents an illustrative collection of seismic-stratigraphy constructs. The upper diagram is a lithostratigraphic cross section prepared from the seismic cross section and available well data. Prominent seismic reflectors are taken to represent stratigraphic horizons. Angular relationships among reflectors record unconformities, off-lap, on-lap, and other stratigraphic features.

The middle diagram in Figure 3.22 recasts the information in a time-stratigraphic cross section. Here, many reflectors of the lithostratigraphic cross section are taken to be time lines. Unconformities are indicated as blank area on the time-stratigraphic cross section; there are no rocks representing that time at that location.

The final seismic-stratigraphy construct is the relative sea-level curve. This curve attempts to portray the relation between the land and the sea through time. In some cases, it is convenient to scale the curve from zero to one on the basis of relative coastal on-lap, with Cretaceous taken as maximum on-lap (1) and mid-Oligocene as maximum off-lap (0). In other cases, the curve can be scaled by measuring coastal aggradation. In either case, the curve should be regarded as a qualitative indication of the net interaction among sedimentation, basic subsidence history, and eustatic sea-level history. With accumulation of global experience with the relative sea-level curve, it may be possible to separate out the global eustatic sea-level component from the varying local components, a point to which we shall return in Chapter 17.

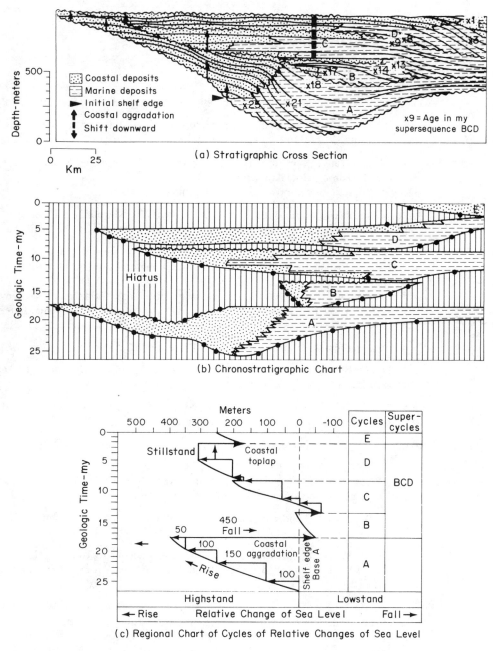

Figure 3.22 Procedure for constructing regional chart of cycles of relative changes in sea level from seismic data. (From Vail *et al.*, 1977.)

Citations and Selected References

ADAMS, JOHN. 1979. Wear of unsound pebbles in river headwaters. Science 203: 171–172.

ALLEN, J. R. L. 1963. The classification of cross-stratified units, with notes on their origin. Sedimentology 2: 93–114.

————. 1968. Current ripples: their relation to patterns of water and sediment motion. North-Holland Publishing Co., Amsterdam. 433 p.
Detailed analysis of the how and why of primary structures.

BERNER, ROBERT A., and GEORGE R. HOLDREN, JR. 1979. Mechanism of feldspar weathering—II: Observations of feldspars from soils. Geochimica et Cosmochimica Acta 43: 1173–1186.

BLATT, H. 1967. Provenance determination and recycling of sediments. J. Sed. Petrology 37: 1031–1044.
An eloquent plea for observational evidence rather than "plausible reasoning" in the discussion of provenance and transportational history of clastic sediments. Extensive bibliography.

————. 1982. Sedimentary Petrology. W. H. Freeman & Co., San Francisco, 564 p.

BLATT, H., G. MIDDLETON, AND R. C. MURRAY. 1980. Origin of sedimentary rocks. 2nd ed. Prentice-Hall, Inc., Englewood Cliffs, N.J. 782 p.

BLOOM, A. L. 1972. Geomorphology of reef complexes. In L. F. Laporte (ed.), Reef complexes in time and space. Soc. Econ. Paleont. Mineral. Spec. Pub. 19.

CAROZZI, A. V. 1960. Microscopic sedimentary petrology. John Wiley & Sons, New York. 485 p.

CUMMINS, W. A. 1962. The greywacke problem. Liverpool and Manchester Geol. J. 3: 51–72. Possible diagenetic origin of clay matrix in these sands.

DICKINSON, WILLIAM R., and CHRISTOPHER A. SUCZEK. 1980. Plate tectonics and sandstone compositions. Bull. Amer. Assoc. Petrol. Geol. 63 (12): 2164–2182.

DUNHAM, R. J. 1962. Classification of carbonate rocks according to depositional texture, p. 108–121. In W. E. Hamm (ed.), Classification of carbonate rocks. Amer. Assoc. Petrol. Geol. Mem. 1.

FOLK, R. L. 1962. Spectral subdivision of limestone types, p. 62–84. In W. E. Hamm (ed.), Classification of carbonate rocks. Amer. Assoc. Petrol. Geol. Mem. 1.
A widely used shorthand system for naming carbonate rocks.

————. 1966. A review of grain size parameters. Sedimentology 6: 73–93.
————. 1980. Petrology of sedimentary rocks. Hemphill Publishing Co., Austin, Tex. 182 p.
A practical, down-to-earth discussion on how to study sedimentary rocks.

FRIEDMAN, G. M. 1967. Dynamic processes and statistical parameters compared for size frequency distribution of beach and river sands. J. Sed. Petrology 37: 327–354.
Application of statistical parameters to a geologic problem, the distinction between beach and river sands collected in bulk, loose, sample.

GARRELS, R. M., and F. T. MACKENZIE. 1971. Evolution of sedimentary rocks. W. W. Norton & Co., New York. 397 p.

GILBERT, C. M. 1954. Sedimentary rocks, p. 251–384. *In* H. Williams, F. J. Turner, and C. M. Gilbert, Petrography. W. H. Freeman and Co., San Francisco. 406 p.

HAMM, W. E. (ed.). 1962. Classification of carbonate rocks. Amer. Assoc. Petrol. Geol. Mem. 1. 279 p.
Symposium volume presenting various approaches to the classification of carbonate rocks.

HJULSTROM, F. 1939. Transportation of detritus by moving water, p. 5–31. *In* P. D. Trask (ed.), Recent marine sediments. Amer. Assoc. Petrol. Geol.

IMBRIE, J. 1964. Factor analytic method in paleoecology, p. 407–422. *In* J. Imbrie and N. D. Newell (eds.), Approaches to paleoecology. John Wiley & Sons, New York.

IMBRIE, J., and H. BUCHANAN. 1965. Sedimentary structures in modern carbonate sands of the Bahamas, p. 149–172. *In* G. V. Middleton (ed.), Primary sedimentary structures and their hydrodynamic interpretation. Soc. Econ. Paleont. Mineral. Spec. Pub. 12.

IMBRIE, J., and E. G. PURDY. 1962. Modern Bahamian carbonate sediments, p. 253–272. *In* W. E. Hamm (ed.), Classification of modern Bahamian carbonate sediments. Amer. Assoc. Petrol. Geol. Mem. 1.

JOPLING, A. V. 1966. Some principles and techniques used in reconstructing the hydraulic parameters of a paleo-flow regime. J. Sed. Petrology 36: 5–49.

KLEIN, G. DEV. 1963. Analysis and review of sandstone classification in the North American geological literature, 1940–1960. Bull. Geol. Soc. Amer. 74: 555–576.

KRUMBEIN, W. C., and L. L. SLOSS. 1963. Stratigraphy and sedimentation. W. H. Freeman and Co., San Francisco. 660 p.

KRYNINE, P. D. 1948. The megascopic study and field classification of sedimentary rocks. J. Geology 56: 130–165.

KUKAL, Z. 1971. Geology of Recent sediments. Academic Press, New York. 490 p.

LAHEE, F. H. 1952. Field geology. McGraw-Hill, New York. 883 p.
Standard reference concerning field methods.

McBRIDE, E. F. 1963. A classification of common sandstones. J. Sed. Petrology 33: 664–669.

McKEE, E. D., and G. W. WEIR. 1953. Terminology for stratification and cross-stratification in sedimentary rocks. Bull. Geol. Soc. Amer. 64: 381–390.

MIDDLETON, G. V. (ed.). 1965. Primary sedimentary structures and their hydro-dynamic interpretation. Soc. Econ. Paleont. Mineral. Spec. Pub. 12. 265 p.
Symposium volume.

PETTIJOHN, F. J. 1954. Classification of sandstones. J. Geology 62: 360–365.

————. 1957. Sedimentary rocks. Harper & Row, New York. 718 p.

PETTIJOHN, F. J., and P. POTTER. 1964. Atlas and glossary of primary sedimentary structures. Springer-Verlag, New York. 370 p.

PINGITORE, N. E., JR., and J. D. SHOTWELL. 1977. Sandstone classification: a view from factor space. N. Jahrbuch f. Geologie u. Paläontologie. Monatshefte 12: 177–188.

PITTMAN, EDWARD D. 1979. Recent advances in sandstone diagenesis. Ann. Rev. Earth Planet. Sci. 7: 39–62.

POTTER, P., and F. J. PETTIJOHN. 1963. Paleocurrents and basin analysis. Springer-Verlag, Berlin. 296 p.

RHOADS, D. C. 1967. Biogenic reworking of intertidal and subtidal sediments in Barnstable Harbor and Buzzards Bay, Massachusetts. J. Geology 75: 461–476.

SCHOLLE, PETER A. 1978. A color illustrated guide to carbonate rock constituents, textures, cements, and porosities. Amer. Assoc. Petrol. Geol. Mem. 27. 248 p.

————. 1979. A color illustrated guide to constituents, textures, cements, and porosities of sandstones and associated rocks. Amer. Assoc. Petrol. Geol. Mem. 28. 201 p.

SHELTON, J. W. 1967. Stratigraphic models and general criteria for recognition of alluvial, barrier-bar, and turbitic-current sand deposits. Bull. Amer. Assoc. Petrol. Geol. 51: 2441–2461.

SIMONS, D. B., E. V. RICHARDSON, and M. L. ALBERTSON. 1961. Flume studies using medium sand (0.45mm). U.S. Geol. Survey Water Supply Paper 1498-A. 76 p.

VAIL, P. R., R. M. MITCHUM, JR., and S. THOMPSON III. 1977. Seismic stratigraphy and global changes of sea level, part 3: relative changes of sea level from coastal onlap, p. 63–81. *In* C. E. Payton (ed.), Seismic stratigraphy—applications to hydrocarbon exploration. Amer. Assoc. Petrol. Geol. Mem. 26.

VISHER, G. S. 1965. Fluvial processes as interpreted from Ancient and Recent fluvial deposits, p. 116–132. *In* G. V. Middleton (ed.), Primary structures and their hydrodynamic interpretation. Soc. Econ. Paleont. Mineral. Spec. Pub. 12.

WALKER, R. G. 1967. Turbidite sedimentary structures and their relationship to proximal and distal depositional environments. J. Sed. Petrology 37: 25–43.

WELLER, J. M. 1960. Stratigraphic principles and practice. Harper & Row, New York. 725 p.

Physical Stratigraphy

4

In this and subsequent chapters, we shall consider the principles and tools that allow us to construct a more detailed understanding of the earth history as recorded in the stratigraphic record. The first step in this direction is to understand the physical relationships among sedimentary rocks within relatively small areas.

The Basic Rationale
of Physical Stratigraphy

Before we discuss the geologic history of an area and the relation of that history to the history of other portions of the earth, we must catalog the contents of that particular area. What rock types are present and what are their relationships to the other rock types in the area?

To answer these questions, we shall be concerned primarily with the physical characteristics of rock units. Time in stratigraphy will be of only secondary concern to us at this level of investigation. Certainly we shall realize that rocks on the bottom of the pile are older than rocks on the top. Similarly, we may recognize times of marine deposition followed by times of continental deposition. The point is that, at this level of investigation, we are not concerned with how these time relationships fit other events on a regional or global scale. For the moment, we want only to make good sense out of the physical events that occurred within a relatively small area. Hopefully, we can quantify the time dimension in subsequent studies.

Rock Units: Group, Formation, and Member. The basic mapping unit of physical stratigraphy is the *formation*. The underlying concept for the definition of a formation is convenience. A formation need only be a mappable rock unit. The word "mappable" clearly suggests that a formation should be defined on the basis of characteristics that are discernible under normal working conditions to the geologist making the map. Gross lithology, color, and, possibly, the general fossil content are common parameters upon which to base the definition of a formation. Alternatively, several lithologies may be grouped as a single formation, provided only that the upper and lower contacts of the formation are conveniently recognized under existing mapping conditions.

The legalisms of stratigraphy require the designation of a type section for each formation. Any stratigraphic section that can be shown to have had physical lithologic continuity with the type section during deposition may be properly assigned to the formation name of the type section.

As mapping progresses, it usually becomes apparent that several formations, existing side by side or stacked one upon the other, may be viewed as a single unit for general purposes. Where one formation thickens, they all thicken; where one formation is absent, they are all absent; and so on. Thus, it may be convenient to have a single designation for all of these formations collectively. Such a collection of formations is referred to as a *group*.

On the other hand, it may become important to keep track of some smaller subdivision of a formation. Once again, the definition of the smaller subdivision is commonly based upon some lithologic characteristic easily recognizable under mapping conditions. Smaller subdivisions of a formation are referred to as *members*.

Thus, the basic mapping units of physical stratigraphy—group, formation, and member—are arbitrary units of convenience. The use of these words will vary widely with the area under consideration and with the field conditions under which the area was first mapped.

In a practical sense, this discussion is more a matter of history than it is a matter of present-day working geology. Few modern stratigraphers will have the opportunity to name new formations. The formations have been defined and named already. We may not always like the definition, but we usually learn to live with it. Often it is convenient to get around original poor definitions of formation boundaries by naming new members within the formation. We shall see such an example in Chapter 12 concerning the Oswego sandstone.

Basement, Unconformities, and Contacts. What constitutes a rational basis for the definition of group, formation, and member? Let us start with the really big items and work our way down. Every area of sediment accumulation has a *basement* of metamorphic or igneous rocks. Clearly, the *contact* between basement and sediment constitutes a starting point for the definition of group and formation. Many stratigraphic sequences contain angular *unconformities*. Sedimentary sequences were deposited over basement; tectonic deformation and erosion occurred; then additional sedimentary sequences were deposited over the erosional surface. Obviously, we should study these naturally occurring discontinuities. We attach group or formation significance to materials above and below the major unconformities.

Major unconformities do not always show angular relationships at every outcrop. A major unconformity may at first be passed over as just another bedding plane (see Figures 3.19 and 3.20, for example). If angular relationships do not exist anywhere within the map area, major unconformities may go unnoticed by the local physical stratigrapher. In such cases, the biostratigrapher may later inform him (or her) that large biostratigraphic time intervals are missing from his single "formation." On the other hand, if angular relationships do exist somewhere within the map area, the physical stratigrapher may recognize his own error early in the mapping process. At all times, the physical stratigrapher must be on the lookout for subtle evidence of a major discontinuity parallel to bedding planes. Conglomerates, regoliths, and solution breccias are but a few examples of the features to watch for.

The Utility of Physical Stratigraphy

We have already emphasized physical stratigraphy as a basic necessity for understanding the earth history of any particular small area. In the next chapter, we shall go on to see how biostratigraphy places this small area into a regional or global time-stratigraphic framework. However, even at the relatively simple level of physical stratigraphy, examination and cataloging of stratigraphic sequences provide some exciting and useful data.

Recognition of General Paleogeography. Glancing ahead for a moment, consider the basic paleogeographic information contained in gross lithostratigraphic data portrayed in Figures 10.14, 10.15, and 10.16. For the moment, disregard the time framework within which these data are cast. Viewed as physical stratigraphy in the absence of any time framework, the major elements of paleography remain intact. Figure 10.14 continues to portray thick clastic sections immediately westward from the present-day crystalline Appalachians. Generalized stratigraphic sequences from Alabama to Pennsylvania (Figure 10.15) still indicate that carbonate sedimentation in the lower portions of the section gives way to clastic input from the east. Recognition of the basic paleogeography throughout the entire length of the Appalachian mountain chain is not dependent upon knowledge of detailed time relationships. The fact that the clastic wedge in Alabama is of Pennsylvania age whereas the clastic wedge of the Catskill Mountains in New York is of Upper Devonian age is an interesting secondary refinement upon our understanding of Appalachian paleogeography. However, the fundamental recognition of the paleogeography need rest only upon physical stratigraphy.

Exploitation and Utilization of Local Areas. Exploration for petroleum best illustrates the utility of approaching sedimentary sequences by means of physical stratigraphy. When the geologist is looking for oil, his view of geology often becomes very localized. Specifically, he wonders whether the hole he is drilling in the ground is going to encounter an oil reservoir. Figure 4.1 schematically summarizes how physical stratigraphy might lead to an oil discovery. Petroleum often accumulates in permeable rocks at the crest anticlines. Consequently, anticlines get drilled early in the exploration history of any oil province. Figure 4.1 shows that

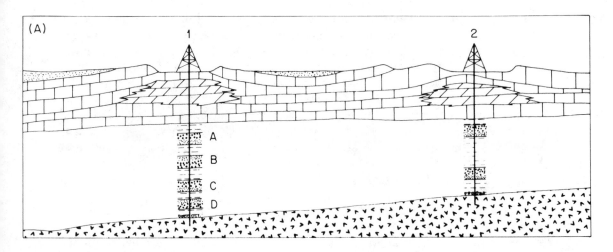

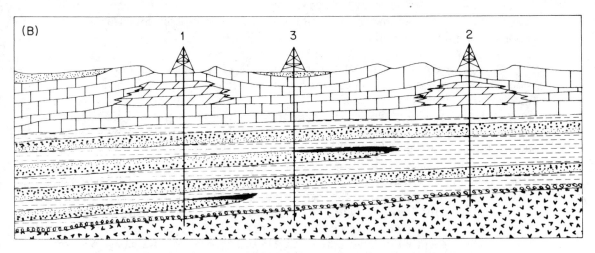

Figure 4.1 Application of physical stratigraphy to a local petroleum exploration problem. As indicated in A, dry holes (1) and (2) were drilled for the wrong reason, but they provide subsurface information from which the physical stratigrapher can begin to make more refined interpretations concerning local geology. As indicated in B, the physical stratigraphy of boreholes (1) and (2) may logically lead to the location of borehole (3). See text for further discussion.

dry holes (1) and (2) were drilled on surface anticlines. It turned out that these anticlines were not structural features but rather were sedimentary drape over reefs located very near to the surface. Both wells were dry, but they provided some interesting physical stratigraphic information. Hole (1) even had an oil show in sandstone (D), although it was not enough to make an oil well.

According to Figure 4.1, both holes begin in limestone and pass abruptly into shale at about the same depth. An unconformity? Perhaps. At least we should call it a formation boundary. Both wells then continue through a marine sandstone-shale sequence until basement is encountered, somewhat deeper in hole (1) than in hole (2). It is a well-established generality that marine clastic sequences tend to thicken basinward. Thus, we may begin to suspect that the transgressions and regressions responsible for the alternating sands and shales came from a more persistent basin to the west. Note further that borehole (1) encounters four sandstone units whereas borehole (2) encounters only two. Closer examination of the lithologic properties of the sandstone units in hole (1) indicate that sandstones (A) and (C) are finer-grained at the bottom and become coarser upward to the point where they are overlain by sharp contact with shale. In contrast, sands (B) and (D) tend to become fine upward and pass gradationally into the overlying shale. At this point, we probably begin to get excited. Sands (A) and (C) display classical marginal-marine, regressive features (coarsening-upward), whereas (B) and (D) appear transgressive. Could it be that rapid sea-level rise drowned the shorelines represented by sands (B) and (D) and left them with porosity pinchouts somewhere between dry holes (1) and (2)? A quick glance at samples from borehole (2) confirms the correlation of these sands with (A) and (C) of borehole (1), and a new wildcat well goes down to seek petroleum accumulations at the pinchout of sands (B) and (D). With any luck, hole (3) will be an oil well. Moreover, stratigraphic information gained from it will also suggest more drilling prospects.

Even though this discussion unravels a small piece of local geology, no one really cares how this geology is related to any other geology, either regionally or in a time-stratigraphic framework. Many of the day-to-day details in petroleum exploration are carried out in precisely this manner.

Similarly, the physical stratigraphy around metropolitan areas may be of major local importance. City planners need not be concerned with how local geology fits into the global time-stratigraphic framework, but they must be interested in whether or not effluent from the dump and the sewage disposal plant gets mixed up with the city water supply. Figure 4.2 portrays a rather simple comedy of errors in this

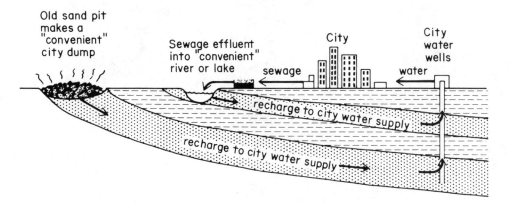

Figure 4.2 A "comedy of errors" concerning physical stratigraphy and city planning.

regard. In this hypothetical example, poor planning has resulted in town dump and sewage treatment plant being situated over the area that recharges the city's aquifer. This example may be exaggerated; yet, the general point is still valid.

Citations and Selected References

AMERICAN COMMISSION ON STRATIGRAPHIC NOMENCLATURE. 1961. Code of stratigraphic nomenclature. Bull. Amer. Assoc. Petrol. Geol. 45: 645–660.
Updated continuously in the AAPG Bulletin.

EICHER, D. L. 1968. Geologic time. Prentice-Hall, Inc., Englewood Cliffs, N.J. 150 p.
Good supplementary reading for Chapters 4 through 7.

KRUMBEIN, W. C., and L. L. SLOSS. 1963. Stratigraphy and sedimentation. W. H. Freeman
and Co., San Francisco. 660 p.
Advanced text.

LAHEE, F. H. 1952. Field geology. McGraw-Hill, New York. 883 p.

WELLER, J. M. 1960. Stratigraphic principles and practice. Harper and Bros., New York.
725 p.

Biostratigraphy:
Introduction to Temporal Correlation

5

In the previous chapter, we noted that rock units are the basic starting point for understanding the sedimentary geology of a small area. As we attempt to expand our discussion to larger and larger areas, we run into two major problems. First, the areal extent of demonstrable physical correlation of rock units is often limited by outcrop patterns and availability of subsurface data. Perhaps the formation under consideration is well exposed in a mountain range, but the next mountain range is 300 kilometers away. Can we really trust lithologic correlation across a 300-kilometer gap? Or perhaps this particular formation has been removed from adjacent areas by erosion. It was there once; we could have walked it out for thousands of kilometers. But now it is gone, and so physical correlation is impossible.

Even if we can establish lithologic correlation over large areas, we run into a second problem. Lithologic units are commonly diachronous on this large a scale. For example, if a sandstone originated as beach deposits prograding westward across a shallow-marine environment, the eastern portion of the lithologic unit will be older than the western portion. In a small mapping area, the fact that beach sand is found to overlie subtidal marine shale portrays a fairly vivid picture of local earth history. Yet, as we view this situation in larger areal extent, we recognize that these lithologic relationships must surely transgress time. Consequently, our view of earth history begins to go out of focus. If we wish to bring large-scale earth history into proper focus, we must pay more serious attention to the dimension of time. It is through biostratigraphy that the geologist traditionally controls this dimension.

Rock Units, Time-Rock Units, and Time Units

Rock units carry with them absolutely no connotation of time. The formation exists as a three-dimensional body of rock. When it was deposited and how it was deposited simply does not enter into the definition of a rock unit. In similar fashion, we accept *a priori* that time existed in the past irrespective of whether or not rocks record the passage of that particular interval of time. For example, we have little doubt that there was "a time"—8 to 10 A.M., March 22, 127,442,361 B.C.—yet, obviously we shall never retrieve the historical record of precisely what happened within that two-hour "time unit." This problem constantly faces us as we attempt to work a time dimension into our understanding of the earth history recorded in sedimentary sequences.

We get around this problem by defining an intermediate time-related statement. Instead of attempting to define absolute time units, stratigraphers have defined a large interwoven system of relative time units called *time-rock units*. A sequence of sediments (the *type section*) is designated to represent a certain unspecified period of time in earth history. Where other sediments can be demonstrated, usually on the basis of fossil content, to have formed at that same time, they are assigned to the same time-rock unit. Ultimately, we seek to relate time-rock units to the absolute time that they represent. Radiometric dating offers a significant potential for the eventual accomplishment of this task. In the meantime, stratigraphers can still understand earth history on a global scale within the framework of time-rock units. Figure 5.1 illustrates the various levels of sophistication concerning our treatment of time in stratigraphy.

No time significance	*Relative time significance*		*Increasingly sophisticated absolute time significance* →
Rock units	Time-rock units	1. Time units	
		Era	2. Radiometric dating of time unit boundaries
Group Formation Member Bed	System – – – –	Period	3. Interpolation between radiometric date points in continuous sedimentation sequences.
	Series – – – –	Epoch	4. Ultimately, "there was a storm the morning of August 23, 467,931 B.C.", etc.; but we cannot hope to really achieve quite that level of sophistication.
	Stage – – – –	Age	

Figure 5.1 Diagram indicating increasing levels of sophistication concerning time in the stratigraphic record. Rock units have no time significance whatsoever. Time-rock units convey relative relationships such as "older than" or "younger than," but they carry with them the implicit recognition that the stratigraphic record of time may have large gaps in it. Time units are the next conceptual step above time-rock units; they designate a continuum of time between two events in the history of the earth. Radiometric dating allows absolute determination of the age of these events within a margin of error. Interpolation between events that have well-known radiometric ages allows still more detailed discussion of absolute time in the stratigraphic record of earth history.

Fossils and Time-Rock Units

Fossil content of the sediments provides the basis for large-scale time-rock correlation. Life on earth has constantly undergone change. By recognizing the changes and putting them in their temporal sequence, we can correlate time and stratigraphy.

If a life-form is to serve as a good biostratigraphic indicator of relative time, it should have four attributes. First, it must produce preservable hard parts; there must be something left for us to find in the stratigraphic record. Second, these hard parts should make up a significant portion of the sediment. The more remains there are, the more easily they are found and used by the biostratigrapher. Third, the organism must have had a life style or life cycle that allowed rapid dispersal around the world or at least over very large areas. Pelagic organisms are useful, but many benthonic organisms have a pelagic larval stage and are therefore just as useful. Finally, this life-form will hopefully have had a rather abrupt beginning and a rather abrupt demise that were not too widely separated in time.

The basic bookkeeping unit of biostratigraphy is the range of the species. What are the oldest strata in which the species first appears, and what are the youngest strata in which the species is last found? Beyond this simple level, zones of maximum abundance, zones containing assemblages of various species, and so on, all have time-stratigraphic significance under various circumstances.

Because of the preservability of fossil remains and because of the distribution of major rock types, shallow-marine invertebrate faunas are by far the most widely used biostratigraphic tools. Although the following discussions will deal solely with shallow-marine invertebrate faunas, the principles are equally applicable to pelagic faunas, terrestrial faunas, and so on.

A Biostratigraphic Overview of Evolution

Let us now investigate why fossils should work as indicators of equivalent time. We shall develop the topic of evolution in a very broad fashion and shall point out generalities that will be applicable both at the species level and at the ecosystem level. To us, *evolution* will simply mean change in the character of life on earth as a function of time. Sometimes this change has occurred within a single phylogenetic lineage. For example, an interbreeding population of molluscs may undergo mutation in such a way that individuals become morphologically distinct from their predecessors. In these situations, our use of the word "evolution" is close to the biologist's use of the word.

But we shall be equally interested in the evolution of community structure within specific environments, regardless of the phyla that may be involved. Consider molluscs, as an example. Today they dominate many of the niches that were dominated by brachiopods in the Paleozoic. "Successful" brachiopods did not evolve into "successful" molluscs. The brachiopods instead lost a long-term competitive battle for these niches to molluscs, which independently evolved a more successful life style.

Diversity of Life. Two aspects of physiological diversity are important for our studies: (1) diversity among the phyla and (2) diversity within single species. As indicated in Figure 5.2, highly organized invertebrate phyla have existed on the earth for at least the last 550 million years. Because the geologic record of life on earth is heavily biased in favor of those organisms that generate preservable hard parts, we must recognize that complicated communities of highly developed organisms existed even prior to many of the "first appearances" indicated in Figure 5.2.

The existence of a complicated community structure including numerous phyla carries with it a certain guarantee for the continued existence of life within the environment occupied by the community. It is highly unlikely that changing conditions on the face of the earth would prove simultaneously unfavorable to all species within the community. If any one species were to undergo decline in response to unfavorable changes in the environment, it is likely that another species would undergo expansion as it exploited at least a portion of the niche vacated by the declining species.

The second aspect of diversity that interests us is the physiologic and morphologic variability within a single species. A *species* is generally defined as a collection of organisms that interbreed in nature to produce fertile offspring more or less resembling themselves. There commonly exists considerable diversity within a single species. Such diversity can usually be expressed in terms of a normal distribution curve, such as Figure 5.3. No matter what characteristic we choose to measure (shape, heat tolerance, etc.), the majority of the population falls within a certain range of variation. Yet, to either side of this dominant range, there are individuals that may still be an interbreeding part of the population. The gene pool of the total population, therefore, includes characteristics lying beyond the norm of the species. If changing conditions of the environment render these abnormal characteristics valuable, then the (+) tail of the normal distribution curve will continue to reproduce, whereas individuals on the (−) tail will tend to die without reproducing. Thus, the shape and position of the normal distribution curve with respect to any characteristic may shift with time as various individuals contribute input to the continuity of a gene pool. We shall discuss this point in greater detail later under selection pressure.

Ecological Limits of Species and Communities. Given steady conditions, communities will approach a dynamic equilibrium with their environment. Availability of nutrients will limit the size of the biomass. Predator-prey relationships and physical factors, such as temperature, salinity, and substrate, will determine the relative abundance of the various species within the community. Both total biomass and abundance of individual species will tend to expand to the limit of the average conditions of the environment. As environmental conditions change, the dynamic equilibrium may be disrupted, and species and communities may undergo modifications that will be interesting as biostratigraphic markers. The following discussion particularly concerns shallow-marine benthonic communities and the changes that they may undergo in response to variations in water depth, living space, and temperature.

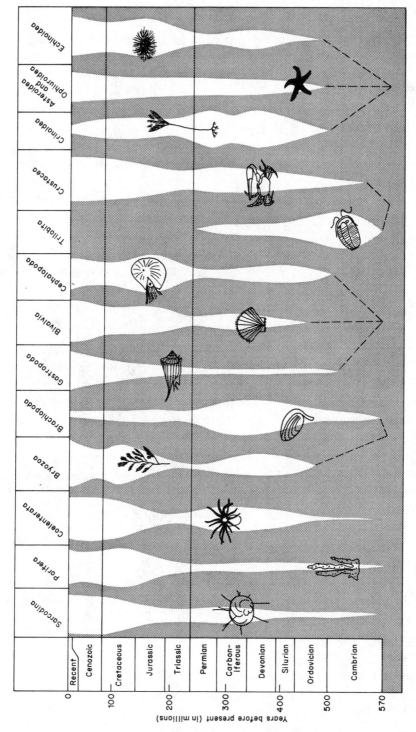

Figure 5.2 Geologic range of major groups of marine invertebrates. Preservable hard parts of most groups of marine invertebrates have existed since the Cambrian. (After McAlester, 1968.)

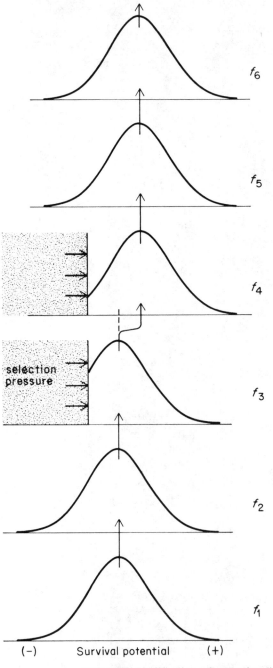

And so on --- return to f_2 and begin the story o over again; perhaps with the same attribute, perhaps with some other attribute important to survival.

With the return of normal conditions, the population continues to reproduce the new mode.

With continuing unfavorable conditions, offspring with low survival potential die before reproducing; mode of the population shifts.

With the onset of unfavorable conditions individuals with low survival potential are killed.

Under good conditions, even individuals with low survival potential are able to survive and reproduce; the population acquires a large standard deviation.

Begin with a population normally distributed with respect to some attribute important to survival.

f_6

f_5

f_4

selection pressure

f_3

f_2

f_1

(−) Survival potential (+)

Figure 5.3 Modification of a species distribution function by selection pressure. Time passes from f_1 to f_6.

Numerous environmental parameters vary as a function of water depth. Wave agitation may provide food supply for filter feeders. It may also produce a firm sandy substrate that contrasts sharply with the soupy mud substrate of deeper water environments. Shallow, clear water allows sufficient light penetration for luxuriant development of a benthonic flora. If living space is available in suitable environments, communities will expand to fill the space. Finally, each species has its own range of preferred temperatures. Some range more widely than others, but all species have their limits. If any of these conditions are altered, we must expect the dynamic equilibrium of the community to respond accordingly—be it a positive response or a negative response.

Changing Conditions on the Earth's Surface. In our review of the dynamics of the earth's surface today (Chapters 8 and 9) and of the general geology of North America (Chapter 10), we shall observe gross variation in all three of the general parameters just cited (water depth, living space, and temperature). The shallow seas that formerly existed over the Interior Lowlands of the United States surely provided more living space for shallow-marine communities than exists on the North American continent today. In addition, note that Pleistocence glaciers existed adjacent to shallow-marine environments off New England. Undoubtedly, the shallow-marine environments of this area were much colder during the Pleistocene than they are today.

Lacking strong evidence to the contrary, we must assume that variable conditions have been the general rule throughout the geologic record. When a change was favorable to a species or a community, it exploited that change by expanding its population. When a change was unfavorable, the population of the species or of the community underwent alteration in response to the selection pressure.

Selection Pressure. If a population is existing at optimum environmental conditions and some aspect of the environment undergoes adverse change, a portion of the population may be adversely affected. In Figure 5.3, we portray the distribution of the generalized variable: *survival potential.* The precise identification of this variable is left open and would presumably vary with the type of stress that the population is about to face. It may be something as simple as temperature tolerance during a three-day cold snap, salinity tolerance when a bay becomes flooded by an unusual amount of fresh water, food-gathering efficiency during times of scarce supply, or perhaps just a willingness to reproduce even during a generally unromantic mating season. Alternatively, survival potential may be an exceedingly complicated collage of all of these and other factors affecting the general maintenance of the population. Indeed, factors contributing to survival potential will change from one crisis situation to the next.

The adverse shift in the environment may be regarded as a selection pressure that will cause the early demise of those individuals having the lowest survival potential. The more serious the environmental crisis, the further into the bell-shaped curve the selection pressure will operate. Thus, selection pressure can substantially modify the gene pool of successive generations. In Figure 5.3, phylogenetic evolution has occurred between f_3 and f_4. If some aspect of this modified distribution shows up in the preservable hard parts of this organism, we may be able to use the shift in population characteristics as a biostratigraphic indicator of time.

Note also how the potential for selection pressure operates among different species. Consider, for example, marine invertebrates *A* and *B,* both with preservable hard parts. They have separate and distinct feeding habits. *A* feeds on *C* and *B* feeds on *D, C* and *D* being species with no preservable hard parts. At this point, *A* and *B* are not competing with each other for food. Then, for reasons that will never be known to us, species *D* simply vanishes from the environment. *A* and *B* now seek to feed on the single species *C.* Species that were formerly not competing with one another are now brought into competition for food and, therefore, for survival. *B,* being the more aggressive food gatherer, continues to thrive. *A,* being the less aggressive food gatherer, undergoes severe selection pressure and is ultimately starved out of the area entirely. Once again, we see the generation of a biostratigraphic shift that must certainly have at least local time-stratigraphic significance; that is, beds containing *A* and *B* give way to beds containing only *B.*

Sea-Level Fluctuation as a Selection Pressure on Shallow-Marine Benthonic Invertebrates. Fluctuating sea level can produce large variations in the amount of subtidal shallow-marine living space available on or near continents. The Recent epoch provides us with an excellent example of this problem in the Great Bahama Banks (Figure 13.1). Shallow subtidal environments cover the top of the banks today and did so numerous times in the past. During lowstands associated with glaciation, however, these banks stood high and dry, forcing subtidal-marine organisms to try to eke out survival within the narrow band between the exposed platform and the deep water surrounding the platform.

Numerous models can be set up to portray the possible interaction between shallow-marine invertebrate communities and fluctuating sea level. Let us work through one of these models just to see how things might fit together. Keep in mind that the possible combinations are more or less infinite.

Figure 5.4 depicts a platform on which highstand living space grossly exceeds lowstand living space. You may think of this area as being on the scale of carbonate platform or on the scale of a continent; the principles are similar.

Let us begin with four lowstand populations situated on the various side of the platform. These populations are isolated from each other by geographic barriers. With rising sea level, the available living space for these communities expands greatly. At this time, there is indeed a great deal of positive environmental shift for the species. Space is available for the population increase; and, as an additional bonus, previously isolated gene pools are now united once again into a single gene pool. Because selective pressure is low, nearly all combinations of genes produce offspring that will survive. Thus, the population grows not only in numbers but also in diversity.

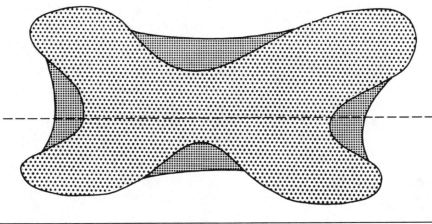

Map

Cross section

high stand
low stand

Figure 5.4 Sea-level fluctuation as a control of living space and, therefore, of selection pressure. During lowstand, shallow-marine benthonic communities are forced to occupy the four small areas of lowstand living space. With highstand of the sea, these four isolated communities can expand onto the much larger area of the platform.

But then the sea level begins to fall, and so the platform becomes emergent once more. Reduction in living space places severe pressure on the species. It must once again split into four isolated communities around the margins of the platform.

Furthermore, each isolated lowstand population endures its own separate selection pressures. With the highstand, benefits gained by selection pressure acting upon the isolated populations are contributed to a single highstand gene pool. For example, if selection pressure has caused the west coast population to acquire a beneficial feeding habit, this information can be transmitted to all populations via the next highstand gene pool. Alternatively, selection pressure on the lowstand west coast community might be so intense as to produce a population that will not interbreed with the other populations during the next highstand. The stage is set for speciation. With the next lowstand, some of the west coast variant will undoubtedly end in several of the lowstand platform-margin havens. If they continue to survive and if they continue to breed only among themselves, speciation has occurred.

As another interesting variation on this model, consider the invasion of one of the lowstand havens by a totally different organism (*B*), which possesses superior adaptation to the environment in question. So long as the platform remains emergent, the new superior organism is excluded from the other lowstand havens by geographic barriers. Although this new organism has completely taken over the southern haven, for example, the west, east, and north havens remain untouched. With the next highstand, however, events move rather rapidly. Organism *B* gets its first look at all that platform area with shallow water over it. Those lowstand populations of organism *A* hardly see what hit them. *B* takes over the whole place, bringing about the rapid demise of *A*. In this fashion, biostratigraphic time horizons of large areal extent can be generated. Note that the example involves replacement of a species by a competitor rather than replacement of a species by its phylogenetic progeny.

The Complications of Niche and Biogeography

The preceding discussion outlines the potential for biostratigraphic horizons to be good time horizons. Now we must recognize some of the pitfalls of these generalities.

Variation in Depositional Environments Perpendicular to Shoreline. All of the various major depositional environments indicated in Figure 5.5 are capable of accumulating a stratigraphic record of a time interval. Yet the organisms that live in these environments may be quite different from each other and may not occur within the same outcrop.

Two considerations are important to us here. First, each of these environments

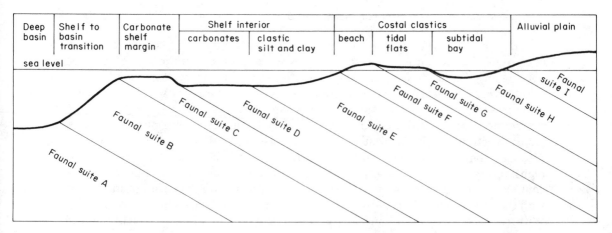

Figure 5.5 Separate and distinct faunal suites as a function of variation in the depositional environment perpendicular to the shoreline. Variations in water depth, substrate, turbidity, and salinity cause different suites of animals to occupy the various environments within a single time plane. Inasmuch as suite A contains no species in common with suites D through I, demonstration that suites A through I are indeed time correlative can be a sizable biostratigraphic task.

has its own set of communities that respond independently to ecologic stress. The equation for survival potential within the deep basin need not have any *a priori* relationship to the equation for survival potential in carbonate shelf-interior environments. For example, sea-level fluctuations may have a profound effect on carbonate, shelf-interior, benthonic environments, yet they need have no effect whatsoever on deep-basin pelagic environments.

Secondly, the first appearance of a fauna in a stratigraphic sequence may record local sea-level fluctuation rather than an evolutionary biostratigraphic event. For example, the first appearance of deep-basin fauna over clastic bay fauna means one of two things: Either the deep-basin fauna has suddenly adapted itself to a very successful life in the clastic bay environment, or the clastic bay environment has been replaced by deep water. The former would be an evolutionary event of potential worldwide importance; the latter may record only local submergence. In most cases, we suspect that the latter is true.

Variation in Depositional Environment Parallel to Shoreline. If a shallow-marine environment covers a sufficiently large area, the temperature varies significantly over its areal extent. Organisms sensitive to temperature may thus be restricted to certain portions of this time plane.

Figure 5.6(A) depicts the distribution of communities along the eastern and southern margins of a continental land mass. Tropical, temperate, and cool communities are recognized. Except for the temperature factor, these communities are occupying the same depositional environments. Because of the temperature factor, however, the tropical and the cool communities do not have a single species in common. The temperate community likely will be less well defined. A few tropical species may be more tolerant of cold than the majority of the tropical community. Similarly, some members of the cold community may extend into the temperate area. Thus, we face a very involved correlation problem if we are to demonstrate contemporaneity of cool and tropical communities. Even if both communities are undergoing evolutionary shifts that will be of regional biostratigraphic importance, numerous sections will be required in between to relate biostratigraphic horizons in the tropical area to biostratigraphic horizons in the cool area.

Time-Transgressive First Appearance in Response to Changing Climate. Given the distribution of tropical, temperate, and cold-water species portrayed in Figure 5.6(A), consider next the stratigraphy generated by sea-level fluctuations during a time of general climatic cooling [Figures 5.6(B) and (C)]. With each highstand, sediments are laid down that record the areal distribution of climatic conditions during the highstand.

To make the model even more illustrative, allow a new species *X* to be introduced to the cold area at time (a) by invasion from another continental shoreline (Figures 5.6 and 5.7). The deep water between the two continents had kept *X* out of this area for millions of years; but once it was introduced to the area by chance migration, it proved to be far superior to the local inhabitants of its niche and

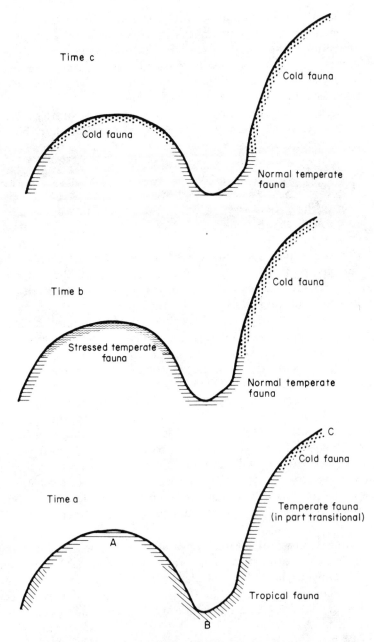

Figure 5.6 Separate and distinct faunal suites as a function of variation in the depositional environment parallel to the shoreline. As indicated at time (a), temperature variation among otherwise similar environments allows faunas that do not resemble each other to occupy similar sediment types within a single time plane. Diagrams (B) and (C) trace some of the biostratigraphic complications that can arise as changing climate interacts with geographic barriers. At time (b), cooling has allowed the cold fauna to migrate considerably south-

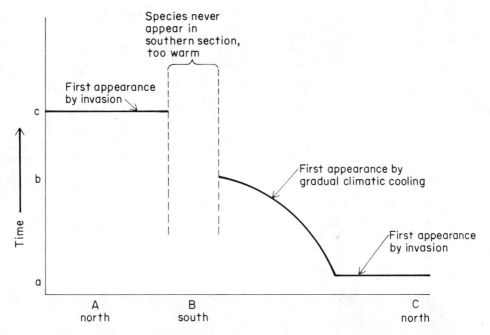

Figure 5.7 Varying significance of the first appearance of species *X* within Figure 5.6. Locations (A), (B), (C) and times (a), (b), (c) are keyed to Figure 5.6. Because of interaction between earth history (climatic change), niche, and geographic barrier, the first appearance of species *X* may be strongly diachronous and of changing stratigraphic values.

proliferated rapidly throughout the cold community. Thus, the first appearance of *X* provides an excellent time line throughout an area dominated by the cold community.

As our model continues, the general cooling trend introduces species *X* into stratigraphic sections farther and farther to the south. Thus, the first appearance of *X* transgresses time as we move southward.

Crossing Geographic Barriers. Our model can become further complicated by the introduction of a geographic barrier. In Figure 5.6(B), cold climatic conditions have pushed down into areas that were temperate during time (a). Observe, however, that the continental mass is a geographic barrier to the migration of species *X* in the western portion of our map area. Ecological conditions are right for *X*, but it cannot get there; it can neither cross the land nor go around the land because the water to the south is too warm.

---

ward, yet the geographic barrier excludes the cold fauna from the northern portions of area (A). At time (c), a chance migration or a brief interconnection between the two areas allows the cold fauna to occupy the northwestern area from which it was previously excluded by a geographic barrier. Thus, the first appearance of the cold fauna is time transgressive from north to south along the east coast, but it is a good time plane where the cold-water fauna invades the northwestern area (A). Figure 5.7 continues this discussion.

Allow a chance migration or a brief interconnection between the two cold-water masses, and X will take over the western area very rapidly. Thus, the first appearance of X becomes an excellent regional time line throughout the western portion of our map area. But note that this first appearance may be greatly separated in time from the first appearance in the northeast portion of the map area (see Figure 5.7).

In looking at this model, we have presumed to know a great many things that the geologist in the field does not know as he or she begins to work. Identifications of tropical, temperate, and cold faunas might be quite difficult in stratigraphic situations. We really know very little about global climatic patterns during the Paleozoic era, for example. Faunal differences that we have assigned to tropical, temperate, and cold regions might easily be mistaken for evolutionary sequences by the biostratigrapher first undertaking temporal correlation within these stratigraphic sequences. The search for the best time correlation is open-ended. We must remain flexible on this point.

Biostratigraphy as Time-Rock Correlation

In spite of all the difficulties that may be anticipated, biostratigraphy is the principal tool of regional and global time-rock correlation. Two approaches are outlined below. It is important to understand the traditional approach because so much of what we think we know is rooted in this method. The graphic correlation method is more logically satisfying but has not achieved wide usage in the literature.

Traditional Biostratigraphy. The preceding discussion views the prospect of biostratigraphic correlation from a clairvoyant perspective. In fact, the roots of biostratigraphy are deeply embedded in geologist's early, dim understanding concerning the history of the earth. Traditionally, the task was begun as the zonation of a particular section of rock. If one could establish that certain molluscs, for example, occurred at certain places within a thick sequence of sediments, then one had established *local range zones* for that particular *type section*. It was then off to some other nearby section to see if the same zonation might be applicable. If the local range zone appeared to be applicable over a wider and wider area, then the zone and its type section took on greater and greater significance in the mind of the biostratigrapher. "Good" type sections might then be zoned on the basis of some other group of organisms; "bad" type sections would fall into disuse, and so on. Through all of these activities, a rather rigid formalism was developed which is still with us today (see Hedberg, 1976, for a convenient, modern summary of the formalities).

If one undertakes to establish a biostratigraphic zonation of a particular local section, the principal data are most commonly local first appearance and local last appearance of various taxa. Given these building blocks, the simplest concept is a *range zone*. This would define that interval of rock that does indeed contain a particular fossil. The range of a taxa is commonly too long to be of biostratigraphic utility. This leads to the more complicated concept of the *concurrent range zone,* where the zone is defined by concurrent occurrence of two or more taxa. Finally, there is the *interval* (or *"gap"*) *zone.* This is conveniently defined as an interval between two "better defined" zones. All of these zones begin as *local* zones based on a particular fossil group; global significance is *inferred* only as the work proceeds.

Note well that all of this is born of utility rather than concept. As one begins this process, one has no way of knowing that the top of a range of taxa in any particular section represents the global extinction of that taxa. That last appearance could just as well represent a shift in environment that forced the taxa to move elsewhere. Thus, when one goes off to stratigraphic sections elsewhere, one may continue to find a local top to the same taxa. Continued utilization of that top as part of the definition of a biostratigraphic zone may be a totally correct application of biostratigraphic formality while at the same time having no time-stratigraphic significance whatsoever. Surely, there is a better way to go at all of this.

Graphic Correlation. One sensible approach to this problem is to consciously avoid *a priori* judgements concerning which fossils will be useful for biostratigraphic correlation or which stratigraphic sections contain "the true" evolutionary base or evolutionary top of a particular taxa. Alternatively, we might demand of ourselves systematic attention to all species present within the stratigraphic interval under consideration, and we might anticipate at least small problems with each and every stratigraphic section under consideration.

Shaw (1964) has developed an extremely useful graphic correlation approach to biostratigraphic correlation. For a more recent summary of the methodology, see also Miller (1977). The graphic correlation technique relies almost exclusively on the base and the top of the range of individual taxa within the various sections under consideration. Note especially that the graphic correlation method does not bring forward complicated concepts such as concurrent range zone or gap zone in the early stages of the exercise. If such concepts have validity, that will fall out at the end of the graphic correlation exercise. Note also that the graphic correlation method can take into account such physical time markers as bentonites or chemical and isotopic time-series signatures.

Figure 5.8 provides several examples of the power of the graphic correlation technique in establishing correlation among two sections. Section (A) is plotted on the *Y*-axis and section (B) on the *X*-axis. For each taxa that the two sections have in common, the base of the range (circles) and the top of the range (crosses) are plotted as *X-Y* coordinates. In Figure 5.8 (A), the majority of the bases and tops cluster along a straight line. The *X* and *Y* coordinates of any point along this line constitute precise biostratigraphic correlation between the two sections. Spurious bases and tops result from species that are strongly facies-dependent or from species that suddenly cross biogeographic barriers between the two sections.

Figures 5.8(B) and (C) provide further examples of the power of this graphic technique. In Figure 5.8(B), the offset in the correlation line clearly indicates missing

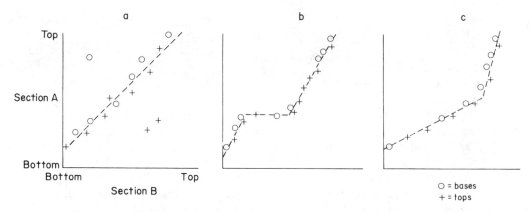

Figure 5.8 Examples of the biostratigraphic correlation technique of Shaw (1964). Data points are the first (○) and the last (+) occurrences of species that the measured sections (A) and (B) have in common. The generally good linear fit in diagram (A) suggests precise correlation between the two sections. In diagram (B), the pronounced offset in the otherwise linear trend line suggests that a section is missing out of (A). In diagram (C), pronounced change in the slope of the correlation line suggests change in relative sedimentation rates.

section in section (A). A fault or an unconformity could produce such a relationship. In Figure 5.8(C), change in slope of the correlation line suggests abrupt change in relative sedimentation rate among the two sections.

While the graphic correlation technique is useful for comparing one section to another, the real power of the method comes in the compilation of a *composite standard reference section.* Having obtained some experience with the entire data set by plotting one section against another, such as in Figure 5.8, one chooses a section that appears to involve the fewest problems and begins to compile information upon that standard reference section. The standard reference section becomes the yardstick of relative time. Precise statements of time correlation will be expressed in "composite standard reference section equivalent depth units." A strong argument can be made that this is as close as the stratigrapher ever needs to come to discussing time. We can state relative time in equivalent depth units to the nearest micrometer if we so desire. We shall never know absolute time to such precision.

Figure 5.9(A) indicates an intermediate step in the process of construction of a composite standard reference section. Using the composite standard reference section as the yardstick of time, the object of the exercise is to establish the *composite standard range* for each taxa by pushing all bases to as low as they occur in any section under consideration and by pushing all tops to as high as they occur.

By way of example, note in Figure 5.9(A) that section (Y) brings some new information to the composite standard reference section. Various taxa are identified by the numbers (1) through (12). Bases are indicated by circles and tops are indicated by crosses. The solid line represents the proposed correlation between section (Y) and composite standard reference section (X). If bases fall on or above the corre-

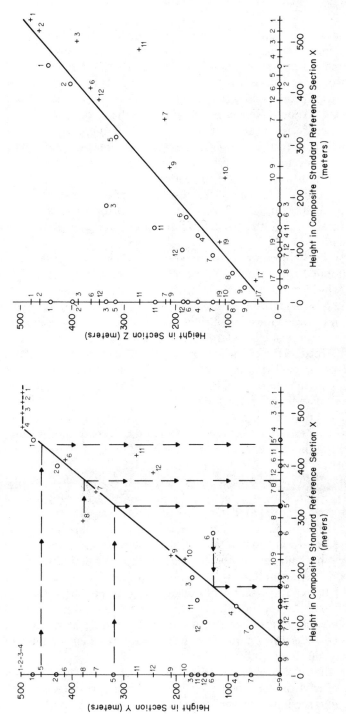

Figure 5.9 Graphic plots illustrating development of a composite standard reference section. Figure 5.9(A) correlates section (Y) with a relatively immature composite standard reference section. Note that section (Y) contributes some new information to the composite. Base (○) (6) is moved to (6′), top (+) (8) to (8′), and taxa (5) is added to the composite for the first time. In Figure 5.9(B), note that section (Z) contributes no new information to the composite; all bases (○) are lower in the composite than in section (Z), and all tops (+) are higher in the composite. The line separating all tops from all bases is the best correlation of (Z) to the composite standard.

lation line, the data in section (Y) contributes nothing to the composite standard reference section. These bases are already known to occur lower in the composite standard reference section than they occur in section (Y). Note, however, that base (6) occurs lower in section (Y) than in section (X). This is new information with regards to the composite standard reference section. As indicated by the arrows, the base of taxa (6) should have occurred at (6') in the composite standard reference section. Likewise, if tops occur below the correlation line, section (Y) does not add any new information to the composite standard reference section. These tops are already known to occur higher in the reference section than they occur in section (Y). Note that the top of taxa (8) in section (Y) is new information. Top (8) is, therefore, moved to (8') to record this new information. Finally, note that taxa (5) does not occur in section (X). Thus, both base and top of taxa (5) in section (Y) are new information and are projected into composite standard reference section as base (5') and top (5').

And so on goes the compilation process! Each new section is compared to the previously compiled composite standard reference section. The concept of the base of a taxa is solidly defined as evolutionary first appearance of taxa. A base can occur higher in a local section because of environmental or biogeographic problems, but a base cannot occur lower than the evolutionary first appearance that is being compiled on the composite standard reference section. Similarly, the top of a taxa is rigorously defined as the extinction of that taxa from the face of the earth. It can be excluded from a local section by environmental or biogeographic problems, but it cannot occur higher than the time of global extinction (discounting for the moment problems of sediment reworking).

Figure 5.9(B) plots a new section against a relatively mature composite standard reference section. Note that all bases lie above the correlation line and all tops lie below the correlation line. That is to say, section (Z) contributes no new information to the composite standard reference section.

In summary, the key concept to keep in mind in graphic correlation is that there can only be one time of evolutionary first appearance of a taxa and one time of global extinction of a taxa. These events have time significance. All other exclusions of a taxa from a particular local section are either environmental or biogeographic. These exclusions do not necessarily have time-stratigraphic significance.

Experience has shown that the correlation line among individual sections and the mature composite standard reference section tends to be a straight line. This means that the thickness of sediment added in any and all sections within a unit of time tends to remain proportional. It is not that sedimentation rate is constant, but only that sedimentation rate varies proportionally among numerous sections. It is interesting to consider this empirical observation in the light of the concept of the equilibrium profile (Figure 11.27).

If the profile of a stream, a shoreline, or a shelf is truly at equilibrium, the vast majority of environments beneath the equilibrium profile are not accumulating new stratigraphic sections. Rather, sediments are simply being transported across the equilibrium profile, which is a zero accumulation rate surface. If the equilibrium

Table 5.1 Common Time-Rock Names of North America and Europe

			NORTH AMERICA	EUROPE		
C	Quaternary	Recent		Calabrian		C
E		Pleistocene		Astian		E
N		Pliocene		Zanclian		N
O				Messinian		O
Z				Tortonian		Z
O	TERTIARY	Miocene		Langhian	NEOGENE	O
I				Burdigalian		I
C				Aquitanian		C
		Oligocene		Chattian		
				Rupelian		
				Tongrian		
		Eocene		Ludian		
			Jacksonian	Bartonian		
				Auversian	PALEOGENE	
			Claibornian	Lutetian		
				Cusian		
			Wilcoxian	Ypresian		
		Paleocene		Thanetian		
			Midwayan	Montian		
				Danian		
C	CRETACEOUS		Laramian	Maestrichtian (Senonian)		M
				Campanian (Senonian)		E
			Montanan	Santonian (Senonian)		S
				Coniacian (Senonian)		O
		Upper Cretaceous	Coloradoan	Turonian	CRETACEOUS	Z
			Dakotan	Cenomanian		O
			Washitan	Albian		I
		Lower Cretaceous	Fredericksburgian	Aptian		C
			Trinitian	Barremian		
				Neocomian		
O	JURASSIC	Upper Jurassic		Portlandian		
Z				Kimmeridgian		
O				Oxfordian		
I				Callovian		
C		Middle Jurassic		Bathonian	JURASSIC	
				Bajocian		
				Aalenian		
				Toarcian		
				Pliensbachian		
		Lower Jurassic		Sinemurian		
				Hettangian		
M	TRIASSIC	Upper Triassic		Rhaetian		
E				Norian		
S				Keuper		
O		Middle Triassic		Karnian	TRIASSIC	
Z				Ladinian		
O				Muschelkalk		
I		Lower Triassic		Anisian		
C				Scythian		
				Buntsandstein		

Table 5.1 (continued)   *NORTH AMERICA*   *EUROPE*

PALEOZOIC

System	Series	North America	Europe	European System
PERMIAN	Upper Permian	Ochoan	Chideruan	PERMIAN
			Kazanian	
		Guadalupian	Kungurian	
	Lower Permian	Leonardian	Artinskian	
		Wolfcampian	Sakmarian	
PENNSYL-VANIAN	Upper Pennsylvanian	Virgilian	Uralian	CARBON-IFEROUS
		Missourian	Gshelian	
		Desmoinesian		
	Middle Pennsylvanian	Atokan	Moscovian	
		Morrowan		
	Lower Pennsylvanian	Springeran	Namurian	
MISSISSIPPIAN	Upper Mississippian	Chesteran	Viséan	
		Meramecian		
		Osagian	Tournaisian	
	Lower Mississippian	Kinderhookian	Etroeungtian	
DEVONIAN	Upper Devonian	Conewangoan	Famennian	DEVONIAN
		Cassadagan		
		Chemungian		
		Fingerlakesian	Frasnian	
	Middle Devonian	Taghanican		
		Tiouchniogan	Givetian	
		Cazenovian	Eifelian	
	Lower Devonian	Onesquethawan		
		Deerparkian	Coblenzian	
		Helderbergian	Gedinnian	
SILURIAN	Upper Silurian (Cayugan)	Keyseran	Dowtonian	GOTHLANDIAN
		Tonolowayan		
		Salinan	Ludlovian	
	Middle Silurian (Niagaran)	Lockportian		
		Cliftonian	Wenlockian	
		Clintonian		
	Lower Silurian	Alexandrian	Llandoverian	SILURIAN
ORDOVICIAN	Upper Ordovician (Cincinnatian)	Richmondian	Ashgillian	
		Maysvillian		
		Edenian	Caradocian	
		Trentonian		
	Middle Ordovician (Mohawkian)	Blackriveran		ORDOVICIAN
		Chazyan	Llandeilian	
			Skiddavian	
	Lower Ordovician	Canadian	Tremadocian	
CAMBRIAN	Upper Cambrian (Croixian)	Trempealeauan		CAMBRIAN
			Potsdamian	
		Franconian	(Lingula Flags)	
		Dresbachian		
	Middle Cambrian	Albertan	Menevian	
			Acadian	
			Solvan	
	Lower Cambrian	Waucoban	Comleyan	
			Georgian	

PALEOZOIC

profile is raised for any reason, all environments under the influence of the equilibrium profile suddenly accumulate new stratigraphic sequence to attain a new equilibrium. Thus, proportional accumulation of stratigraphic thickness should be the rule. Relatively large perturbations of the equilibrium profile should lead to relatively large accumulations of new sediment regardless of sedimentary facies. Relatively small perturbations of the equilibrium profile should likewise lead to relatively small accumulations of new stratigraphic thickness.

Biostratigraphic Nomenclature

The foregoing discussion gives some idea of the complexity involved in the establishment of reliable biostratigraphic correlation. Such efforts have provided us with a list of recognized names for time-rock units. Table 5.1 is a convenient scorecard containing many of the commonly used names of time-rock units.

Nevertheless, we must still face the task of placing these time-rock units into a framework of absolute time. Absolute chronology in geologic history rests heavily upon radiometric dating and is the subject of Chapter 7.

Citations and Selected References

BERGGREN, W. A., and J. A. VAN COUVERING. 1974. Late Neogene biostratigraphy, geochronology, and paleoclimatology of last 15 million years in marine and continental sequences. Paleogeo. Paleoclimat. Paleoecol. 16: 1–216.

BERGGREN, W. A., and B. U. HAG. 1976. The Andalusian stage (Late Miocene): biostratigraphy, biochronology, and paleoecology. Paleogeog. Paleoclimat. Paleoecol. 20: 67–129.

BRETSKY, P. W. 1969. Central Appalachian, late Ordovician communities. Bull. Geol. Soc. Amer. 80: 193–212.

GIGNOUX, M. 1955. Stratigraphic geology. W. H. Freeman and Co., San Francisco. 682 p.
Classical reference to European stratigraphy. Strongly biostratigraphic.

HALLAM, A. (ed.). 1973. Atlas of palaeobiogeography. Elsevier, Amsterdam. 531 p.

HALLOCK, P. 1982. Evolution and extinction in larger Foraminifera. Third North American Paleontological Convention Proceedings, 1: 221–225.
Evolutionary radiation on transgressions; extinctions on regressions.

HAZEL, J. E. 1970. Binary co-efficients and clustering in biostratigraphy. Bull. Geol. Soc. Amer. 81: 3237–3252. Biostratigraphic zonation by statistical techniques.

HEDBERG, H. D. (ed.). 1976. International stratigraphic guide. John Wiley & Sons, New York. 200 p.
Comparatively easy reading concerning the legalistic formalisms of stratigraphy.

KAUFFMAN, E. G., and J. E. HAZEL. (eds.). 1977. Concepts and methods of biostratigraphy. Dowden, Hutchinson & Ross, Inc., Stroudsburg, Penn. 658 p.
A collection of modern papers concerning various aspects of biostratigraphy.

KUMMEL, B., and C. TEICHERT. 1971. Stratigraphic boundary problems: Permian and Triassic of West Pakistan. Univ. Press of Kansas. 480 p.

LAPORTE, L. F. 1968. Ancient environments. Prentice-Hall, Inc., Englewood Cliffs, N.J. 115 p.

MAYR, ERNST. 1976. Evolution and the diversity of life: selected essays. Belknap Press of Harvard Univ. Press, Cambridge, Mass. 722 p.

MCALESTER, A. L. 1968. The history of life. Prentice-Hall, Inc., Englewood Cliffs, N.J. 151 p.

MILLER, F. X. 1977. The graphic correlation method in biostratigraphy, p. 165–186. *In* E. G. Kauffman and J. E. Hazel (eds.), Concepts and methods of biostratigraphy. Dowden, Hutchinson & Ross, Inc., Stroudsburg, Penn.

MOORE, T. C., JR., T. H. VAN ANDEL, C. SANCETTA and N. PISIAS. 1978. Cenozoic hiatuses in pelagic sediments. Micropaleontology 24 (2): 113–138.
Defines stratigraphic section to be missing on the basis of the absence of one or more biostratigraphic zones.

SHAW, A. B. 1964. Time in stratigraphy. McGraw-Hill, New York. 365 p.
An excellent and scholarly discussion concerning the recognition of precise time relationships within stratigraphic sequences.

STANLEY, STEVEN M. 1968. Post-Paleozoic adaptive radiation of infaunal bivalve molluscs: a consequence of mantle fusion and siphon formation. J. Paleont. 42: 214–229.
Adaptation to living *in* the sediment instead of *on* the sediment afforded competitive advantages to these molluscs.

STANLEY, STEVEN M., WARREN O. ADDICOTT, and KIYOTAKA CHINZEI. 1980. Lyellian curves in paleontology: possibilities and limitations. Geology 8(9): 422–426.

STEBBINS, G. LEDYARD, and FRANCISCO J. AYALA. 1981. Is a new evolutionary synthesis necessary? Science 213 (4511): 967–971.

VAIL, P. R., R. M. MITCHUM, JR., and S. THOMPSON III. 1977. Seismic stratigraphic and global changes of sea level, part 4: global cycles of relative changes of sea level, p. 83–97. *In* C. E. Payton (ed.), Seismic stratigraphy—applications to hydrocarbon exploration. Amer. Assoc. Petrol. Geol. Tulsa, Okla.
Convenient summary of biostratigraphic and geochronologic units within the context of global sea-level history.

WALKER, K. R., and L. F. LAPORTE. 1970. Congruent fossil communities of Ordovician and Devonian carbonates, New York. J. Paleont. 44: 928–944.

WILSON, E. O., and W. H. BOSSERT. 1971. A primer of population biology. Sinauer Associates, Inc., Stamford, Conn. 92 p.

Time-Stratigraphic Correlation Based on Physical Events of "Short" Duration

6

The search for the best time-stratigraphic correlation is unending. We use all our knowledge but still cannot be absolutely certain that we have events worked out correctly. We must become accustomed to bringing as many lines of evidence as possible to bear on our time-correlation problems. We can develop a real confidence in our correlations only when several lines of evidence begin to agree.

Many physical events in the history of the earth have left us with hints of time-stratigraphic correlation. If we can learn to use these hints, they can help us write the precise history of the earth. Some of the events in the following discussion are global in scope; others are of an extremely local nature. Some of the events may have occurred in a matter of hours or days; others may have occurred over a period of thousands or tens of thousands of years. Whereas worldwide correlation may sound conceptually simple, it is in fact a gigantic undertaking. A geologist may feel that two years of his life were well spent if he has genuinely demonstrated precise time correlation between two sedimentary sections 100 kilometers apart.

Magnetic Stratigraphy

From time to time, the polarity of the earth's magnetic field has become reversed. Precisely why this reversal has happened is a rather difficult problem that we shall leave to the geophysicist. Instead, we shall just use this phenomenon as a time-stratigraphic tool.

Magnetic minerals in rocks preserve a magnetic vector indicative of the orientation and intensity of the global magnetic field within which the rock formed. In the case of igneous rocks, this vector is acquired as the cooling rock passes through the Curie point. In the case of sediments, detrital magnetic minerals may take on a preferred orientation during or shortly after sedimentation, or diagenetic magnetic minerals may record the earth's magnetic field at the time of diagenesis.

Both polarity and intensity of the magnetic field are potentially useful data for time-stratigraphic correlation purposes. To date, most of the emphasis has been placed on correlation of polarity time intervals, and black and white time-stratigraphic diagrams, such as Figure 6.1, have resulted. Clearly, there are certain difficulties when we attempt to relate a new section of rock to one of these diagrams. Numerous times in the history of the earth, the magnetic field has been normal, and numerous times it has been reversed. Correlation can be based only on similarities concerning the spacing of magnetic events. Because this graphic technique does not assign any particular identifying characteristic to any one magnetic interval, magnetic stratigraphy has found its greatest application in relatively young rocks and sediments.

Volcanic Rocks. The fundamental documentation of the earth's magnetic history over the last 4.5 million years has been carried out in volcanic sequences of ocean islands and continental areas. Lavas are particularly good for magnetic work because they have formed hard rock before their temperature lowers through the Curie point of the magnetic minerals. Furthermore, these rocks are relatively easy to date by radiometric methods. Thus, we can get good magnetic data on rocks that can be easily placed within an absolute time-stratigraphy.

The one serious disadvantage with developing a magnetic stratigraphy based upon volcanic rocks is the fact that we cannot be certain that volcanism has been continuous through time. Even if we sample every lava flow on a volcanic island— a physical impossibility in itself—we cannot be sure that a complete record of earth history is represented by the suite of samples collected.

Figure 6.1 summarizes the state of our knowledge in 1969 concerning the polarity history of the earth over the last 4.5 million years. It is important to bear in mind that this area of research is rapidly evolving. As indicated in Figure 6.2, the more data we gather, the more magnetic events we recognize.

Oceanic Sediments. Fine-grained, deep-ocean sediments also record a magnetic history. Detrital magnetic minerals tend to take on a preferred orientation that is, at least in some statistical sense, related to the magnetic field in which the sediment is ultimately deposited. Reaching this final sedimentation is a more complicated process than it might at first appear. The major problem is the continuing activity of burrowing organisms. As a particle of detrital magnetic mineral first adheres to the sediment-water interface, it is likely very well oriented with respect to the existing magnetic field. Then, along come the burrowing organisms. They ingest the sedi-

---

Figure 6.1 (*opposite*) Magnetic polarity time scale for the last 4.5 million years. Compiled from magnetic data on volcanic rocks that have been dated by the potassium-argon method. [After A. Cox, "Geomagnetic Reversals," *Science,* **163,** 237–245 (17 January 1969).] Copyright 1969 by the American Association for the Advancement of Science.

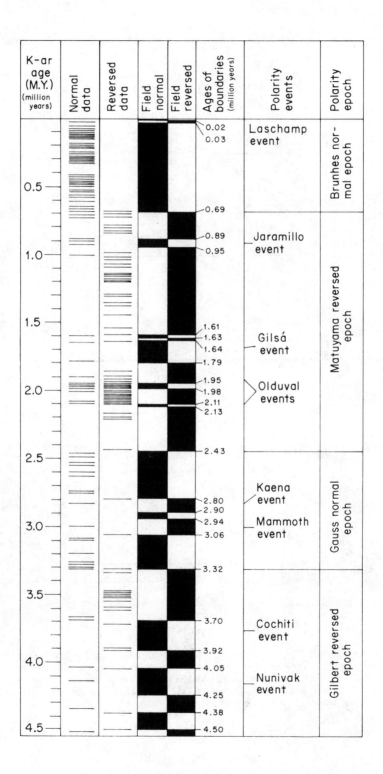

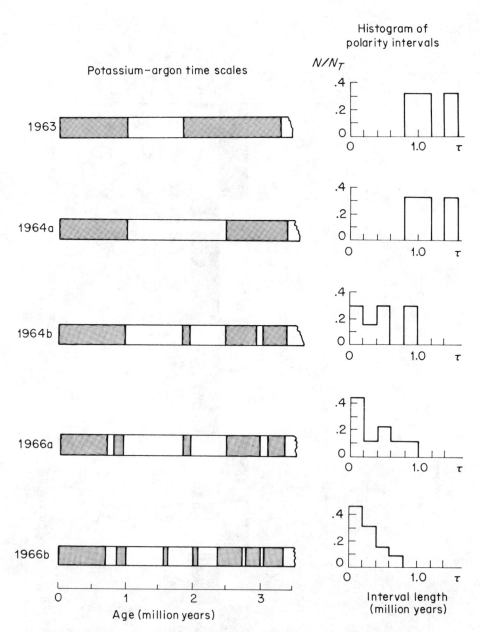

Figure 6.2 Successive refinements of the magnetic polarity time scale. Polarity is either normal or reversed. As more and more data are accumulated, more and more reversed events are found in normal epochs and normal events in reversed epochs. [After A. Cox. "Geomagnetic Reversals," *Science,* **163**, 237–245 (17 January 1969).] Copyright 1969 by the American Association for the Advancement of Science.

ment to extract organic matter and excrete the mineral residue. The precise effect of this activity upon the orientation of magnetic minerals is not well understood. Nevertheless, two points seem very clear: (1) Any given sedimentary particle does not achieve its ultimate orientation within the sediment until after sediment accumulation has placed that particle out of the reach of burrowing organisms. Clearly, this fact will tend to blur the precise stratigraphic positioning of magnetic events in the deep-sea cores. (2) It must be anticipated that magnetic events of short duration may be lost from the sedimentary record by homogenization of the sediment because of burrowing activity.

Thus, both volcanic rocks and oceanic sediments offer certain assets and certain liabilities concerning the precise recording of the earth's magnetic history. Whereas volcanic rocks afford excellent magnetic data and can be dated readily by radiometric methods, they do not record a continuous history, owing to the periodic nature of volcanism. In contrast, oceanic sediments offer a continuous record of the earth's magnetic history; but the record is somewhat blurred by burrowing activity, and so precise dating by radiometric methods is often difficult. Where the ultimate goal of scientific research is the establishment of the complete and detailed magnetic history of the earth, scientists move back and forth among magnetic data from volcanic rocks, deep-sea sediment, and mid-ocean ridges. In our present discussion, the main concern is using magnetic reversals as physical events that allow time-stratigraphic correlation.

Figure 6.3 depicts correlation of seven Antarctic deep-sea cores on the basis of their magnetic stratigraphy and relates this correlation to the standard magnetic stratigraphy composite section. Note that the magnetic stratigraphy provides rather convincing correlation among these cores, even though they contain varying types of sediment and are physically separated by as much as 2000 kilometers.

Of further interest to stratigraphers is the fact that biostratigraphic events, such as beginnings and terminations of ranges, appear in some cases to be closely associated with magnetic events (see Hays and Opdyke, 1967; and Phillips *et al.,* 1968, for example). This fact offers the stratigrapher the possibility that a single event in the history of the earth may be expressed in both *physical* (magnetic properties) and *biological* (termination or beginning of a range) data. The event, therefore, becomes a concept that allows correlation among various types of data.

Continental Iron-Bearing Sediments. Alluvial sediments deposited in an oxidizing environment may also record the magnetic history of the earth. The usual problems of detrital versus authogenic origin of the magnetic minerals must be taken into account in these sediments. Furthermore, interaction between stream transport and gravity may produce depositional fabrics unrelated to the magnetic field. Thus, cross-bedded, channel-sand deposits may be a poor choice for magnetic study, whereas associated muddy, flood-plain deposits may yield reliable magnetic data.

Soil zones may likewise contain significant magnetic minerals to be suitable for study. The activity of burrowing organisms in the soil and the movement of sedimentary particles by root growth tend to disrupt primary depositional fabrics

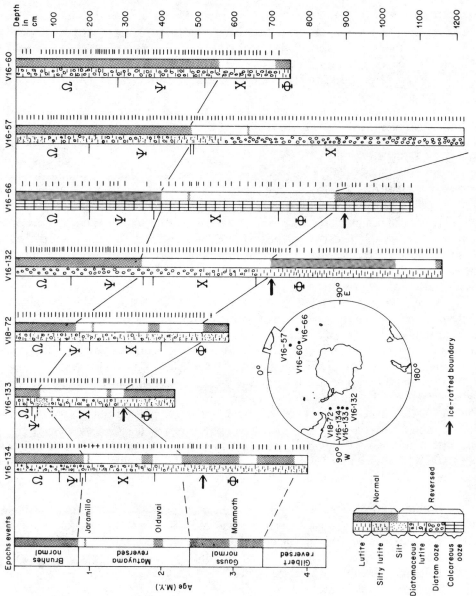

Figure 6.3 Correlation of Antarctic cores by magnetic stratigraphy. Greek letters indicate radiolarian biostratigraphic zones. Note that biostratigraphic boundaries often occur near magnetic events. [After N. D. Opdyke et al., "Paleomagnetic Study of Antarctic Deep-Sea Cores," *Science*, **154**, 349–357 (21 October 1966).] Copyright 1966 by the American Association for the Advancement of Science.

resulting from particle shape and gravity. As homogenization removes these strong primary fabrics, the magnetic mineral may take on a weak, but statistically significant, preferred orientation with respect to the magnetic field.

Time-Stratigraphic Correlation Based on Climatic Fluctuations

Physical events, such as magnetic reversals, provide actual *physical* basis for time-stratigraphic correlation. Physical events may also provide *conceptual* basis for time-stratigraphic correlation. In this case, the physical event is set forward as the unifying concept that allows patterns within stratigraphic time-series data to be assigned time-stratigraphic significance.

Environmental Stratigraphy. Variation in climate commonly induces variation in the distribution pattern of organisms. Even if the organisms involved in these shifting patterns remain the same over large intervals of time, changes in their distribution pattern may produce a stratigraphy in which climatic fluctuations provide the conceptual basis for time-stratigraphic correlation. A generalized example of this concept as it might apply to deep-sea planktic organisms is provided in Figure 6.4. Note that time-stratigraphic correlation is based on ecological shifts among populations of organisms rather than on first appearances or last appearances of the various taxa.

Occasional extreme shifts of environment can provide useful end-members of this correlation scheme. For example, it is common in the lagoonal environment to

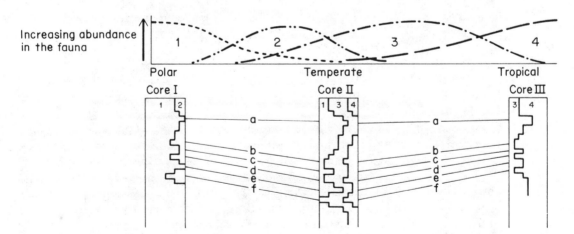

Figure 6.4 Schematic example of time-stratigraphic correlation based on shifts of fauna in response to climatic change. Faunas (1), (2), (3), and (4) occur predominantly in polar, subpolar, subtropical, and tropical regions, respectively. Even though the same organisms may exist throughout a long time interval, variations in their distribution pattern may provide a basis for event correlation. In this fashion, horizon *a* through *f* correlate shifting patterns of "warmer" to "cooler" fluctuations down each of three cores.

find particular beds extremely rich in the remains of certain organisms that are not otherwise present within the stratigraphic sequence. Presumably, environmental conditions throughout the lagoon were suitable for this organism for a brief period of time.

$\delta^{18}O$ Variation in Biogenic Calcite. The abundance of oxygen-18 in a calcite is normally reported in the standard "del" notation, where:

$$\delta^{18}O \; (°/_{oo}) = \left(\frac{^{18}O/^{16}O \; (sample)}{^{18}O/^{16}O \; (standard)} - 1 \right) \times 1000$$

The customary standard for comparison among calcites is the University of Chicago PDB standard (being a particular belemnite from the Pee Dee Formation of South Carolina). Isotopic composition of natural waters is commonly referred to SMOW (being Standard Mean Ocean Water).

It has long been recognized that the isotopic composition of biogenic calcite reflects the temperature and isotopic composition of the water from which the calcite grew (Epstein *et al.,* 1953). A commonly cited equation expressing this relationship is as follows:

$$T = 16.5 - 4.3 \, (\delta_c - \delta_w) + 0.14 \, (\delta_c - \delta_w)^2$$

where: δ_c = equilibrium oxygen isotopic composition of the calcite; δ_w = the isotopic composition of the water from which the calcite was precipitated; and T = temperature of the water from which the calcite was precipitated (°C).

It is easy to measure the $\delta^{18}O_{calcite}$ for fossils such as molluscs and foraminifers. Early interest in this measurement centered around the potential for estimating paleotemperatures. More recently, attention has become focused on estimation of variation in $\delta^{18}O_{water}$ as a measure of global ice volume (Shackleton and Opdyke, 1973, for example). This relationship between seawater $\delta^{18}O_{water}$ and the volume of ice requires further explanation.

The fractionation of heavy oxygen during the hydrologic cycle is depicted in Figure 6.5. When water evaporates from the ocean into the atmosphere, there is a fractionation of water containing oxygen-16 and oxygen-18. Statistically, water containing heavy oxygen tends to be left in the ocean because the molecule is heavier. Likewise, when atmospheric moisture returns to the surface as precipitation, it is the heavier water that will tend to precipitate first. Thus, precipitation to the interior of continents or in polar regions is significantly depleted in oxygen-18 when compared to seawater. It follows then that the relative abundance of oxygen-18 in ocean water is dependent upon the amount of ice impounded in continental glaciers. Thus, to the extent that temperature can be constrained to remain constant, the measurement of $\delta^{18}O_{calcite}$ in organisms within stratigraphic sequence provides a record of changes in $\delta^{18}O_{water}$ and thereby changes in global ice volume. The concept that the $\delta^{18}O_{calcite}$ in stratigraphic sequence is varying in response to the global phenomenon of ice-volume fluctuation encourages us to think of $\delta^{18}O_{calcite}$ correlation among stratigraphic sequences as time-stratigraphic correlation.

As indicated in Figure 6.5, an intercalibration can be obtained among $\delta^{18}O_{water}$, ice volume (and thereby changing sea level), and position of the shoreline. Derivation of this calibration is treated in detail in Chapter 16. For the present time, simply note that the concept of fluctuation in global ice volume is a unifying concept

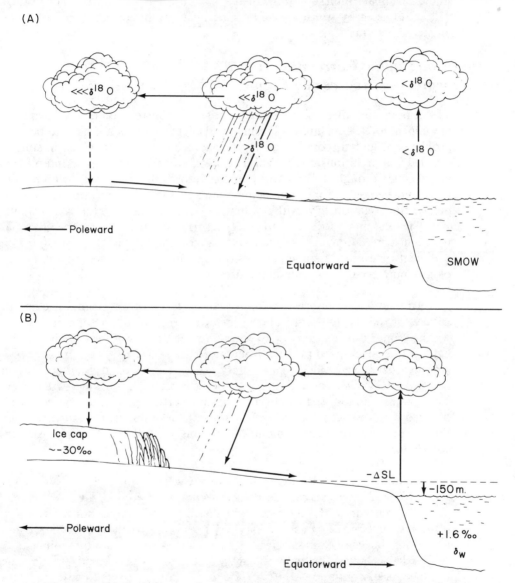

(A)

$<<<\delta^{18}O$ $<<\delta^{18}O$ $<\delta^{18}O$

$>\delta^{18}O$ $<\delta^{18}O$

Poleward →

Equatorward →   SMOW

(B)

Ice cap
~-30‰

Poleward →

$-\Delta SL$

-150 m.

Equatorward →

+1.6‰

δ_W

Figure 6.5 Schematic cross section indicating oxygen-18/oxygen-16 fractionation during hydrologic cycle. Note that storage of relatively light water in continental ice sheets causes the average isotopic composition of the ocean to become more heavy. Further, fluctuating ice volume on the continents causes sea level and shorelines to fluctuate.

that should manifest itself in different ways in different stratigraphic sequences. In the pelagic realm, we have the opportunity to measure this fluctuation as the $\delta^{18}O_{calcite}$ signal in planktic organisms. In bank-margin carbonates, we have the opportunity to measure sea-level lowering in the diagenetic overprint of fresh water upon unstable carbonate mineralogy. In the coastal clastic regions, fluctuating ice volume will be reflected by transgressive and regressive events measuring hundreds of kilometers in lateral dimension.

Time-Stratigraphic Correlation Based on Transgressive-Regressive Events

At this point, our discussion of time-stratigraphic correlation drops down to a much more local level. Our interest in time in stratigraphy does not have to be limited to problems of global scope. A geologist may become quite involved in studying the detailed time relationships within a relatively small stratigraphic interval and not give a single thought to how this area ties into global time-rock or time units. The petroleum geologist, for example, may find a stratigraphic trap simply by unraveling precise time relationships within a limited stratigraphic interval over a small area.

The relation of sediments to sea level forms another important conceptual basis for event correlation. As the sea level changes, the sedimentation patterns change. Sediment types may assume new positions within the paleogeography, or new lithologies may occur over widespread areas.

Symmetrical Transgressive-Regressive Cycles. Lithologic and paleontologic attributes of rock types usually allow generalized reconstruction of a paleogeographic profile from basin margin to basin center. Let us view a measured section in the light of this generalized profile. When more basinward rocks occur over more marginward rocks, we can see that a transgression has occurred. Similarly, when marginward rocks occur over basinward rocks, a regression has occurred. Where we can demonstrate that sediment types within a vertical profile are symmetrically distributed with respect to the most basinward facies present, that surface connecting the most basinward facies among the various vertical profiles approximates time correlation.

Asymmetrical Transgressive-Regressive Cycles. Submergence followed by progradation and aggradation commonly results in an asymmetrical cycle. Submergence often occurs with such rapidity that sedimentation does not record the orderly transgression from one facies to another. Rather, we frequently find extremely shallow-water facies or even continental facies abruptly overlain by sediments that indicate deeper water or more open marine conditions. Rapid submergence has produced a discontinuous shift, or *kickback,* in the regional configuration of the generalized paleogeographic profile. In such a case, time correlation can be made just above the discontinuity surface. All lithologies in that stratigraphic position existed almost immediately after the kickback event. Those sediments overlying the kickback surface record the synchronous reestablishment of a paleogeographic profile immediately following the transgressive event. From this point onward, progradation and aggradation may produce a complicated facies mosaic that defies precise time correlation.

Thin Limestone Beds in Clastic Marine Sequences. Clastic sedimentation in shallow-marine environments is closely related to sediment supply from the land area. Because of the characteristics of meandering-stream sedimentation and deltaic sedimentation (discussed in later chapters), transgressive pulses may cause the clastic sediment supply to be momentarily trapped in alluvial and deltaic environments. "Momentarily" is used here in the geologic context; probably thousands of years would be involved. Such disruption of clastic sediment supply to shallow-marine environments may allow carbonate sedimentation to occupy large areas for a short time; thus, there is opportunity for a special case of kickback correlation. The subsequent return of normal conditions of clastic sediment supply will cause the limestones to be replaced in time by renewed clastic sedimentation. The net result of this overall sequence of events will be the deposition of thin beds of limestone throughout areas that are normally dominated by clastic sedimentation. Such limestones have been recognized historically as valuable marker beds for *physical* stratigraphic correlation. By tracing their relationship to transgressive events, we have a conceptual basis for considering their significance in *time*-stratigraphic correlation.

Rapid Depositional Events

Once again, we return to our opening remarks: The search for the best time-stratigraphic correlation is unending. Some sedimentary events have long been acknowledged as useful in time-stratigraphic correlations; other events appear quite reasonable but have not been widely used.

The impact of a large meteorite colliding with the earth would surely be a rapid and spectacular event. The resulting dust clouds would not only significantly alter global climate but also might settle out as an identifiable clay horizon over large areas. The search for such events requires rather sophisticated chemical analysis of clay minerals (Alvarez *et al.,* 1980, for example). On a somewhat less spectacular scale, volcanic ashfalls (called *bentonites* by stratigraphers) are valuable aids in time-stratigraphic correlation studies. Major eruptions of volcanoes are discrete events of relatively short duration and widely separated in time. The ash from such an event may blanket large areas and therefore provide a time marker within the sedimentary sequence.

Turbidity current deposits may also be catastrophic events that blanket large areas. Other examples are sandstorms, which may lavishly deposit petrographically or mineralogically identifiable material. The possibilities are numerous. The ultimate completeness of a list of such events depends only on keen observation and a mind sufficiently alert to make sense out of what has been observed.

Citations and Selected References

BUREK, P. J. 1970. Magnetic reversals: their application to stratigraphic problems. Bull. Amer. Assoc. Petrol. Geol. 54: 1120–1139.
Application of magnetic stratigraphy to the problem of defining the Paleozoic-Mesozoic boundary.

CLUBE, S. V. M., and W. M. NAPIER. 1982. The role of episodic bombardment in geophysics. Earth Planet. Sci. Lett. 57, 251–262.
Good review of the meteorite impact hypothesis.

COX, A. 1969. Geomagnetic reversals. Science 163: 237–245.
Excellent summary paper.

COX, A., G. B. DALRYMPLE, and R. R. DOELL. 1967. Reversals of the earth's magnetic field. Sci. American 216: 44–61.
A popular summary including a discussion about methods of measurement.

CURRY, W. B., and R. K. MATTHEWS. 1981. Paleo-oceanographic utility of oxygen isotopic measurements on planktic foramin-ifera: Indian Ocean core-top evidence. Palaeogeog. Palaeoclimat. Palaeoecol. 33: 173–191.
Some planktic foraminifers record sea-surface conditions; others do not.

EMILIANI, C. 1966. Isotopic paleotemperatures. Science 154: 851–857.
Temperature interpretation of $^{18}O_{calcite}$ record. See also Shackleton and Opdyke (1973).

EPSTEIN, S., R. BUCHSBAUM, H. A. LOWENSTAM and H. C. UREY. 1953. Revised carbonate-water isotopic temperature scale. Bull. Geol. Soc. Amer. 64: 1315–1326.

GLASS, B. P., and M. J. ZWART. 1979. North American microtektites in Deep Sea Drilling Project cores from the Caribbean Sea and Gulf of Mexico. Bull. Geol. Soc. Amer. 90 (6): 595–602.
Recognition of microtectite horizons allows time-stratigraphic correlation over a large area.

HARRISON, C. G. A., IAN MCDOUGALL, and N. D. WATKINS. 1979. A geomagnetic field reversal time scale back to 13.0 million years. Earth Planet. Sci. Lett. 42: 143–152.

HAYS, J. D. 1971. Faunal extinctions and reversals of the earth's magnetic field. Bull. Geol. Soc. Amer. 82: 2433–2448.

HAYS, J. D., and N. D. OPDYKE. 1967. Antarctic radiolaria, magnetic reversals, and climatic change. Science 158: 1001–1011.
Relates the disappearance of some radiolaria to magnetic events.

HAYS, J. D., JOHN IMBRIE, and N. J. SHACKLETON. 1976. Variations in the earth's orbit: pacemaker of the ice ages. Science 194: 1121–1132.
A classic paper timing stratigraphic variations to global climate change.

IMBRIE, J. and N. G. KIPP. 1971. A new micropaleontological method for quantitative paleoclimatology: application to a late Pleistocene Caribbean core, p. 71–181. *In* K. Turekian (ed.), The Late Cenozoic glacial ages. Yale Univ. Press, New Haven, Conn.

IRVING, E. 1964. Paleomagnetism and its application to geological and geophysical problems. John Wiley & Sons, New York. 399 p.

IRWIN, M. L. 1965. General theory of epeiric clear water sedimentation. Bull. Amer. Assoc. Petrol. Geol. 49: 445–459.
A classic paper concerning the recognition of time-stratigraphic correlation that cuts across lithostratigraphic correlation.

ISRAELSKY, M. C. 1949. Oscillation chart. Bull. Amer. Assoc. Petrol. Geol. 33: 92–98. Proposes a symmetrical cycle for transgression and regression based on benthonic foraminifera assemblages.

MALKUS, W. V. R. 1968. Precession of the earth as the cause of geomagnetism. Science 160: 259–264.

MCELHINNY, M. W. 1971. Geomagnetic reversals during the Phanerozoic. Science 172: 157–159.
Generalized magnetic history of the earth from the Cambrian to the Recent.

NAGATA, T. 1961. Rock magnetism. Plenum Press, New York. 350 p.

OPDYKE, N. D., B. GLASS, J. D. HAYS, and J. FOSTER. 1966. Paleomagnetic study of Antarctic deep-sea cores. Science 154: 349–357.

PHILLIPS, J. D., W. A. BERGGREN, A. BERTELS, and D. WALL. 1968. Paleomagnetic stratigraphy and micropaleontology of three deep-sea cores from the central North Atlantic ocean. Earth Planet. Sci. Lett. 4: 118–130.
Relates the evolutionary transition from one foraminifera species to another during the Olduvai event.

POLLACK, J. B., O. B. TOON, T. P. ACKERMAN, C. P. MCKAY, and R. P. TURCO. 1983. Environmental effects of an impact-generated dust cloud: implications for the Cretaceous-Tertiary extinctions. Science 219: 287–289.
Spectacular dust cloud would result from large meteorite impact.

SHACKLETON, N. J., and N. D. OPDYKE. 1973. Oxygen isotope and palaeomagnetic stratigraphy of equatorial Pacific core V28-238: oxygen isotope temperatures and ice volumes on a 10^5 year and 10^6 year scale. Quaternary Res. 3: 39–55.
Ice volume interpretation of the $\delta^{18}O_{calcite}$ signal. See also Emiliani (1966).

WATKINS, N. D., and A. ABDEL-MONEM. 1971. Detection of the Gilsa geomagnetic polarity event on the islands of Madeira. Bull. Geol. Soc. Amer. 82: 191–198.

Absolute Time
in the Stratigraphic Record

7

In previous chapters, we have explored the methods that the stratigrapher uses to place rocks in their proper sequential relationships within the history of the earth. The ultimate conclusions drawn from these exercises are time-rock statements: These rocks are the same age as those rocks; these rocks are younger than those rocks, and so on. Now we shall look at the ways in which we can obtain estimates of absolute age within the time-rock sequence.

Radioactive Isotopes:
The Clocks of Geologic Time

Radiometric dating techniques have been a highly significant fallout of human interest in the exploitation of nuclear energy. Radioactive isotopes undergo systematic change with time by gaining or losing subatomic particles. All these reactions proceed as an exponential function of time, which can be characterized by the half-life of the reaction. In the case of decay reactions, for example, after one half-life, the abundance of the parent nuclei is one-half what it was at the beginning; after two half-lives, one-quarter the original amount; after three half-lives, one-eighth; after four half-lives, one-sixteenth; and so on. Table 7.1 summarizes some pertinent information concerning the utility of various reactions for radiometric dating. Details of applying these methods are discussed elsewhere (Hurley, 1959; Faul, 1966; and Eicher, 1968, for example).

In the following pages, we shall address ourselves to the stratigraphic aspects of radiometric dating. We assume that the geochemist is smart enough to give us a reliable date if we are smart enough to give him or her a sample that will have genuine stratigraphic significance once it has been dated.

Table 7.1 The Chief Methods of Radiometric Age Determination

Parent Nuclide	Half-life (in years)	Useful Age Ranges (years B.P.)	Daughter Nuclide	Minerals and Rocks Commonly Dated
Carbon-14	5730	< 25,000	Carbon-12	Wood, peat, $CaCO_3$
Uranium-235	——	< 150,000	Protactinium-231[a]	Aragonite corals
Uranium-234	——	< 250,000	Thorium-230[a]	Deep-sea sediment Aragonite corals
Uranium series	——	200,000 to tens of millions	Helium 4	Deep-sea sediment Aragonite corals
Uranium-238	4,510 million	> 5 million[b]	Lead-206	Zircon Uraninite Pitchblende
Uranium-235	713 million	> 60 million[b]	Lead-207	Zircon Uraninite Pitchblende
Potassium-40	1,300 million	> 50,000[b]	Argon-40	Muscovite Biotite Hornblende Glauconite Sanidine Whole volcanic rock
Rubidium-87	47,000 million	> 5 million[b]	Strontium-87	Muscovite Biotite Lepidolite Microcline Glauconite Whole metamorphic rock

(a) Intermediate daughter products.
(b) These methods may give useful information below the minimum indicated, but the methods become increasingly subject to serious error with decreasing age.

Radiometric Dating
of the Stratigraphic Record

A glance at Table 7.1 confirms that we may have some difficulties when we attempt to apply radiometric dating methods to the stratigraphic record. The rocks and minerals on our list are, in general, igneous and metamorphic. Indeed, there are relatively few situations in which we can directly tell the age of a sediment. More often, we must determine the age of associated igneous rocks and estimate the age of the sediments by stratigraphic inference based on the relationship of the sediments to the igneous rocks.

Direct Dating of the Sediment. Carbon-14 dating on materials in the sediment has proven invaluable in the study of the last 25,000 years of earth history. The dating is usually performed on wood, mummified organic matter, peat, or the skeletal remains of carbonate-secreting organisms. Although the analytical equipment used in C-14 determinations is capable of measuring apparent ages as old as 40,000 years, the possibility of contamination by Recent carbon makes ages beyond 25,000 years highly suspect.

Various methods involving thorium-230 and protactinium-231 are directly applicable to sediments in the time range of 1000 to 200,000 B.P. Once again, we note that these sediments are relatively young. Yet, a precise understanding of this interval of earth history is extremely important to our understanding of earth history as a whole. Thorium-230 and protactinium-231 are not detectable in seawater, whereas their parent nuclides, uranium-238 and U-235, are relatively abundant. Thus, thorium-230 or protactinium-231 growth in the skeletons of such marine organisms as corals provides an excellent clock, which can be easily read. We need only determine the abundance of either pair of nuclides to estimate the age of an aragonite coral sample. Because the daughter products were not present in the seawater in which the coral grew, their presence in the coral skeleton records only the decay of the parent after the coral skeleton formed. Thorium-230 and protactinium-231 undergo decay themselves. Thus, the ratios of these intermediate daughter products to their parents approach an equilibrium value with the continuing passage of time. Thorium-230 growth reaches steady state after approximately 250,000 years; protactinium-231 reaches steady state after 150,000 years.

Because thorium-230 and protactinium-231 are produced by their parents at different rates, the ratio of thorium-230 to protactinium-231 changes systematically with time. The ratio method can be applied to deep-sea sediments that initially contained measurable amounts of detrital thorium and protactinium. If we can assume constant isotopic composition of the detrital input to the sediment, the age of sediment from deep-sea cores can be calculated from the analytical data. In some cases, this assumption seems valid; in other cases, the assumption is almost certainly invalid. Thus, the ratio method on deep-sea cores yields age estimates that are inferior to estimates based on thorium-230 and protactinium-231 growth methods on corals.

The growth of helium concentration in aragonite coral skeletons provides a dating method potentially applicable to sediments as much as tens of millions of years old. Uranium-series nuclides produce helium as a decay product. The helium is trapped within the aragonite skeleton, and its concentration provides an estimate concerning the age of the coral.

All these methods for dating calcium carbonate require unaltered skeletal material. If the skeleton has undergone recrystallization, the system has been opened at some unspecified time and under some unspecified environmental conditions. The apparent date on such material will be invalid. Inasmuch as carbonate skeletons commonly undergo recrystallization, the specification of unaltered skeletal material places serious limitations upon the application of these techniques to the stratigraphic record.

In certain special cases, potassium-argon dates on clay minerals provide direct evidence for the time at which the sedimentary rock was deposited. However, we must approach the direct dating of clay minerals with extreme caution. For example, many "illites" in the stratigraphic record are nothing more than finely divided detrital grains of igneous or metamorphic muscovite. Other common clay minerals may have a complex origin involving chemical alteration of igneous and metamorphic materials followed by interaction with the water chemistry of the depositional environment. Thus, the dates for these minerals may not record the time at which the sediment was deposited but may indicate instead the time of origin for the igneous or metamorphic rock from which the detrital sediment was derived. Such a date provides information concerning the maximum possible age of the sediment. The sediment, however, may be much younger.

We can easily infer that some clay minerals are indeed the authogenic product of the marine environment in which the sediment was deposited. Potassium-argon dates on the mineral glauconite, for example, are commonly in fair agreement with other lines of evidence concerning the precise age of the sediments containing them.

Contemporaneous Volcanics. In general, volcanics are relatively easy to date, particularly by the potassium-argon method. Occasionally, we may be so fortunate as to find volcanics interbedded with otherwise continuous sedimentary sequences. Ashfalls, or bentonites, are particularly valuable because the observant stratigrapher can often demonstrate conclusively that the ashfall and the surrounding sediments are truly contemporaneous.

If the ashfall occurred within the sedimentary environment, then the date on the ash is a date on the biostratigraphic interval containing it. If, on the other hand, the ash simply fell upon an exposed rock surface of some ancient paleogeography, then the date on the ash simply tells us that everything below the ash is older and everything above the ash is younger than the data on the ash.

There are many things we can look for to infer the contemporaneity of the ash and the surrounding sediment. First and foremost, biostratigraphic zones are the main working tool of the stratigrapher. Are the rocks above and below the ash assignable to the same biostratigraphic zone? Furthermore, can it be demonstrated that the ash fell on soft sediment? Burrowing organisms frequently carry portions of the ash bed down into the underlying sediment. They also may carry overlying sediment down into the ash layer. Distribution of the ash layer with respect to the paleogeography may provide another important clue. Within shallow-marine environments, for example, the ash should accumulate within quiet-water environments, but it would presumably be disrupted and removed from agitated environments. Thus, lagoons and offshore areas should have the ash, whereas beach deposits should not. If the ash does cut across lagoons, beaches and offshore areas, we should suspect that the underlying paleogeography was not the site of active, shallow-marine sedimentation at the time the ash was laid down. Perhaps, for example, the entire paleogeography was lithified rock sitting well above sea level at the time the ash fell.

Lavas interbedded with sediments provide another opportunity to date volcanic material that may be synchronous with the interbedded sediment. The deposition of lavas into a stratigraphic sequence is a much more violent process than the deposition of ashfalls. Contact relationships between the sediments and the lavas will not be so conclusive as the evidence concerning ashfalls. In addition, we must be careful to distinguish lava flows from injected sills, a problem that we shall take up shortly.

Age Brackets on Sedimentary Sequences from Radiometric Dates on Related Igneous Rocks. Radiometric dating techniques have much more general application to igneous and metamorphic rocks. If we can demonstrate a relationship between sediments and nearby datable igneous rocks, then we have information concerning the absolute time of deposition of the sediment. Usually this information takes the form of an "older than" or "younger than" statement. If a sediment unconformably overlies metamorphic rocks and the sediment itself shows no signs of metamorphism, a date on the metamorphic rocks clearly tells us that the sediments are younger than that absolute age. At this point, we have no idea how much younger; the age of the metamorphic rocks simply puts a lower limit on our estimation of the age of the sediment.

If we continue to be concerned about the precise age of these sediments, we must find stratigraphic relationships with datable materials to provide an "older than" statement. For example, if we can demonstrate igneous intrusion into these same sedimentary beds, then the date of igneous intrusion tells us that the sediments are older than that age. Thus, in this example, two dates on unrelated igneous and metamorphic materials yield absolute ages that much bracket the age of the sediment.

Note that we are obliged to take very special care in documenting contact relationships between the sediments and associated igneous or metamorphic rocks that are to be dated. Incorrect interpretation of stratigraphic relationships results in patently wrong inference concerning the age of the sediment. If the sediment is either "older than" or "younger than" a given date, there is clearly no room to be halfway right.

The Correlation Web and Conflicting Radiomtric Dates. The quest for "a number" can become an overriding obsession. At some point, scientists must ask themselves what they really need to know and what may remain unknown. For example, a glauconite "age date" on marine sediments may be linked by a complex web of biostratigraphic and magnetostratigraphic correlation to a volcanic "age date" that is 5 million years older. All would agree that 5 million years is a significant amount of geologic time, but do we really have to resolve this problem in the next year? For most stratigraphic purposes, the answer will be "No!" As discussed under "Graphic Correlation" in Chapter 5, the stratigraphic goals are usually best served by establishing correlation as precisely as possible and by acknowledging that the exact age is beyond our requirements.

Table 7.2 Geologic Time Scale

Era	Period	Epoch	Duration in Millions of Years (Approx.)	Million of Years Ago (Approx.)
Cenozoic	Quaternary	Recent	Approx. last 5,000 years	
		Pleistocene	2.5	2.5
	Tertiary	Pliocene	4.5	7
		Miocene	19	26
		Oligocene	12	38
		Eocene	16	54
		Paleocene	11	65
Mesozoic	Cretaceous		71	136
	Jurassic		54	190
	Triassic		35	225
Paleozoic	Permian		55	280
	Carboniferous — Pennsylvanian		45	325
	Carboniferous — Mississippian		20	345
	Devonian		50	395
	Silurian		35	430
	Ordovician		70	500
	Cambrian		70	570
Precambrian				

Millions of years

The Geologic Time Scale

The various techniques discussed in the last three chapters have been applied for approximately 100 years. The result has been the continuing development and improvement of the geologic time scale. Tables 5.1 and 7.2 are a convenient summary of the words and absolute ages that are used today in discussing the history of the earth. Nevertheless, we should anticipate further modification and addition.

Citations and Selected References

BENDER, M. L., R. G. FAIRBANKS, F. W. TAYLOR, R. K. MATTHEWS, and J. G. GODDARD. 1979. Uranium-series dating of the Pleistocene reef tracts of Barbados, West Indies. Geol. Soc. Amer. Bull. 90 (pt. 1): 577–594.
Applies helium growth method to corals. Also demonstrates some pitfalls of the thorium-230 method.

BERGGREN, W. A., M. McKENNA, J. HARDENEN, and T. OBRADOVICH. 1978. Revised Paleogene polarity time scale. Geology 86 (1): 67–81. Berggren summaries generally well received.

DALRYMPLE, G. B. 1979. Research note: critical tables for conversion of K-Ar ages from old to new constants. Geology 7 (11): 558–560.
A seldom-discussed problem for those who really want to know precise age: How good are the constants in the equations?

DALRYMPLE, G. B., and M. A. LAMPHERE. 1969. Potassium-argon dating. W. H. Freeman and Co., San Francisco. 258 p.

EICHER, D. L. 1968. Geologic time. Prentice-Hall, Inc., Englewood Cliffs, N.J. 150 p.
Good supplementary reading for Chapter 5 through 7.

EMERY, K. O., and A. S. MERRILL. 1979. Relict oysters on the United States Atlantic continental shelf: a reconsideration of their usefulness in understanding late Quaternary sea-level history: discussion and reply. Bull. Geol. Soc. Amer. 90 (pt. 1): 689–694.
A long-smoldering question: Are 30,000-year carbon-14 dates real or just contaminated old shells?

EVERNDEN, J. F., and G. T. JAMES. 1964. Potassium-argon dates and the Tertiary floras of North America. Amer. J. Sci. 262: 945–974.

EVERNDEN, J. F., D. E. SAVAGE, G. H. CURTIS, and G. T. JAMES. 1964. Potassium-argon dates and the Cenozoic mammalian chronology of North America. Amer. J. Sci. 262: 145–198.

FAUL, H. 1966. Ages of rocks, planets, and stars. McGraw-Hill, New York. 109 p. Convenient discussion of radiometric dating methods.

HAMILTON, E. I., and R. M. FARQUHAR. 1968. Radiometric dating for geologists. Wiley-Interscience, New York. 506 p.
Collection of papers at the research level.

HARMON, R. S., T.-L. KU, R. K. MATTHEWS, and P. L. SMART. 1979. Limits of U-series analysis: phase 1 results of the Uranium-Series Intercomparison Project. Geology 7 (8): 405–409.
Interlab calibrations are always fun!

HART, S. R. 1964. The petrology and isotopic-mineral age relations of a contact zone in the Front Range, Colorado. J. Geology 72: 493–525.
Metamorphism affects radiometric clocks to varying degrees.

HURLEY, P. M. 1959. How old is the earth? Doubleday and Co., Inc., Anchor Books, New York. 160 p.
Colorful easy reading concerning radiometric dating and the history of the earth.

KULP, J. L. 1960. Geologic time scale. Science 133: 1105–1114.
Absolute ages of the periods.

LABRECQUE, JOHN L., DENNIS V. KENT, and STEVEN C. CANDE. 1977. Revised magnetic polarity time scale for late Cretaceous and Cenozoic time. Geology 5: 330–335.

LIBBY, W. F. 1955. Radiocarbon dating. 2nd ed. Univ. of Chicago Press, Chicago. 124 p.

MANKINEN, EDWARD A., and G. BRENT DALRYMPLE. 1979. Revised geomagnetic polarity time scale for the interval 0–5 m.y. B.P. J. Geophys. Res. 84 (B2): 615–626.

VAN HINTE, J. E. 1976. A Cretaceous time scale. Bull. Amer. Assoc. Petrol. Geol. 60 (4): 498–516.

———. 1976. A Jurassic time scale. Bull. Amer. Assoc. Petrol. Geol. 60 (4): 489–497.

Geodynamic Context
Of Sediment Accumulation

III

Bridging the gap from examination of Recent sediments to understanding of similar environments in the stratigraphic record requires a conceptual framework concerning large-scale dynamical processes that are constantly affecting the lithosphere upon which our sediments are accumulating. For example, we recognize that depositional environments impart properties to sediments that can easily be preserved in the stratigraphic record. Waves, tides, and currents of the beach environment commonly produce sediments that are well-rounded, well-sorted sands with abraded marine fossils and well-developed, low-angle cross-stratification. If we examine a hand specimen and thin section of a well-lithified sandstone possessing these properties, we are likely to deduce that the sandstone was once beach sediment. So far, so good.

Now, suppose that the hand specimen came from strata that included 200 meters of that lithology. The modern beach environment occupies a position within no more than a few meters of mean sea level. How could the geologic record stack up 200 meters of this lithology? Clearly, we are dealing with a problem involving dynamics. One logical explanation for the observed stratigraphic thickness would be that this portion of the earth's surface subsided during the time when beach sedimentation was occurring upon it. Thus, relationships between the beach environment and sea level remain constant on a day-to-day basis, while significant thickness of beach sediments accumulated over the long time interval.

Indeed, there would be no sedimentary record if it were not for dynamical changes in lithosphere—sea-level relationship. If the lithosphere were not in motion and if sea level did not rise and fall, an equilibrium profile would develop by which material eroded from the continents would be simply passed along down the equilibrium profile to ultimate sedimentation in the deep sea. While there would be sedimentation all along this path, there would be no net accumulation of new stratigraphic record. Everything coming in at the beginning of the system (erosion) would simply be passed through all the various sedimentary environments along the way with no net deposition until the sediment reached its final resting place in the deep ocean.

Fortunately for the stratigrapher, the earth is a dynamic place. Lithosphere plates do move around; sea level does go up and down. These generalities have been recognized in vague, qualitative terms by the stratigrapher for over 100 years. However, it is the geophysicists who have gathered a wealth of quantitative information concerning the ever-changing surface of the earth. They have studied earthquakes, plate tectonics, magnetics, gravity, and isostasy, to mention a few examples.

Until fairly recent years, stratigraphers have not dealt precisely with change or rate of change in the stratigraphic record. They did not have adequate control on absolute time in stratigraphy, and they did not have truly credible geophysical models for the changes that should be expected to be represented in the stratigraphic record. Now, stratigraphy and geophysics have developed to the point that we should be able to relate stratigraphic observations of change to the dynamics quantitatively described by geophysicists.

What the stratigrapher described qualitatively as *orogeny,* the geophysicist now quantifies as plate boundary tectonics; *epeirogeny,* lithospheric reheating; *eustasy,* ice-volume fluctuations and changing spreading rates. Interaction among these three quantifiable elements of earth dynamics will largely dictate facies geometry within the stratigraphic record.

If we are truly to consider "the present as key to the past," we must include dynamics in our view of the present. Just as modern beach sedimentation is the key to recognizing beach deposits in the stratigraphic record, modern dynamics of the earth surface are surely the key to understanding change and rate of change in the stratigraphic record. Let us therefore examine the generalities of the geophysicists' view of the earth and attempt to relate this view to what we might expect to see in the stratigraphic record.

Plate Tectonics
and the Stratigraphic Record

8

For a number of years, it has been recognized that the earth's magnetic field from time to time was reversed from its present polarity. Scientists have found it particularly convenient to study this phenomenon in layered sequences of volcanic rocks. As lava cools, it takes on magnetic properties indicative of the magnetic field in which it is crystallizing. Thus, normal and reversed magnetic fields are easily discernible in volcanic stratigraphic sequences. Furthermore, volcanic rocks are datable by conventional radiometric methods. On this basis, a late Cenozoic magnetic stratigraphy has been developed as indicated in Figure 6.1.

Concurrent with the development of radiometrically dated magnetic stratigraphy in volcanic stratigraphic sequences, it became apparent that certain portions of the sea floor had extremely systematic variation in magnetic properties that were quite hard to explain by prevailing paradigms. While other scientists raced out to gather more data on this important observation, it was a young graduate student named Vine who first put two and two together to get the elegantly simple answer that now forms a cornerstone of modern geology. Vine and Matthews (1963) proposed that the same patterns of normal and reversed magnetization, well known from volcanic stratigraphic sequences, explain the magnetic properties of oceanic crust observed along mid-ocean ridges (see Figures 1.2 and 8.1). If new formation of oceanic crust takes place along mid-ocean ridges, then the new crust should take on magnetic properties dictated by orientation and intensity of the magnetic field at the time of formation. With continuous generation of new crust beneath mid-ocean ridges, the ocean floor spreads apart, producing a horizontal stratigraphy of the earth's magnetic history that can be correlated with the vertical stratigraphy developed in layered volcanic rocks on the land. This hypothesis has become the cornerstone for one of the most important unifying concepts in modern geology.

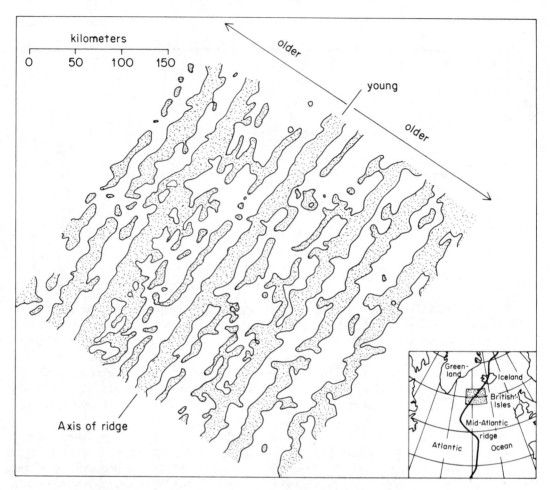

Figure 8.1 Magnetic anomaly pattern over the Reykjanes Ridge. The intensity of the present-day magnetic field is anomalously high over the stippled areas and anomalously low over the white areas. This results from the fact that oceanic basalt beneath the stippled areas has normal magnetization whereas oceanic basalt beneath the white areas has reversed magnetization because it solidified during a reversed magnetic epoch.

As indicated in Figure 8.2, new lithosphere is being generated beneath mid-ocean ridges and old lithosphere is being consumed beneath the deep-sea trenches.

Plate tectonics begins as the story of oceanic lithosphere, yet it is the stratigraphy of the continents that is of primary economic importance. In the following pages, we shall find that plate tectonics causes the continents to move laterally, to go up and down, to break apart, and to come together. Many of these phenomenon have been described quantitatively and can be demonstrated to have time frames of millions to tens of millions of years. They are an important element of the concep-

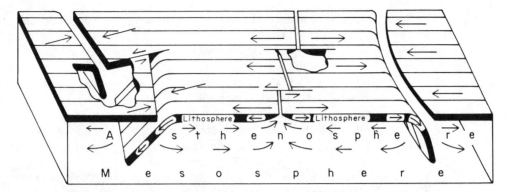

Figure 8.2 Schematic diagram indicating the major elements of plate tectonics. New oceanic lithosphere is generated along mid-ocean ridges and old oceanic lithosphere is consumed beneath trenches. The distribution of major ridges and trenches is indicated in Figure 8.4. [After B. Isacks *et al.*, *Jour. Geophysical Research*, **73**, No. 18, 5857 (1968).] Courtesy American Geophysical Union.

tual framework within which we shall consider the accumulation of stratigraphic sequences.

Continental and Oceanic Lithosphere

The continents and the ocean basins constitute the first-order geologic features of our globe. Several aspects of continents within the context of plate tectonics will be of special interest to us.

Compositional Topography. The upper 20 to 50 kilometers of continental lithosphere is composed of rocks (notably granite and sediments) that are less dense than oceanic basalt. Because of these compositional variations, it takes a thicker section of continental lithosphere to be at isostatic equilibrium with nearby oceanic lithosphere. Thus, compositional variation provides us with the first-order topography of our globe: continents and ocean basins.

The principal continental masses of today's world have been a feature of the earth's surface throughout the Phanerozoic. Geometric relationships among the continents will change dramatically, but the continents will always be there, the ocean basins will always be there, and there will be a shoreline somewhere between continent and ocean which will be the site of active sedimentation processes.

Subduction and Continental Lithosphere. In general, formation of new oceanic lithosphere at mid-ocean ridges requires subduction of lithosphere at trenches if the size of the globe is to remain more or less constant (see Figure 8.2). Subduction of oceanic lithosphere is quite well documented. However, when plate motions send continental lithosphere toward a zone of active subduction, buoyancy forces of the relatively low-density continental crust counterbalance the forces driving subduction. The continental crust tends to remain at the surface, and the zone of active subduction is shifted to some new region of oceanic lithosphere.

Figure 8.3 depicts such a situation presently in the making to the east of Japan. Shatsky Rise and Hess Rise have crustal density profiles more like continental lithosphere than oceanic lithosphere. The prediction depicted in Figure 8.3 is that these regions will not go down the Japan Trench, but rather will be accreted onto the Asian continent when they reach the present site of the Japan Trench. This phenomenon of accumulation of small continental masses onto larger continental masses has been going on throughout the Phanerozoic. In Figure 8.4, note especially the numerous continental blocks of central and southern Asia. The zone between two continental blocks is referred to as a suture zone and is characterized by intensive deformation, metamorphism, and uplift.

Paleomagnetic Data on Continental Blocks. The plate tectonics history of individual continental blocks can be studied by gathering paleomagnetic data on the rocks of that continent or subcontinent. Following substantially the same methodology as magnetic stratigraphy (Chapter 7), we may obtain a quantitative estimate of the apparent magnetic pole position responsible for the magnetic properties of various igneous or sedimentary rocks found within the continents. Given radiometric or biostratigraphic time control on these rocks, we can construct maps of apparent pole position through time for the various continents. Figure 8.5 presents such a map for North America, and Figure 8.6 presents one for Europe.

Many paleomagnetic studies precede the modern concept of plate tectonics. Indeed, early workers thought they were studying something rather unusual about the history of the earth's magnetic pole. The concept was that for some reason (which would be very interesting to know!) the earth's magnetic pole was at times in the past not coincident with the earth's present pole of rotation; hence, the study of "polar wandering."

In retrospect, we see that it is not the poles that have wandered, but rather the continents. Note, for example, that the polar wandering path for North America (Figure 8.5) is quite systematically different from the polar wandering path of Europe (Figure 8.6). The modern interpretation of these data is that the magnetic pole has remained more or less coincident with the pole of rotation and that the data reflect the motion of the continental blocks in response to plate tectonics. In Part V, we shall make extensive use of this relationship to reconstruct the positions of the major continents throughout the Phanerozoic.

Thermal Topography and Related Vertical Motions

Once again, we shall find that phenomenon well documented in oceanic lithosphere can be generalized to the more complicated regions of continental lithosphere and continental margin.

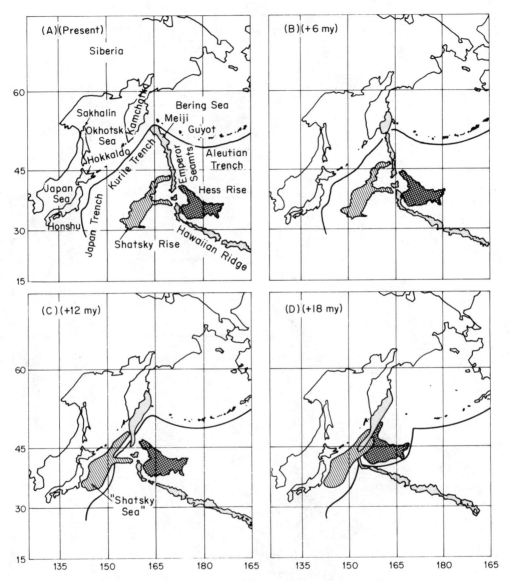

Figure 8.3 Series of sketch maps indicating possible future continental accretion events in the northwest Pacific. Shatsky Rise and Hess Rise are comprised of crust that is less dense than typical oceanic crust. Whereas typical oceanic lithosphere tends to be subducted beneath arc trench systems, less dense crust tends to be accreted onto the upper slab and a new subduction zone develops basinward. Thus, a continent may be a complex collage of microcontinents, separated from one another by highly deformed suture zones. A global perspective of microcontinents and suture zones is presented in Figure 8.4 [After Z. Ben-Avraham *et al.*, "Continental Accretion: From Oceanic Plateaus to Allochtonous Terranes," *Science,* **213,** 47–54 (3 July 1981).] Copyright 1981 by the American Association for the Advancement of Science.

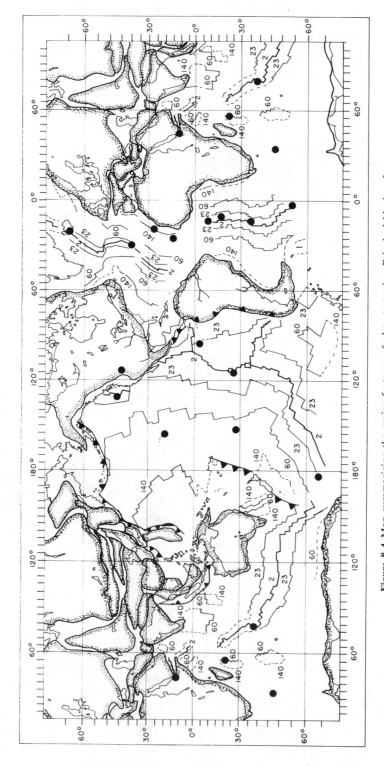

Figure 8.4 Map summarizing the major features of plate tectonics. Principal blocks of continental lithosphere are indicated by stipple pattern. The major elements of the mid-ocean ridge system are indicated by the heavy solid line labeled 2 million years. Position of 23-million-year-old lithosphere (Miocene–Oligocene boundary), 60-million-year-old lithosphere (Paleocene), and 140-million-year-old lithosphere (late Jurassic) are also indicated. The principal circum-Pacific arc-trench systems are indicated by solid lines with triangular flags pointing down the subduction zone. Closed circles indicate modern mantle "hot spots." Numerous other subduction zones are recognized within the Alpine–Himalayan tectonic belts. (Compiled from various sources including Scotese *et al.*, 1979.)

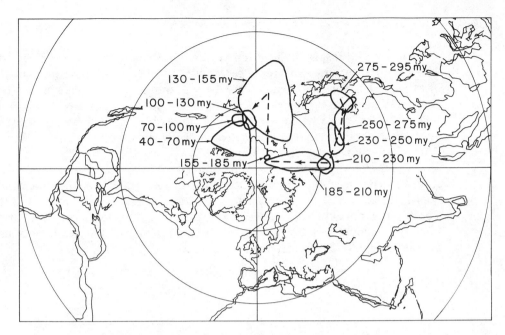

Figure 8.5 Apparent position of North American magnetic pole position throughout the last 300 million years. Magnetic properties of rocks within the North American continent allow measurement of an "apparent pole position" for the time represented by the rocks. Many such studies predate formulation of modern plate tectonics concepts. Thus, this field of investigation carries with it today the colloquial title of "polar wandering." The modern view is that it is not the poles that have wandered, but rather the continents.

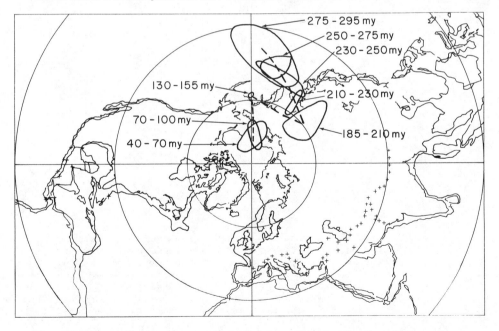

Figure 8.6 Map indicating apparent magnetic pole position as a function of time as determined from European rocks. Determination of paleolatitude from such data is extremely important to reconstruction of position of the continents throughout Mesozoic and Paleozoic times.

Oceanic Lithosphere Elevation versus Age Relationship. It is an empirical observation that water depth over oceanic lithosphere varies as a function of age of lithosphere (Parsons and Sclater, 1977, for example). Mid-ocean ridges, where new oceanic lithosphere is being formed today, have an average water depth of approximately 2800 meters whereas abyssal plains floored by Mesoszoic oceanic lithosphere generally have more than 5 kilometers of water plus sediment over oceanic crust. This relationship is depicted in Figure 8.7.

The thermal structure of oceanic lithosphere of various ages appears sufficient to explain much of the observed variation in topography. As oceanic lithosphere cools, the surface of the lithosphere subsides. As indicated in Figure 8.7, the relationship is exponential as a function of age. Young lithosphere subsides at approximately 10 centimeters per 1000 years; 80-million-year-old lithosphere is subsiding at something like 1 centimeter per 1000 years.

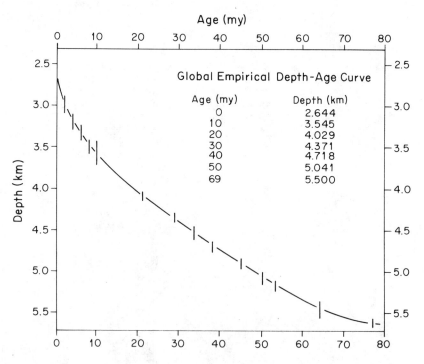

Figure 8.7 Global empirical age-versus-depth relationship for the top of oceanic crustal basalt. Young, hot lithosphere becomes more dense as it cools. Cooling of oceanic lithosphere produces the depth-versus-age relationship depicted here. Subsidence rates during the first 10 million years of lithospheric cooling are approximately 10 centimeters per 1000 years, whereas subsidence rates beyond 70 million years are only about 1 centimeter per 1000 years. This well-known subsidence curve can be taken as an approximation for regional subsidence of continental lithosphere associated with the opening of ocean basins. (From Pitman, 1978.)

Continental Lithosphere in High-Heatflow Areas. Mid-ocean ridges have high heatflow because they are situated over upwelling limbs of major convection cells within the asthenosphere. If a major continental block were to become situated over regions of major upward convection in the asthenosphere, that portion of the continent should become hotter and therefore thermally expanded and thinned as compared to the continent away from the high-heatflow region or to the continent before it became situated over the high-heatflow region.

Continents are very complex places compared to oceanic lithosphere, but this phenomenon is probably at least in part responsible for the general tendency to increased elevation in the western United States as compared to the eastern half of the North American continent. On the average, a dissected peneplain surface in the desert of the southwestern United States (New Mexico, Arizona, Nevada, Utah, and Colorado) stands about 2 kilometers higher than the Canadian Shield. Stratigraphic similarities between the Paleozoic of the Colorado Plateau and that of the eastern portions of the North American continent clearly indicate these regions were at comparable elevation at an earlier time in the history of the continent. The model of Bird (1979) places major uplift of the Colorado Plateau around 25 to 30 million years ago and indicates an uplift rate on the order of 20 centimeters per 1000 years.

Rifted Continents and Passive-Margin Subsidence. The next stage in the continuing saga of continental lithosphere situated over well-organized upward convection in the asthenosphere is for the continent to become broken apart (rifted) and for new oceanic lithosphere to form over the axis of upward convection. We shall see later in Part V that this phenomenon occurred on a grand scale during Mesozoic and early Cenozoic time when North America became rifted from West Africa, South America from central and southern Africa, and so on. On a lesser scale, the phenomenon appears to be occurring today within the region of the Red Sea.

Figure 8.8 presents a map and cross section of the Red Sea region. Thermal expansion of continental lithosphere preceded rifting of the Arabian subcontinent from the African continent. This is clearly indicated by large areas of elevated Precambrian basement to either side of the Red Sea rift. Note in Figure 8.8 that present elevation of Precambrian basement in eastern Egypt and Sudan and the Arabian Shield area is about 3 kilometers above sea level. Rifting and initiation of formation of oceanic crust in the Red Sea appears to be a late Tertiary phenomenon.

The East Coast of North America provides an example of a rifted continental margin considerably further along in its history. As rifting occurred in the early Mesozoic and North Atlantic oceanic crust began to form at the Mid-Atlantic Ridge, a lithospheric cooling-subsidence process was set in motion which was not unlike that depicted for oceanic lithosphere today (Figure 8.7). As the North American continental margin subsided, the weight of accumulating shallow-marine sediments further accentuated the subsidence process (discussed further in Chapter 10 under isostatic adjustment to new load). As indicated in Figure 8.9, the result is the ac-

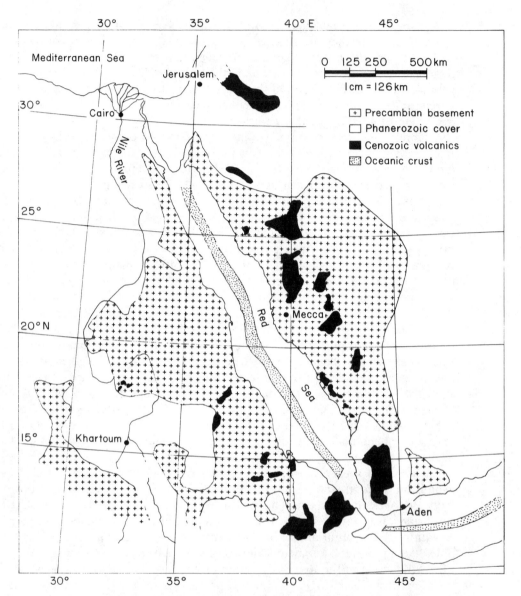

Figure 8.8 Sketch map indicating major geological features of the Red Sea rift. Geological evolution of the present situation began with uplift of Precambrian basement rocks to an elevation kilometers above sea level as a result of subcrustal heating. Subsequently, crustal lithosphere failed and emplacement of oceanic lithosphere commenced along the length of the present Red Sea. Thus, the present situation can be generally portrayed as generation of new oceanic lithosphere flanked on both sides by Precambrian continental crust that stands several kilometers above sea level. If sea-floor spreading should continue to occur within the Red Sea, one can imagine the continental margins subsiding along a thermal subsidence curve similar to that depicted in Figure 8.7. (Compiled with modification from Brown, 1970; Ghuma and Rogers, 1978; and Gass, 1970.)

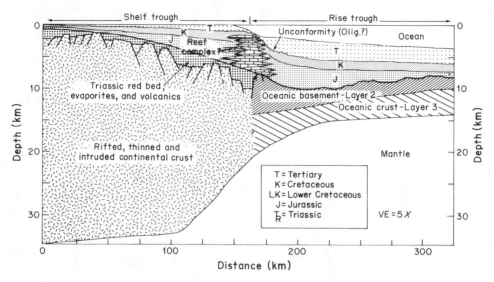

Figure 8.9 Schematic cross section of the North American continental margin east of New York. Note analogy between block-faulted basement, Triassic red beds, evaporites, and volcanics depicted here and the modern continental margins of the Red Sea. With continued opening of the North Atlantic, thermal subsidence has accommodated the deposition of approximately 8 kilometers of Jurassic through Tertiary continental margin sediments. (After Folger *et al.,* 1979.)

cumulation of some 14 kilometers of Mesozoic and Cenozoic sediments along the North American eastern continental margin.

Aborted Pull-Aparts. The presentation above takes a logical progression from thermally activated continental lithosphere (western U.S.) to rifted continents (the Red Sea rift) to passive-margin subsidence (East Coast of North America). An important variation on this general theme is the aborted pull-apart basin. The aborted pull-apart, also referred to as an *aulacogen,* can be thought of as a continental rifting system in which sea-floor spreading proceeded for a while and then ceased. Aborted pull-aparts most commonly occur at places in continental lithosphere where three directions of sea-floor spreading are tending to occur at the same time. The common occurrence is for two of these directions to become predominant and the third axis of sea-floor spreading to become an aborted pull-apart basin.

As indicated in Figure 8.10, aborted pull-apart basins are well known as being associated with the Mesozoic opening of the Atlantic Ocean. The North Sea graben, the Benue trough of Nigeria, and the Amazon basin are clearly failed limbs of old three-way spreading centers. The failed limb becomes the pull-apart basin and the other two limbs of the spreading system continue to spread to become the Atlantic Ocean (Dewey and Burke, 1974). Older basins are likewise associated with still older openings of paleo-oceans. The southern Oklahoma aulacogen (Burke and Wilson,

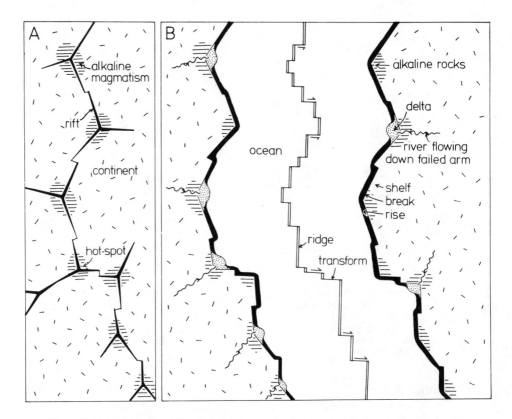

Figure 8.10 Sketch maps indicating development of aborted pull-apart basins along the failed limbs of former rift-rift-rift triple junctions in continental lithosphere. Thermal doming of continental lithosphere is envisioned to occur much as observed in Colorado Plateau and prerifting Egyptian-Arabian Shield areas (discussed above). Radius of the area affected by thermal doming is typically 500 kilometers. As propagating rifts begin to hook up to delineate the newly forming mid-ocean ridge, rifts that do not hook up become aborted pull-apart basins. These basins then undergo a thermal subsidence history consistent with the rest of the passive margin of the newly formed ocean basin. (After Dewey and Burke, 1974.)

1976b, for example) appears to be an earliest Paleozoic aborted pull-apart associated with a previous opening of the Atlantic. Hoffman, Dewey, and Burke (1974) develop the aulacogen model for Proterozoic geology of the Great Slave Lake region of Canada. The Mississippi embayment of central North America likewise shows many of the characteristics of an aborted pull-apart, but this feature is not clearly related to any well-defined triple junction.

Aborted pull-apart basins marginal to an opening ocean share many of the features cited above for passive continental margins. The pull-apart begins in thermally expanded continental lithosphere, a certain amount of oceanic lithosphere is formed in the middle of the pull-apart structure, and the region will undergo a very systematic subsidence history once the pull-apart is aborted and the feature is moved away from the high-heatflow ridge crest.

Hot-Spot Volcanism and Absolute Plate Motions. The hot-spot concept was originally invoked to explain volcanism that did not otherwise fit the plate tectonics model. It is proposed that there are spots in the earth where molten material persistently rises through the asthenosphere and invades the overlying lithosphere (Morgan, 1971; Crough *et al.*, 1980). By far the most spectacular example of this concept is the Hawaiian Islands—Emperor Seamount chain of the central North Pacific (Figure 8.11). As indicated in Figure 8.11, the basic data is that the age of major accumulations of basaltic material (the Hawaiian Islands and the seamount chains to the northwest) become progressively older away from the island of Hawaii. The hot-spot hypothesis is that there exists a persistent source of basalt rising through the asthenosphere beneath the present site of Hawaii. As the Pacific plate has moved westward and northward through the last 70 million years, this persistent source of basaltic magma has left a chain of extinct volcanoes which we now see as the Emperor Seamount chain and the Hawaiian Islands.

For any particular hot spot, we could equally argue that the position of the hot spot within the asthenosphere has migrated laterally with time. However, consistency among the traces of numerous hot spots tends to favor the concept that the hot spots are fixed within the asthenosphere and the traces of hot spots preserved as seamount chains reflect motion of lithosphere plates across fixed hot spots within the asthenosphere.

This concept is extremely attractive with regard to reconstruction of plate positions back through time. Whereas knowledge concerning the age of basaltic oceanic crust allows us to reconstruct the relation of one plate to another back through time, these data alone do not fix the plates with regard to the earth's rotational axis. By assuming that hot spots maintain a constant relationship with the earth's rotational axis, we can thereby track the motion of plates in a fixed coordinate system of latitude and longitude. A number of generally accepted hot spots are indicated in Figure 8.4.

Reheating and Thinning of Oceanic Lithosphere Passing over Hot Spots. There is commonly a broad swell extending hundreds of kilometers to either side of volcanic islands or seamount chains associated with hot spots. It is proposed that this increase in elevation of the sea floor reflects reheating and thinning of old lithosphere by the hot spot. This phenomenon is especially well known in the central Pacific Ocean basin where lithosphere asthenosphere relationships are especially easy to discern (Figure 8.12).

The net effect of passage of old lithosphere across a hot spot is to reset the thermal subsidence curve of that piece of lithosphere (Crough, 1978, 1979). Thus, a region of oceanic lithosphere that was passing down the age-versus-elevation curve depicted in Figure 8.7 might pass over a hot spot and have its elevation increased dramatically from the infusion of new heat. As the lithosphere moves away from

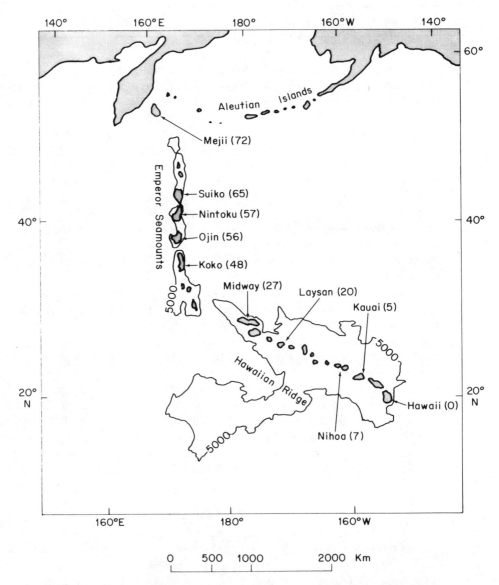

Figure 8.11 Sketch map indicating location of Emperor Seamount–Hawaiian Ridge volcanic chain. This chain of islands and seamounts is taken to record passage of the Pacific plate across a mantle plume (hot spot) presently located beneath Hawaii. The north end of the Emperor Seamount chain is dated at approximately 70 million years. The bend between Emperor Seamount chain and Hawaiian Ridge occurs at approximately 40 million years. Numerous dates on volcanic structures from Midway Island to Hawaii show good linear relation of age to distance from Hawaii, with Midway being approximately 27 million years old. The volcanic structures are located along the axis of the Hawaiian Swell, a broad uplift of oceanic lithosphere approximately 1000 kilometers wide and rising approximately 1 kilometer above surrounding abyssal plains. (After Dalrymple *et al.,* 1981, with substantial modification.)

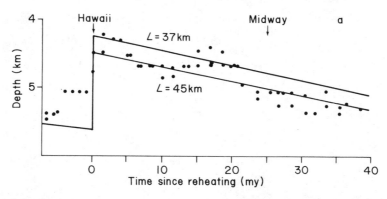

Figure 8.12 Observed and predicted uplift and subsidence of the Hawaiian Swell as the result of passage of oceanic lithosphere across a mantle plume (hot spot). Dots are 1° × 1° averages of depth of the crest of the Hawaiian Swell. Solid curves are uplift and predicted subsidence for lithospheric reheating models where L is the lithospheric thickness after reheating. Note that average depth to the top of oceanic lithosphere southeast of Hawaii is approximately 5.5 kilometers. Note that average depth of reheated lithosphere around Hawaii is around 4.4 kilometers. Note that subsidence models are in fair agreement with the observed depths out to as much as 40 million years after lithospheric reheating. (After Detrick *et al.,* 1981. Reprinted by permission from *Nature,* **292,** 142–143.) Copyright 1981 MacMillan Journals Limited.

the hot spot, subsidence of lithosphere in accordance with the general curve would again resume, but the lithosphere would not follow a cooling curve appropriate to a younger piece of lithosphere because its "apparent thermal age" was reset by the infusion of heat from the hot spot.

This complex history of thermal subsidence and thermal uplift of oceanic lithosphere does little to change the sedimentary record overlying the ocean floor. However, the data from such simple regions allow us to quantify our expectations concerning the interaction of hot spots with continental crust. Effects of such phenomena on continental crust and on resultant sedimentation patterns should be profound.

Reheating and Thinning of Continental Lithosphere by Hot Spots. Volcanic activity occurring within continental lithosphere plates has likewise been attributed to hot spot activity (Suppe *et al.,* 1975; Crough, 1981, for example). We should expect the hot spot phenomena to produce vertical motions in continental lithosphere as a result of lithospheric reheating and thinning such as described above for oceanic lithosphere. The phenomena is especially well displayed with regards to several hot spots that appear to exist beneath North African lithosphere. The fact that the North African plate is presently relatively stationary with regards to the mantle (Burke and Wilson, 1972) allows surficial geology of continent—hot spot interaction to be displayed in classic simplicity.

Geologic expression of several prominent North African hot spots is depicted in Figure 8.13. Note especially the relation between Precambrian "massifs" and Tertiary volcanism in the Hoggar and Tibesti regions of southern Algeria and north-

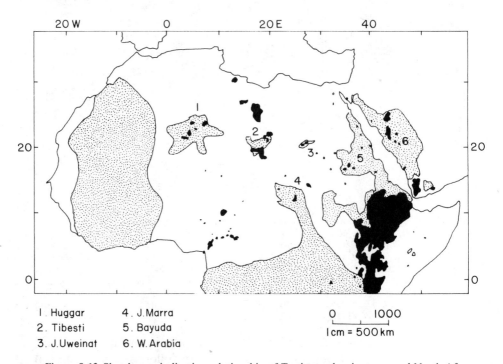

1. Huggar   4. J.Marra
2. Tibesti   5. Bayuda
3. J.Uweinat   6. W. Arabia

0   1000
1 cm = 500 km

Figure 8.13 Sketch map indicating relationship of Tertiary volcanics to several North African "Precambrian massifs." (Precambrian massifs are stippled, and Cenozoic volcanic fields are shaded.) It is proposed that both the high topography on Precambrian basement and the Tertiary volcanics result from the fact that these regions are underlain by mantle hot spots. Whereas the continent of Africa appears to be more or less stationary with regard to these hot spots, it is particularly instructive to consider the possibility of a continent passing across such hot spots or the possibility of intermittent hot-spot activity in the mantle. Either scenario will result in broad-scale regional uplift followed by subsidence in accordance with a relationship similar to that depicted in Figure 8.7. (Map compiled from Gass *et al.,* 1978, and Ghuma and Rogers, 1978, with modification.)

ern Chad. These Precambrian massifs have an average elevation of about 1 kilometer above sea level, an anomaly of several kilometers when compared to the depth of Precambrian basement in basins north and south of these regions. Intimately associated with topographic highs of the Precambrian massifs are Cenozoic volcanics. The interpretation is that the Hoggar and Tibesti massifs owe their topography of reheating and thinning of continental lithosphere by hot spots below. Several kilometers of Paleozoic sediment that existed over these regions prior to the onset of hot-spot—induced epeirogenic movements have been eroded.

It is interesting to speculate concerning the fate of the Hoggar and Tibesti regions if the African plate were to move these sites away from the hot spots or if the hot spots were to become inactive. Presumably, cooling and thickening of continental lithosphere would result in a subsidence-versus-time curve similar to those depicted in Figures 8.7 and 8.12. In this manner, it is reasonable that mid-plate continental topographic prominences underlain by Precambrian rocks might indeed become large-scale intercratonic basins. The time scale for this phenomenal topographic inversion would be measured in several tens of millions of years, a relatively short time in the total history of a continent.

Crough (1979) offers an interesting calculation concerning the probable importance of hot spot-induced epeirogenic uplift of continents throughout geologic time. Given today's distribution of presumed hot spots and today's average velocity of continental lithosphere plate motion, it is calculated that 50% of the surface of the continents has been effected by a hot spot within the past 420 million years. Thus, as a simple rule of thumb, whenever we look at a regional stratigraphic sequence that extends from Cambrian to Recent, we should anticipate that profound epeirogenic complications are lurking somewhere within the strata. These complications will be the rule rather than the exception, and the sorting out of regional epeirogenic effects from global eustatic effects will be a major task for the stratigrapher.

The rate at which a region undergoes epeirogenic uplift associated with passage over a hot spot can be roughly calculated as the maximum anticipated uplift (1 to 3 kilometers) divided by the time required for the continent to traverse 500 kilometers (taken as the distance to which modern hot spots appear to be affecting lithosphere topography). For average continental motions, this yields epeirogenic uplift rates on the order of 20 centimeters per 1000 years. Gass *et al.* (1978) offers a more thorough consideration of this problem. Epeirogenic subsidence rates can be taken to follow driving subsidence curves such as depicted in Figures 8.7 and 8.12. Driving subsidence rates of less than 10 centimeters per 1000 years are indicated. To these effects, we must add isostatic subsidence in response to accumulation of new sedimentary load, a topic to which we shall return in Chapter 9.

Tectonic Topography and Rates of Vertical Motion

In the preceding pages, we have noted some spectacular topographic contrasts (continents and ocean basins, for example), and we have noted large-scale thermal processes that produce uplift and subsidence with kilometers of vertical displacement. However, rates of vertical displacement have thus far been relatively small numbers, persistent in sign over relatively large time intervals, but generally measured in centimeters per 1000 years. To complete our picture of rates of vertical motion on the surface of the earth today, let us direct our attention to those relatively small portions of the earth "where the action is"—the plate boundaries.

The vast majority of plate tectonics literature is written in terms of lateral displacement rates rather than vertical displacement rates. The stratigrapher must view these processes in terms of vertical displacement, for it is subsidence which will allow room for new sediment accumulation. If we understand the depositional environment and paleotopography within a region, we must ask how fast tectonic subsidence might allow this picture to change. In practice, our knowledge of rates of vertical motion in orogenic belts is seriously limited by our need for precise time control if we are to calculate meaningful rates. We shall find over and over again in geology that it is very easy to estimate magnitude of uplift or subsidence, but very difficult to estimate the time represented by that vertical displacement. Thus, it is convenient to divide our discussion according to the method by which we shall estimate time.

The Earthquake Time Interval. Earthquakes provide the most direct opportunity to observe deformation of the earth's surface. Figure 8.14, for example, summarizes vertical displacements that accompanied the magnitude 8.5 Alaska earthquake of 1964 (Plafker, 1965). Local vertical displacements as great as 8 meters and regional vertical displacements of 1 to 2 meters to 75 kilometers to either side of the strike of the seismic zone were observed. Surely, these displacements occurred at the time of the earthquake or within a relatively few days following the quake. Can we estimate a geologically meaningful *average* rate of vertical displacement from these data? There are difficulties.

To begin with, these displacements record *discontinuous* ajdustment of the earth's surface to *continuous* processes that are occurring deep in the earth. While it is useful to the stratigrapher to know that the instantaneous displacements are measured in meters rather than tens or hundreds of meters, the fact that a particular piece of land rises 8 meters overnight is not necessarily useful information concerning the *average* rate at which tectonic uplift is occurring. Indeed, Plafker notes that the general history of Prince William Sound immediately prior to the earthquake was one of general subsidence, as indicated by drowned forests and intertidal peat bogs along the shores. Thus, large-scale tectonic deformation occurred overnight, yet similar rapid deformation is not likely to occur again in the near future.

A magnitude 8.5 earthquake is so large and unusual that we have very poor statistical data concerning the likelihood of recurrence of such a phenomenon within the same region. Furthermore, as we shall note in greater detail in Chapter 9, isostatic adjustment to new load occurs on a time scale of 10^3 to 10^4 years. Thus, a significant portion of the tectonic deformation observable immediately following an earthquake may be smoothed out over time as isostatic response to changing load occurs. All in all, it is difficult to estimate an *average* rate of tectonic deformation over geologically significant time scales from the study of deformation accompanying individual earthquakes.

Carbon-14 Dating and the 10^3-Year Time Interval. Continuing with the 1964 Alaska earthquake example, Plafker also notes a radiocarbon date on driftwood associated with the highest of five uplifted beaches on Middletown Island. This beach deposit dates 4470 B.P. ±250 years and is presently situated 30 meters above sea level. This suggests an average rate of tectonic uplift no greater than 8 meters per 1000 years over the last 5000 years. It is interesting to compare this average rate with the deformation resulting from a single earthquake event. Note in Figure 8.14 that Middletown Island experienced more than 1 meter of uplift during the 1964 earthquake. The radiocarbon dating of raised beaches now indicates that amount

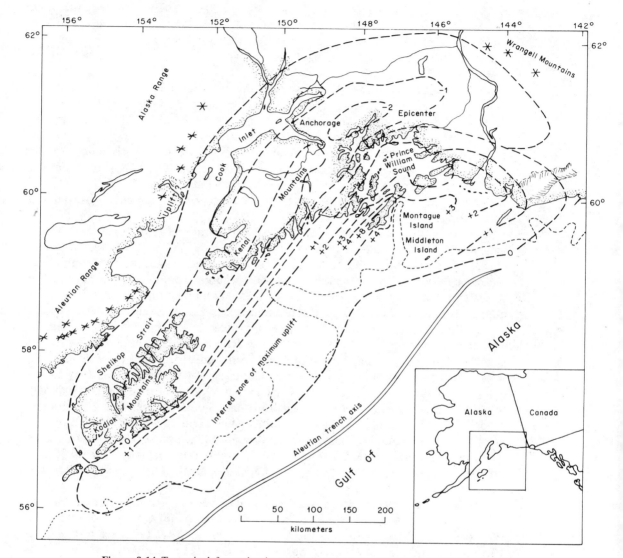

Figure 8.14 Tectonic deformation in south central Alaska following the 1964 earthquake. Contours indicate vertical displacement of the land surface; contour interval is 1 meter. Note that displacements of several meters can occur over very large areas virtually instantaneously. However, earthquakes of this magnitude are not an everyday occurrence. Thus, the displacement rate on the geological time scale is a relatively small number. [After G. Plafker, "Tectonic Deformation Associated with the 1964 Alaska Earthquake," *Science*, **148**, 1675–1687 (25 June 1965), with modifications.] Copyright 1965 by the American Association for the Advancement of Science.

of uplift to occur on the average of about every 200 years. Even so, we must note that 5000 years is an extremely short period of time from a sedimentological and stratigraphic point of view. Let us therefore look for some places where nature has run some longer experiment for us.

Uranium-Series Dating of Corals and the 10^5-Year Time Interval. From the study of geologically stable regions, we know that approximately 125,000 years ago the sea stood a few meters above its present level (Ku *et al.,* 1974; Neumann and Moore, 1975). Where coral reefs grew in association with this sea level, the geologist is provided with both a material (aragonite corals) that can be easily dated (by the thorium-230 growth method) and with a datum plane that closely approximates the 125,000 B.P. sea-level surface. Subsequent deformation of this 125,000 B.P. sea-level datum provides a good measure of the average rates of tectonic uplift or subsidence on the time scale of 10^5 years.

Tectonic uplift would place the 125,000 B.P. coral reefs above sea level, where they would be conspicuous features of the landscape. Tectonic subsidence would place the 125,000 B.P. reefs below sea level, where they would be inevitably recolonized by younger coral reefs and thus be obscured from easy examination. For this reason, data concerning uplift of this 125,000 B.P. datum are better and more numerous than data concerning its subsidence.

Uplifted Pleistocene coral reefs have been dated by the thorium-230 and similar radiometric methods in at least three areas widely separated around the world. The island of Barbados in the West Indies has been studied in detail (Mesolella *et al.,* 1969; Fairbanks and Matthews, 1978; Bender *et al.,* 1979). The case history of this study is discussed further in Chapter 13 as an example of how we can apply sedimentological data to unravel details of earth history. Similar studies in the Pacific are reported for the north coast of New Guinea (Veeh and Chappell, 1970; Bloom *et al.,* 1974) and for southern Japanese islands (Konishi, Schlanger, and Omura, 1970). The Barbados data suggests average tectonic uplift rates on the order of 30 centimeters per 1000 years. The Pacific data indicate average uplift rates up to 2 meters per 1000 years.

In Belize, Caribbean South America, shallow seismic profiling over the modern coral reefs and associated sediments affords an example of tectonic subsidence of Pleistocene coral reefs (Purdy and Matthews, 1964; Purdy, 1974; see also Figures 13.17 and 13.18). In the northern portion of the study area, Pleistocene coral reefs outcrop on land. The Pleistocene surface gradually slopes southward. In the southern portion of the study area, some 150 kilometers south of the subaerial outcrop, the Pleistocene surface is 30 meters below present sea level. An average rate of tectonic subsidence on the order of 30 centimeters per 1000 years is indicated.

Biostratigraphic Datums and the 10^7-Year Time Interval. If we wish to examine the average rates of tectonic deformation over a still longer time scale, we must look beyond reefs, beaches, and radiometric dating techniques. Raised reefs and other records of sea level events older than 125,000 years B.P. are numerous. However, they seldom contain datable materials, owing to recrystallization or chemical contamination of their contained fossils.

Christensen (1965) attempts to estimate vertical displacements in the central California coast ranges and San Joaquin Valley within the Pliocene-Pleistocene biostratigraphic framework. His results are summarized in Figure 8.15. In the southeastern corner of the map area as much as 3000 meters of subsidence has occurred within the last 3 million years. Over broad areas of the San Joaquin Valley, however, 500 meters of subsidence is a more appropriate general figure. Similarly, the coast ranges have been uplifted in places by as much as 1000 meters over the

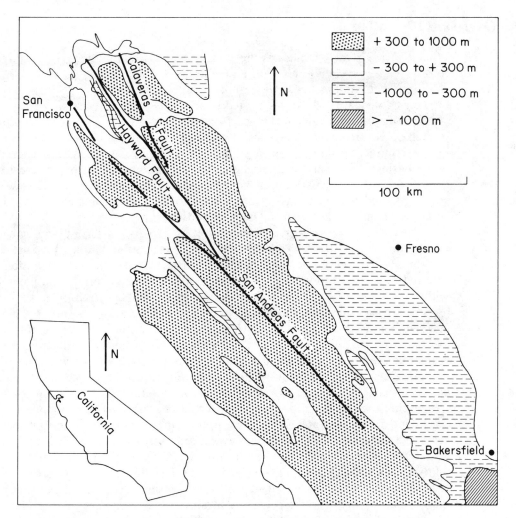

Figure 8.15 Uplift and subsidence in central California during the last 3 million years. (After Christensen, 1965, with simplification.)

past 3 million years. Once again, however, a general estimate of 700 meters uplift during 3 million years is a more realistic description of large areas within the coast ranges. Thus, estimates of the rate of tectonic subsidence range from an average value around 15 or 20 centimeters per 1000 years to as much as 1 meter per 1000 years. Similarly, rates of tectonic uplift range from an average of 20 to 25 centimeters per 1000 years to as much as 35 centimeters per 1000 years.

Matsuda, Nakamura, and Sugimura (1967) present a similar analysis based on the elevation of marine Miocene deposits over a large portion of Japan. They estimate average vertical displacement rates in the range of ±1 meter per 1000 years, although local figures may be as high as 5 meters per 1000 years.

Overview

When we view plate tectonics as the context of the accumulation of a stratigraphic record, it is convenient to recognize three levels of complexity. To begin with, there is an inherent compositional topography (the continents and the ocean basins) to the face of the earth which has existed in about its present form for several billion years. Some of the sediment derived from the ever-present continents is either deposited on continents during times of relatively high eustatic sea level, or is deposited at continental margins during times of relatively low eustatic sea level. The ocean basins continually receive sediment. Throughout this history of erosion and deposition, isostatic balance must be maintained. This relatively simple view of the continents and the oceans as a framework for the stratigraphic record will indeed fit large regions of the earth for large periods of time.

However, we have seen that reheating of preexisting lithosphere can induce vertical motions at the earth's surface that are measured in kilometers. Further, the present distribution of hot spots suggests that these large-scale vertical motions have affected intraplate topography of more than half of the world's continental area within the time frame of the Phanerozoic. Uplift may be accentuated by isostatic response to erosion, and subsidence may be accentuated by isostatic response to new sedimentary load. Rates of uplift associated with lithospheric reheating are not precisely known, but appear to be in the range of 10 to 20 centimeters per 1000 years. Subsidence due to lithospheric cooling appears well known with rates of driving subsidence ranging downward from 10 centimeters per 1000 years. This rate may be accentuated by a factor of two or three by isostatic response to new load.

Within the tectonically active regions that characterize plate boundaries, well-documented average rates of tectonic uplift range from 25 centimeters per 1000 years to 8 meters per 1000 years, and average rates of tectonic subsidence (including isostatic compensation for new load) range from 30 centimeters per 1000 years to 5 meters per 1000 years. This is a fairly wide variation in rate of vertical motion around plate boundaries. What sort of numbers should we expect as we approach the geologic record? Note especially that all the high estimates of tectonic vertical displacement rate come from the major island-arc provinces of the northern and western Pacific. Plate convergence rates in this region are the most rapid on the face of the earth today. Thus, it would appear that high rates of convergence give rise to high rates of vertical motion. For example, it is interesting to observe that the island of Barbados, West Indies, is in a position tectonically similar to that of Middletown Island, Alaska. Both islands are situated on broad tectonic uplifts (accretionary prisms) situated 100 to 200 kilometers in front of the volcanic island arc. Yet, the Aleutian arc-trench system is extremely active seismically whereas the Caribbean arc-trench system is the site of only moderate seismic activity. The respective estimates of average rate of tectonic uplift conform with this contrast: 8 meters per 1000 years for Middletown Island and only 30 centimeters per 1000 years for Barbados.

Thus, in areas of extremely rapid plate convergence, we may expect vertical tectonic displacement rates toward the high end of the spectrum. In all other areas, we shall probably consider a rate of 1 meter per 1000 years up or down, as the maximum rate to be anticipated. Comparison between this number and the average rate of eustatic sea-level fluctuations will be extremely important in later chapters.

Citations and Selected References

ALVAREZ, LUIS W., WALTER ALVAREZ, FRANK ASARO, and HELEN V. MICHEL. 1980. Extraterrestrial cause for the Cretaceous-Tertiary extinction. Science 208 (4448): 1095–1108.
The earth has taken lots of shots from space. Some *may* have profound global effects; the ultimate "event correlation" event!

BALLY, A. W., P. L. BENDER, T. R. MCGETCHIN, and R. I. WALCOTT (eds.). 1980. Dynamics of plate interiors. Amer. Geophys. Union, Geol. Soc. Amer., Geodynamics Series vol. 1, 162 p.

BEN-AVRAHAM, Z., A. NUR, D. JONES, and A. COX. 1981. Continental accretion: from oceanic plateaus to allochthonous terranes. Science 213 (4503): 47–54.

BENDER, MICHAEL L., RICHARD G. FAIRBANKS, R. W. TAYLOR, R. K. MATHEWS, JOHN G. GODDARD, and WALLACE S. BROECKER. 1979. Uranium-series dating of the Pleistocene reef tracts of Barbados, West Indies. Bull. Geol. Soc. Amer. 90 (6): 577–594.

BIRD, P. 1979. Continental delamination and the Colorado Plateau. J. Geophys. Res. 84: 7561–7571.
Process alternative to lithospheric reheating.

BLOOM, A. L., W. S. BROECKER, M. A. CHAPPELL, R. K. MATTHEWS, and K. J. MESOLELLA. 1974. Quaternary sea level fluctuations on a tectonic coast: new ^{230}Th/^{234}U dates from the Huon Peninsula, New Guinea. Quaternary Res. 4: 185–205.

BROWN, G. F. 1970. Eastern margin of the Red Sea and the coastal structures in Saudi Arabia. Phil. Trans. Roy. Soc. London A., 267: 75–87.

BURKE, KEVIN C., and J. TUZO WILSON. 1972. Is the African plate stationary? Nature 239: 387–390.

———. 1976a. Hot spots on the earth's surface. p. 58–69. *In* J. Tuzo Wilson (ed.), Continents adrift and continents aground. W. H. Freeman and Co., San Francisco.

———. 1976b. Hotspots on the earth's surface. Sci. American 235 (2): 46–59.

CHRISTIANSEN, M. N. 1965. Late Cenozoic deformation in the central coast ranges of California. Bull. Geol. Soc. Amer. 76: 1105–1124.

CROUGH, S. THOMAS. 1978. Thermal origin of mid-plate hot-spot swells. Geophys. J. R. Astr. Soc. 55: 451–469.

———. 1979. Hot spot epeirogeny. Tectonophysics 61: 321–333.

———. 1981. Mesozoic hotspot epeirogeny in eastern North America. Geology 9: 2–6.

CROUGH, S. THOMAS, W. JASON MORGAN, and ROBERT B. HARGRAVES. 1980. Kimberlites: their relation to mantle hotspots. Earth Planet. Sci. Lett. 50: 260–274.

DALRYMPLE, G. BRENT, DAVID A. CLAGUE, MICHAEL O. GARCIA, and SEPHEN W. BRIGHT. 1981. Petrology and K-Ar ages of dredged samples from Laysan Island and Northhampton Bank volcanoes, Hawaiian Ridge, and evolution of the Hawaiian-Emperor chain: summary. Bull. Geol. Soc. Amer. 92 (pt. 1, no. 6): 315–318.

DETRICK, R. S., R. P. VAN HERZEN, S. T. CROUGH, E. EPP, and U. FEHN. 1981. Heat flow on the Hawaiian Swell and lithospheric reheating. Nature 292: 142–143.

DEWEY, JOHN F., and KEVIN BURKE. 1974. Hot spots and continental break-up: implications for collisional orogeny. Geology 2: 57–60.

FAIRBANKS, RICHARD G., and R. K. MATTHEWS. 1978. The marine oxygen isotope record in Pleistocene coral, Barbados, West Indies: Quaternary Res. 10: 181–196.

GASS, I. G. 1970. The evolution of volcanism in the junction area of the Red Sea, Gulf of Aden and Ethiopian rifts. Phil. Trans. Roy. Soc. London A. 267: 369–381.

GASS, I. G., D. S. CHAPMAN, H. N. POLLACK, and R. S. THORPE. 1978. Geological and geophysical parameters of mid-plate volcanism. Phil. Trans. Roy. Soc. London A. 288: 581–597.

GHUMA, MOHAMED, A., and JOHN J. W. ROGERS. 1978. Geology, geochemistry, and tectonic setting of the Ben Ghnema batholith, Tibesti massif, southern Libya. Bull. Geol. Soc. Amer. 89 (9): 1351–1358.

GORDON, RICHARD G., and ALLAN COX. 1980. Paleomagnetic test of the early Tertiary plate circuit between the Pacific basin plates and the Indian Plate. Geophys. Res. 85 (B11): 6534–6546.

HAGER, BRADFORD H., and RICHARD J. O'CONNELL. 1981. A simple global model of plate dynamics and mantle convection. J. Geophys. Res. 86 (B6): 4843–4867.

HAXBY, W. F., D. L. TURCOTTE, and J. M. BIRD. 1976. Thermal and mechanical evolution of the Michigan Basin. Tectonophysics 36: 57–75.

HEESTAND, RICHARD LEE, and S. THOMAS CROUGH. 1981. The effect of hot spots on the oceanic age-depth relation. J. Geophys. Res. 86 (B7): 6107–6114.

HOFFMAN, P., J. F. DEWEY, and K. BURKE. 1974. Aulacogens and their genetic relation to geosynclines, with a Proterozoic example from Great Slave Lake, Canada, p. 38–55. *In*

R. H. Dott, Jr., and R. H. Shaver (eds.), Modern and ancient geosynclinal sedimentation. Soc. Econ. Paleont. Mineral. Spec. Pub. 19.

ILLIES, J. H. (ed.). 1981. Mechanism of Graben formation. Tectonophysics, 73 (Special Issue, no. 1-3): 1-266.

ISACKS, B., J. OLIVER, and L. R. SYKES. 1968. Seismology and the new global tectonics. J. Geophys. Res. 73: 5855-5899.
Excellent summary paper concerning seismology and plate tectonics.

KONISHI, A., S. O. SCHLANGER, and A. OMURA. 1970. Neotectonic rates in the Central Ryukyu Islands derived from thorium-230 coral ages. Marine Geology 9: 225-240.

KU, TEH-LUNG, MARGARET A. KIMMEL, WILLIAM H. EASTON, and THOMAS J. O'NEIL. 1974. Eustatic sea level 120,000 years ago on Oahu, Hawaii. Science 183: 959-962.

LE PICHON, XAVIER, and JEAN-CLAUDE SIBUET. 1981. Passive margins: a model of formation. J. Geophys. Res. 86 (B5): 3708-3720.

MATSUDA, T., K. NAKAMURA, and A. SUGIMURA. 1967. Late Cenozoic orogeny in Japan. Tectonophysics. 4: 349-366.

MCELHINNY, M. W. 1973. Palaeomagnetism and plate tectonics. Cambridge Univ. Press, Cambridge.

MCKENZIE, D. P., and F. RICHTER. 1976. Convection currents in the earth's mantle. Sci. American 235 (5): 72-85.

MESOLELLA, K. J., R. K. MATHEWS, W. S. BROECKER, and D. L. THURBER. 1969. The astronomical theory of climatic change: Barbados data. J. Geology 77:250-274.

MOLNAR, PETER, and DALE GRAY. 1979. Subduction of continental lithosphere: some constraints and uncertainties. Geology 7 (1): 58-62.

MORGAN, W. JASON. 1971. Convection plumes in the lower mantle. Nature 230: 42-43.

———. 1972. Plate motions and deep mantle convection; p. 7-21. *In* R. Shagram, *et al.,* Studies in earth and space sciences. Geol. Soc. Amer. Mem. 132.

NEILL, WILLIAM M. 1976. Mesozoic epeirogeny at the South Atlantic margin and the Tristan hot spot. Geology 4: 495-498.

NEUGEBAUER, HORST J., and PAUL TEMME. 1981. Crustal uplift and the propagation of failure zones. Tectonophysics 73: 33-51.

NEUMANN, A. CONRAD, and WILLARD S. MOORE. 1975. Sea level events and Pleistocene coral ages in the northern Bahamas. Quaternary Res. 5: 215-224.

PARSONS, BARRY, and J. G. SCLATER, 1977. An analysis of the variation of ocean floor heat flow and bathymetry with age. J. Geophys. Res. 82: 803-827.

PARSONS, BARRY, and DAN MCKENZIE. 1979. Mantle convection and the thermal structure of the plates. J. Geophys. Res. 83 (B9): 4485-4496.

PITMAN, WALTER C., III. 1978. Relationship between eustacy and stratigraphic sequences of passive margins. Bull. Geol. Soc. Amer. 89 (9): 1389-1403.

PLAFKER, G. 1965. Tectonic deformation associated with the 1964 Alaska earthquake. Science 148: 1675–1687.

PURDY, EDWARD G. 1974. Reef configurations: cause and effect, p. 9–76. *In* Leo F. Laporte (ed.), Reefs in time and space: selected examples from the Recent and Ancient. Soc. Econ. Paleont. Mineral. Spec. Pub. 18.

PURDY, E. G., and R. K. MATTHEWS. 1964. Structural control of Recent calcium carbonate deposition in British Honduras [abstract]. Geol. Soc. Amer. Program with Abstracts, 1964 Miami National Convention, 157.

ROSS, CHARLES A. 1979. Late Paleozoic collision of North and South America. Geology 7 (1): 41–44.

SCOTESE, CHRISTOPHER, R, RICHARD K. BAMBACH, COLLEEN BARTON, R. VAN DER VOO, and ALFRED M. ZIEGLER. 1979. Paleozoic base maps. J. Geology 87 (3): 217–277.

SLEEP, NORMAN, H. 1976. Platform subsidence mechanisms and "eustatic" sea-level changes. Tectonophysics 36: 45–56.

SMITH, R. B., and R. L. CHRISTIANSEN. 1980. Yellowstone Park as a window on the earth's interior. Sci. American 242: 104–117.
Easy reading concerning interaction of the North American plate with a hot spot.

STEINER, J. 1977. An expanding earth on the basis of sea-floor spreading and subduction rates. Geology 5 (5): 313–318.

STEWART, A. D. 1981. The expanding earth. Nature 290 (5808): 627.

SUPPE, J., C. POWELL, and R. BERRY. 1975. Regional topography, seismicity, Quaternary volcanism and the present-day tectonics of the western U.S. Amer. J. Sci. 275A: 397–436.
A classic "big picture" paper.

TAPPONNIER, PAUL, and JEAN FRANCHETEAU. 1978. Necking of the lithosphere and the mechanics of slowly accreting plate boundaries. J. Geophys. Res. 83 (B8): 3955–3970.

TAYLOR, BRIAN. 1979. Bismarck Sea: evolution of a back-arc basin. Geology 7 (4): 171–174.

VEEH, H. H., and J. CHAPPELL. 1970. Astronomical theory of climatic change: support from New Guinea. Science 167: 862–865.

VINE, F. J. 1966. Spreading of the ocean floor: new evidence. Science (3755): 1405–1415.
Plate tectonics: a giant at age three years!

VINE, F. J., and D. H. MATTHEWS, 1963. Magnetic anomalies over oceanic ridges. Nature 199 (4897): 947–949.
The birth of plate tectonics!

Eustasy, Isostasy,
and the Stratigraphic Record

9

There is no doubt about it, plate tectonics is the concept with the most visible attractions! How can anything else compete with all those earthquakes and volcanoes? However, when it comes to understanding the stratigraphic record, it is equally important that we thoroughly understand the concepts of eustasy and isostasy. Eustasy is the noun used to set apart sea-level fluctuations that are taken to be global in nature as opposed to more local "relative sea level" variations that might alternatively be ascribed to local tectonic activity. Isostasy is the noun used for the concept that the rigid outer portion of the earth (lithosphere) can be visualized as "floating" upon a more dense viscous asthenosphere. With regard to sedimentation and stratigraphy, the concept of isostasy thus dictates that lithosphere will rise if material is removed from it (erosion) and that lithosphere will sink if material is added on top of it (sedimentation).

In the following pages, we shall investigate the credibility and consequences of the concepts of eustasy and isostasy as they relate to accumulation of a stratigraphic record. In conclusion, we shall review some of the terminology that stratigraphers have generated to deal with the ambiguities that inevitably arise in the study of the stratigraphic record.

Glacio-Eustatic Sea-Level Fluctuations

By far the best known cause of global, synchronous sea-level fluctuations is variation in the amount of water stored in continental ice sheets. At present, the earth has a considerable volume of water stored in ice sheets over land areas (primarily Antarctica and Greenland). When there are fluctuations in the size of continental ice sheets, there are corresponding fluctuations in sea level. During Wisconsin time (approximately 18,000 B.P.), for example, the sea stood somewhere between 80 and

150 meters below its present level. At the other end of the spectrum, if all of the ice presently on the face of the earth were to melt, sea level would rise to approximately 65 meters above present sea level.

The history of ice volume fluctuation will be treated in greater detail in Part V. For now, we shall develop some rough numbers for rate of sea-level fluctuation in order that we may compare them to rates of vertical motion associated with plate tectonics (Chapter 8) and other eustatic mechanisms.

The Post-Wisconsin Eustatic Sea-Level Rise. Relict sediments associated with previous shorelines allow us to reconstruct the history of sea-level rise following Wisconsin glaciation (approximately 18,000 B.P.). In particular, samples of bottom sediment taken from water depths as great as 100 meters commonly contain peat or certain species of molluscs that are known to accumulate only under shallow-marine or brackish-water conditions. The presence of these materials *in situ* dictates that the sea was at that level. Thus, carbon-14 dates on relict shoreline deposits of the present-day continental shelf allow the reconstruction of a history of post-Wisconsin sea-level rise.

Figure 9.1 summarizes carbon-14 data pertaining to the history of sea-level rise over the past 17,000 years. From 15,000 B.P. to 6,000 B.P., it would appear that the average rate of sea-level rise was approximately 8 meters per 1000 years. It will be developed in Chapter 16 that the $\delta^{18}O$ variation in deep-sea cores is likewise related to fluctuations in global continental ice volume. For the present discussion, suffice it to say the shape of the curves for oxygen-18 variation in deep-sea cores (Figure 16.4, for example) suggests that other sea-level rises during the Pleistocene occurred at similar rates.

Rate of Glacio-Eustatic Sea-Level Lowering. The history of sea-level lowering into the main Wisconsin glaciation lies beyond the range of carbon-14 dating and has been further obscured by sediment reworking associated with the post-Wisconsin sea-level rise. We do know that sea level stood a few meters above present levels approximately 125,000 years ago and that it stood at least 80 meters below the present level approximately 18,000 years ago. These data constitute a minimum rate of sea-level lowering during the last glacial-interglacial cycle: 80 centimeters per 1000 years. However, the deep-sea oxygen isotope record suggests the transition from interglacial to glacial sea level was discontinuous and interrupted by numerous minor sea-level rises. Over the short term, rates of sea-level lowering in the range of 2 to 8 meters per 1000 years are to be anticipated.

Other Possible Causes of Eustatic Sea-Level Fluctuations

Glacio-eustasy is by far the dominant factor in late Cenozoic sea-level fluctuations. Its effects mask other possible effects that might contribute to eustasy. Nevertheless, calculations can be made concerning several physical phenomena that might cause

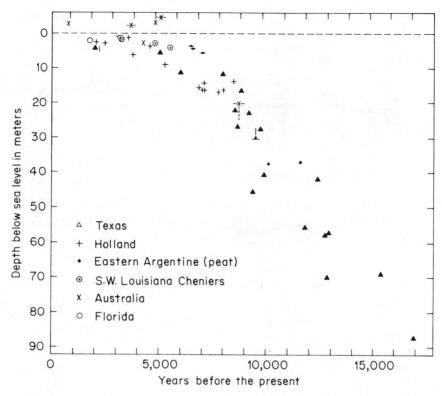

Figure 9.1 Sea-level rise over the last 17,000 years, as determined by carbon-14 dating of peat deposits and brackish-water molluscs. (After Sheppard, 1960.)

eustatic sea-level fluctuations. Pitman (1978) provides an especially useful summary of a number of these phenomena. Some of the more straightforward concepts are discussed below.

Changing Volume of the Mid-Ocean Ridge System. Global synthesis of data concerning age of oceanic lithosphere suggests that the rate of sea-floor spreading and resultant production of new oceanic lithosphere at ridge crests is not constant. It appears especially well demonstrated that late Cretaceous spreading rates were faster than modern spreading rates. Thus, during late Cretaceous time, a larger portion of the global ocean basin was underlain by relatively young lithosphere. As indicated in Figure 8.7, relatively young lithosphere commonly has around 3 kilometers water depth over it whereas relatively old lithosphere commonly has around 5.5 kilometers water depth over it. Thus, an ocean basin that is underlain by relatively young lithosphere must displace water upwards, a global eustatic high sea level by definition. The magnitude of this effect is proposed to be approximately 500 meter sea-level fluctuation. The rate at which this phenomenon might drive sea-level fluctuations is calculated to be approximately 1 centimeters per 1000 years.

Changing Ratio of Continental to Oceanic Lithosphere. Plate tectonics can have the effect of creating areas of new ocean basin at the expense of the area of continents. The most clear-cut example of this phenomenon is the compression and uplift of the Himalalyas where the Indian subcontinent has been driven into the Asian continent over the past 20 million years. Thickening of continental crust beneath the Himalayas has substantially reduced the area of continental crust within this region. This reduction in area of continental lithosphere is substantially offset by creation of new oceanic lithosphere south of India. A rough calculation would indicate an increase in global ocean basin volume sufficient to accommodate a 40-meter lowering of global sea level. An average rate of this sea-level lowering would be approximately 0.2 centimeters per 1000 years.

Sedimentation into the Global Ocean Basin. The converse of continental crustal thickening is clearly that the new high continent will simply deliver more sediment back to the ocean and thereby cause sea level to rise. Consideration of global average erosion rates suggests a sea-level rise from this mechanism might have a rate as high as 0.2 centimeters per 1000 years. However, it is extremely difficult to estimate the possible amplitude of sea-level fluctuations resulting from the combination of continental ice sheets. The uniqueness of the glacio-eustatic sea-level rise rate will become especially important for us in consideration of the existence of deep-water sediments over shallow-water or continental deposits. Sedimentation rates for most sedimentary environments are sufficiently rapid to keep pace with any other mechanism that might tend to cause relative sea-level rise. If sea level rises slowly onto a beach, beach sedimentation will simply perpetuate the beach upward and outward into rising water. Glacio-eustatic sea-level fluctuations are sufficiently rapid to overcome this tendency.

Isostasy and the Stratigraphic Record

The interior of the earth is viscous and the lithosphere is sufficiently weak that it would collapse under its own weight if it were not supported from below. Thus, the major topography of the earth's surface can be modeled as a collection of less dense blocks floating upon a more dense liquid (see Figure 9.2). If all the blocks have the same density, thicker blocks will produce higher topography. If the blocks are of differing density, less dense blocks will produce high topography.

In fact, the lithosphere is a complicated combination of these two hypotheses. Further, the lithosphere does have some strength. Thus, inhomogeneous distributions of mass can be supported by surrounding regions out to some measurable distance and for up to some measurable length of time. Erosion and sedimentation is basically the removal of material from a high block and the placing of that material on a low block. This redistribution of mass disturbs the previous isostatic equilibrium and the system must adjust to the new distribution of load by viscous flow beneath the lithosphere.

Airy hypothesis

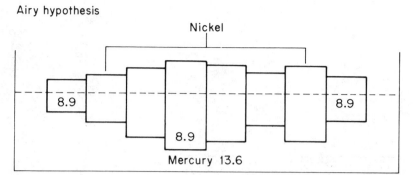

Pratt hypothesis

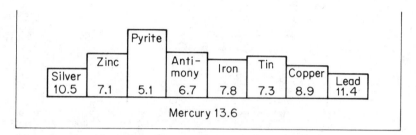

Figure 9.2 Basic concepts of isostasy. Both the Airy and the Pratt hypothesis seek to explain the topography of the earth with disconnected blocks floating in a heavy liquid. The Airy hypothesis suggests that topographic highs are compensated by unusual thicknesses of the same material, whereas the Pratt hypothesis proposes that topographic highs are composed of less dense material than topographic lows. The truth concerning the real earth is a combination of these two concepts. (After Leet and Judson, 1971.)

Isostasy and Basin Filling. Consider first the thickness of sediments beneath an alluvial fan adjacent to high topography. For the present discussion, assume that local isostatic equilibrium is maintained. We shall give further consideration to that assumption later. Presume that the surface of the alluvial fan now stands some 300 meters above the valley floor. How did things get this way and what lies beneath the alluvial fan?

Figure 9.3 develops the present-day situation in a stepwise fashion. Add the first 300 meters of alluvial fan sediment to the low block and what happens? Sediment of density 2.3 grams per cubic centimeter displaces mantle material of density 3.3 grams per cubic centimeter. Thus, isostatic readjustment will depress the surface of the alluvial fan by 70 percent of the thickness of sediment added. After 300 meters of alluvial fan has been deposited, the alluvial fan has only 90 meters of topographic relief. By the time the alluvial fan has 300 meters of topographic relief, the total thickness of sediment is approximately 1000 meters.

Another common situation is the deposition of sediments into a water-filled

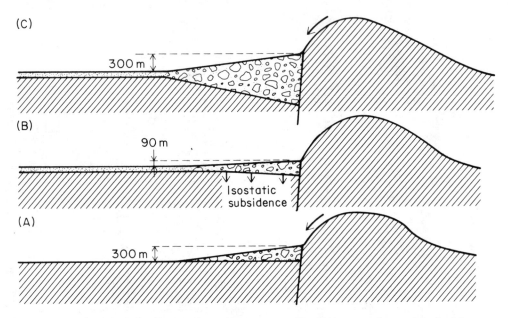

Figure 9.3 Schematic cross section indicating isostatic subsidence accompanying alluvial fan sedimentation. Steps (A) through (C) indicate how an alluvial fan with 300 meters of topographic relief may actually accommodate 1000 meters of total sediment thickness because of isostatic subsidence accompanying the sedimentation.

basin. For example, this process is occurring at the north end of the Gulf of Mexico, where the Mississippi River has been depositing large volumes of sediment for the last 50 to 70 million years. The stepwise calculation of the effects of isostatic subsidence are esssentially the same as in the preceding model with the exception that a lower sediment density will be more typical of such regions. For this calculation, let us assume a sediment density of 2.0 grams per cubic centimeter.

Figure 9.4 develops the present-day situation in a stepwise fashion for sedimentation into one end of a basin that is initially 4000 meters deep and filled with water. As sediment becomes deposited in the basin, water of density 1.0 grams per cubic centimeter is displaced by sediment of density 2.0 grams per cubic centimeter. Thus, the net addition of load onto the basin floor is 1.0 grams per cubic centimeter of sediment deposited. This new load displaces mantle material of density 3.3 grams per cubic centimeter. Therefore, each unit of thickness of new sediment deposited will cause basin subsidence amounting to approximately 30 percent of the thickness of the new sediment. Thus, the first 4000 meters of sediment do not fill the basin; instead, isostatic subsidence has generated 1200 meters of new space above the sediment-water interface. The weight of this new water must itself be isostatically compensated, making the total new space 1700 meters. This process continues until approximately 7000 meters of sediment will be required to fill the original 4000-meter depth of basin.

In reality, this situation can become much more complicated by compaction

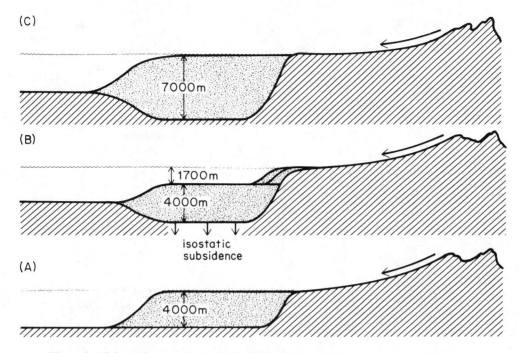

Figure 9.4 Schematic cross section indicating isostatic subsidence accompanying marine sedimentation into a basin initially filled with 4000 meters of water. Steps (A) through (C) indicate how isostatic subsidence accompanying sedimentation may allow an ultimate accumulation of sediment thickness that is nearly twice as great as the initial depth of the water column.

and diagenesis of the sediment with increasing burial. Thus, the ratio of total sediment thickness to initial basin depth may be considerably greater than indicated in Figure 9.4.

Isostatic Response to a Single Eustatic Sea-Level Rise. In the study of flat-lying strata to the interior of continents, it is common to find marine sand sediments sandwiched in among nonmarine deposits. As we stand on an outcrop and look at such a marine unit, we inevitably wish to know the magnitude of the transgression that was responsible for these rocks. Let us, therefore, consider the relationship between the height of a sea-level rise and the thickness of marine strata they may accumulate in response to that sea-level rise, taking into account isostatic compensation for new load.

Figure 9.5 summarizes events that would follow a 100-meter sea-level rise onto a low-lying continental area. Isostatic subsidence of the continental area due to the weight of new water load would result in a water depth of 143 meters over the former continental surface. One hundred meters of this depth is the result of the sea-level rise; 43 meters is the integrated isostatic subsidence of the continental area in response to the new load of water. The water is then displaced by sediment of

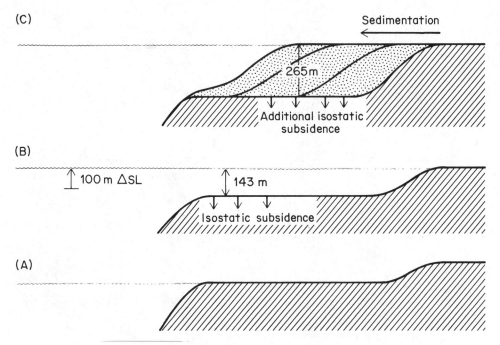

Figure 9.5 Estimation of sediment thickness accommodated by a 100-meter eustatis sea-level rise. Steps (A) through (C) summarize the interaction between sea-level rise, isostatic subsidence, and sedimentation.

density 2.0 grams per cubic centimeter. The total accumulation of sediment required to produce a sediment surface at the new high sea level will be approximately 265 meters (ignoring effects of compaction and lithification). Thus, the sea-level rise required to explain any given thickness of shallow-marine sediments over a continental area is approximately 30% to 40% of the total sediment thickness observed. The remaining 60% to 70% of the sediment thickness is accommodated by isostatic subsidence.

Isostatic Response to New Load Accompanying Driving Subsidence. We have noted that tectonic and thermal effects can cause continental lithosphere to subside, thus creating new room for shallow-marine sedimentation. It is convenient to think of such subsidence as consisting of two components, a driving subsidence and an isostatic response to new load accompanying the subsidence. The situation works out to be identical to that described in Figure 9.5 except that the continental area subsides relative to sea level rather than sea level rising relative to the continent. Either way, net result is creation of new space for shallow-marine sediments where there was only continent-air interface before. If shallow-marine deposition keeps pace with driving subsidence, the thickness of sediment thus accommodated per unit time will be approximately 2.5 to 3.0 times the actual amount of driving subsidence. That is, approximately one-third of the subsidence results from tectonic or thermal processes acting upon the region and approximately two-thirds of the observed sediment thickness accumulates in response to isostatic adjustment to new load.

Relaxation Times Resulting from Viscosity of the Mantle. In the preceding discussion, we have talked about flow of mantle material induced by loading of new sediment onto portions of the lithosphere. Inasmuch as mantle material is quite viscous, flow cannot instantaneously achieve a new isostatic equilibrium. The half-life of the process of going from one state of isostatic equilibrium to a new state of isostatic equilibrium is referred to as the relaxation time.

Classic geophysical investigations of relaxation time deal with the rebound of continental areas that were previously occupied by Pleistocene glaciers (see McConnell, 1968, for example). Pleistocene glaciation has provided the geophysicist with a natural experiment. Some 12,000 to 18,000 years ago, massive ice sheets existed over Scandinavia and Canada. If we assume that the ice was more or less at isostatic equilibrium, then the weight of the ice must have depressed the lithosphere. As the ice melted, the lithosphere should have risen to achieve new isostatic equilibrium.

Figure 9.6 indicates present elevation of a 5000-year-old Scandinavian shoreline. Five thousand years ago, this surface was at or near sea level. Today, isostatic rebound in response to removal of the Scandinavian ice sheet has deformed the 5000 B.P. datum plane by some 100 meters. Gravity data indicate that an additional 200 meters of uplift is to be anticipated before isostatic equilibrium is achieved. From such data, estimates can be made concerning the time required to achieve new isostatic equilibrium following a rapid loading or unloading of the lithosphere. Relaxation times on the order of 10^3 years are to be anticipated.

Flexural Rigidity of the Lithosphere. Following the general concept of isostacy, we have thus far considered that the lithosphere is essentially floating on viscous mantle material. To make this picture more reasonable, we must now consider the structural properties of the lithosphere itself. Is there local isostatic compensation for each small region of lithosphere, or do large portions of the lithosphere have sufficient rigidity to behave as single units? If a new load is placed over a portion of the lithosphere, will isostatic equilibrium obtained directly be below that load, or will compensation of the load be distributed over a broad area?

Once again, the geophysicist seeks areas where nature has run an experiment for him (see Walcott, 1970, for example). Proglacial lakes once contained a depth of water that can be estimated, and these lakes had shorelines that were approximately level. Deformation of these shorelines provides information about the area over which the load was distributed. Similarly, volcanic islands commonly have a closely associated bathymetric depression, suggesting that the load represented by the volcanic island is distributed over the surrounding lithosphere. Resulting estimates of the apparent flexural rigidity of the lithosphere are summarized in Figure 9.7.

The data suggest that the lithosphere behaves like a Maxwell solid. With rapid application of load ("rapid" being 10^3 to 10^4 years!), the lithosphere has high ap-

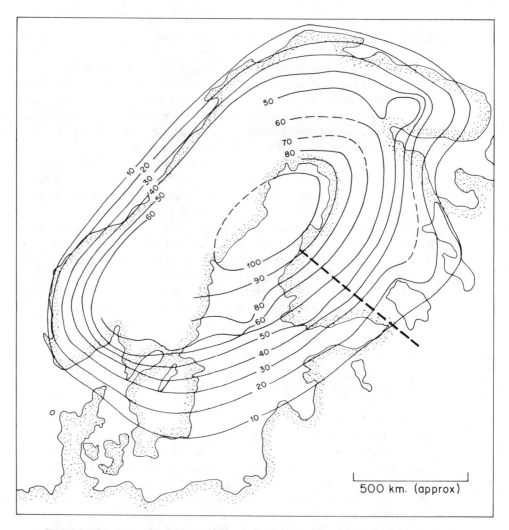

Figure 9.6 Isostatic rebound of Scandinavia as shown by the present elevation of the 5000 B.P. shoreline. [After R. K. McConnell, Jr., *Jour. Geophysical Research,* **73,** No. 22, 7090 (1968).] Courtesy American Geophysical Union.

parent flexural rigidity and behaves elastically. With continued application of load (10^6 to 10^8 years), apparent flexural rigidity decreases by one or more orders of magnitude and viscous deformation of the lithosphere occurs.

The viscosity of the asthenosphere regulates the rate at which isostatic equilibrium is attained, but it is the apparent flexural rigidity of the lithosphere that determines the area over which the new load will become compensated. Figure 9.8 indicates the effect of flexural rigidity upon the distance to which a new load may be isostatically compensated. Our interest in this phenomenon centers around various permutations of the following question: Can rapid application of new load (sedimentation or tectonic overthrusting, for example) cause subsidence over broad areas adjacent to the new load? If we look back at Figure 9.3, for example, we would ask whether the deposition of the alluvial fan would cause isostatic subsidence only beneath the fan or throughout the entire valley. Note that the viscosity of the mantle dictates that isostatic response to new load will occur within 10^3 to 10^4 years, whereas apparent flexural rigidity of the lithosphere does not undergo significant deterioration until 10^5 to 10^7 years after application of new load. Thus, if it can be demonstrated that new load was applied within 10^3 to 10^4 years, we would expect apparent flexural rigidity of the lithosphere to be rather high. Regional downwarping because of the new load should extend approximately 500 kilometers outward from the load. If, on the other hand, it could be demonstrated that the load was applied gradually over a long time interval (perhaps 10^7 to 10^8 years), then apparent flexural rigidity of the lithosphere would be much lower and isostatic compensation would occur on a more local basis.

Isostatic Response to Small Loads. How small a load will induce geologically significant isostatic response? The geophysicists have worked extensively with the unloading of continental glaciers. Glacial retreat involves the unloading of at least 150 bars. Such unloading is clearly sufficient to initiate measurable isostatic response during a reasonably short time. Can we expect to observe response generated by a significantly smaller load?

Bloom (1967) recognized yet another simple experiment that nature has run for us over the past few thousand years. It is generally agreed that post-Wisconsin sea level has been rising rather slowly throughout the last 5000 years. However, various locations give differing estimates as to the shape of the sea-level rise curve (see Figure 9.9). Bloom proposed that these discrepancies are a function of isostatic response of the continental shelf to new water load; the thicker the average depth of water near these shorelines, the greater the isostatic subsidence of that shoreline sea-level record. To test this hypothesis, Bloom undertook to estimate average water depth within 50 kilometers (a convenient arbitrary dimension) of the various sea-level data points under consideration. Figure 9.10 summarizes a portion of his results. The data strongly suggest that isostatic response to a load change of only a few bars has indeed occurred.

Terminology for Description of the Stratigraphic Record of Earth Dynamics

The foregoing discussion outlines the various ways in which the relationship between land and sea level may be altered. The roots of the study of stratigraphy go back much further than the modern concepts presented above. Further, it is not always possible to make the stratigraphic observation that would fit cleanly into our conceptual framework. For both of these reasons, there are descriptive words which

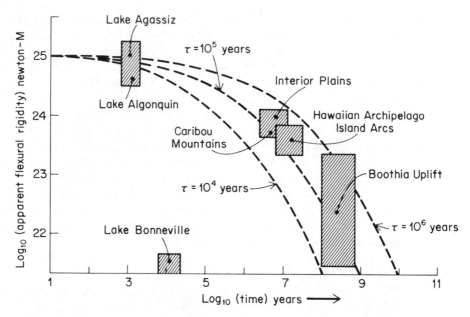

Figure 9.7 Apparent flexural rigidity of the lithosphere as a function of the duration of the naturally occurring experiment from which the flexural rigidity estimates were made. The low flexural rigidity of the late Bonneville area is probably due to extreme fracturing of the crust over broad areas of the western United States. [After R. I. Walcott, *Jour. Geophysical Research,* **75,** No. 20, 3951 (1970).] Courtesy American Geophysical Union.

simply must be precisely memorized. Three sets of such terminology are in common usage.

A long-standing stratigraphic concept is that of *transgression* and *regression.* Using the "center of the basin" as a reference point, one speaks of a transgression when the *shoreline* moves outward (onto the "continent") and regression when the *shoreline* moves inward (toward the center of the basin) (Krumbein and Sloss, 1963). Note that neither term indicates the cause of this dynamic interaction between water and the land. The words "transgression" and "regression" used alone carry absolutely no connotation concerning the tectonic, eustatic, isostatic, or sedimentological nature of the dynamic event. In this book, we shall use these two words sparingly and only as "inclusive terms of ignorance." If nonmarine rocks are overlain by marine rocks, there rather clearly has been a "transgression." However, let's not stop here. What caused this relationship? Was the cause tectonic, eustatic, or isostatic?

The word "regression" is particularly difficult. In addition to possible tectonic, eustatic, or isostatic causal mechanisms, a "regression" may simply record that the basin has become increasingly filled with sediment *regardless* of what other mechanisms may be operating.

To get around these problems, we must employ new words. It would be senseless to redefine "transgression" and "regression"; they are too ingrained into the

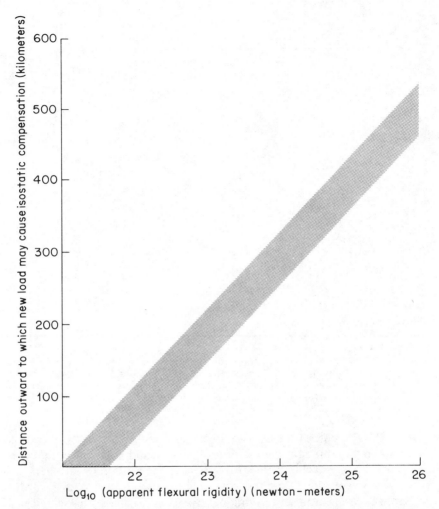

Figure 9.8 High flexural rigidity allows isostatic compensation to be spread over a large area; low flexural rigidity demands local isostatic compensation for new load. The indicated relationship is approximate. [Compiled from data and calculations contained in R. I. Walcott, *Jour. Geophysical Research,* **75,** No. 20, 3951 (1970).] Courtesy American Geophysical Union.

stratigraphic literature. Instead, the science has placed more precise definitions on the terms "submergence," "emergence" "progradation," and "aggradation."

Submergence and *emergence* refer to changes in the *vertical* relationship between sea level and a fixed spot on or beneath subtidal sediments. If the thickness of water plus sediments over that spot increases, the area is undergoing submergence. Similarly, a decrease will be referred to as emergence. If the causal mechanism is known, a modifier is appropriate, such as "glacio-eustatic submergence."

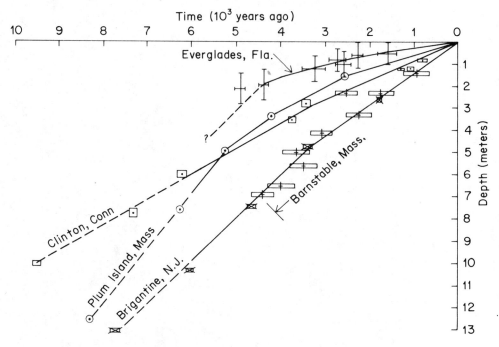

Figure 9.9 Submergence history of five sites along the eastern coast of the United States. Data points are carbon-14 ages determined on molluscs or peat deposits that accumulated at or near the sea level existing at that time. Each curve records the interaction between rising sea level and local subsidence or emergence. (After Bloom, 1967.)

Progradation occurs as sediment accumulation builds outward laterally. For example, continued accumulation of beach sand may cause the beach to prograde seaward. *Aggradation* refers to vertical accumulation of sediment. For example, a coastal swamp might aggrade if sea level rose slowly. With regards to the topset, foreset, and bottomset terminology for bedding planes presented in Chapter 12, it is convenient to think of aggradation as occurring in topset and bottomset strata whereas progradation occurs in foreset strata. Note well that the terms "progradation" and "aggradation" are substantially independent of the terms "submergence" and "emergence." For example, there is no reason why progradation cannot occur under conditions of general submergence. Sea level may be rising, but there is sufficient supply of sediment to continue seaward progradation. These are points upon which we shall dwell at some length in later chapters.

Finally, we must note the convenience of using the terms *on-lap* and *off-lap* as a general description of stratigraphic relationship among bedding planes and unconformities regardless of the sedimentary facies involved. As noted in Chapter 3, it is particularly common in the study of seismic stratigraphy that we are able to map the geometry of various layers within the sediments without having precise knowledge of the sediment types involved. In the vast majority of cases, "on-lap" is

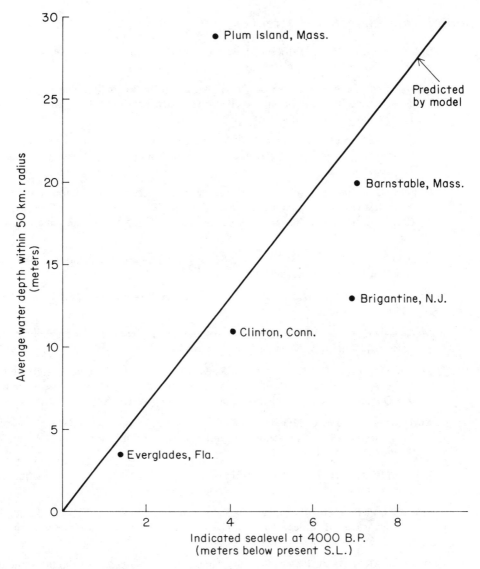

Figure 9.10 Isostatic response to new water load on continental shelves produces a systematic relationship between apparent sea level at 4000 B.P. and average water depth within a 50-kilometer radius of the area where the sea-level estimate was obtained. The solid line is the relationship predicted by the model. (Compiled from data presented in Bloom, 1967.)

conceptually compatible with "submergence" and "off-lap" with "emergence." However, definition of these different sets of terminology are rooted in totally different stratigraphic observations. Thus, they cannot be used interchangeably but must be used to convey the nature of the observation upon which the stratigrapher is basing his or her discussion.

Citations and Selected References

BLOOM, A. L., 1967. Pleistocene shoreline: a new test of isostacy. Bull. Geol. Soc. Amer. 78: 1477–1494.

———. 1980. Late Quaternary sea level change on South Pacific coasts: study in tectonic diversity, p. 505–516. *In* Nils-Axel Mörner (ed.), Earth rheology, isostacy and eustasy. John Wiley & Sons, New York.

CLARK, J. A., and C. S. LINGLE. 1979. Predicted relative sea-level changes (18,000 years B.P. to present) caused by late-glacial retreat of the Antarctic ice sheet. Quaternary Res. 11: 279–298.

CURRAY, J. R. 1960. Sediments and history of Holocene transgression, continental shelf, northwest Gulf of Mexico, p. 221–266. *In* Francis P. Shepard, Fred B. Phleger, and Tjeerd H. van Andel (eds.), Recent sediments, northwest Gulf of Mexico. Amer. Assoc. Petrol. Geol.

KRUMBEIN, W. C., and L. L. SLOSS. 1963. Stratigraphy and sedimentation. W. H. Freeman and Co., San Francisco, 660 p.

LEET, L. D., and S. JUDSON, 1971. Physical geology. Prentice-Hall, Englewood Cliffs, N.J., 687 p.

McCONNELL, R. K., JR. 1968. Viscosity of the mantle from relaxation time spectra of isostatic adjustment. J. Geophys. Res. 73: 7089–7105.

MACINTYRE, I. G., O. H. PILKEY, and R. STUCKENRATH. 1978. Relict oysters on the United States Atlantic continental shelf: a reconsideration of their usefulness in understanding late Quaternary sea-level history. Bull. Geol. Soc. Amer. 89: 277–282.

MATTHEWS, J. L., B. C. HEEZEN, R. CATALANO, A. COOGAN, M. THARP, J. NATLAND, and M. RAWSON. 1974. Cretaceous drowning of reefs on mid-Pacific and Japanese guyots. Science 184: 462–464.

PITMAN, W. C., III. 1978. Relationship between eustacy and stratigraphic sequences of passive margins. Bull. Geol. Soc. Amer. 89: 1389–1403.

SHEPARD, F. P. 1960. Rise of sea level along northwest Gulf of Mexico, p. 338–344. *In* F. P. Shepard, F. B. Phleger, and T. H. van Andel (eds.) Recent sediment, northwest Gulf of Mexico. Amer. Assoc. Petrol. Geol., Tulsa, Okla.

WALCOTT, R. I. 1970. Flexural rigidity, thickness, and viscosity of the lithosphere. J. Geophys. Res. 75: 3941–3954.

WATTS, A. B., and W. B. F. RYAN. 1976. Flexure of the lithosphere and continental margin basins. Tectonophysics 36: 25–44.

Geologic Framework
of Sediment Accumulation

10

In previous chapters, we have looked at the variations among sedimentary rocks and at the dynamics of the earth. Now, we must begin to put these two sets of information into the context of earth history. Where have dynamic events happened in the past, and what is the stratigraphic record of those events? In this chapter, we shall approach these questions on a very general level. Our purpose here is to establish the broad context in which we shall examine the details of small areas in Parts IV and V.

The following discussion divides the sedimentary geology of a continent into three general categories: (1) the craton, (2) the passive continental margin, and (3) the active continental margin. To these three regional concepts, we must add the possibility that the tectonic context of a region may change with time. Our discussion of the architecture of a continent will center around the geology of central North America (see Figure 10.1).

The Craton

The fundamental property of this geologic province is that it is underlain at relatively shallow depth by Precambrian basement rock. A relatively flat-lying unconformity separates the Precambrian basement complex from flat-lying sedimentary rocks above.

General Characteristics of the Interior Lowlands of the United States. The Interior Lowlands is the sediment-covered area which lies south or west of the Canadian Shield, west of the Appalachian Mountains, east of the Cordilleran Mountains, and north of the Ouachita Mountains.

The sediments are predominantly Paleozoic and Mesozoic shallow-marine de-

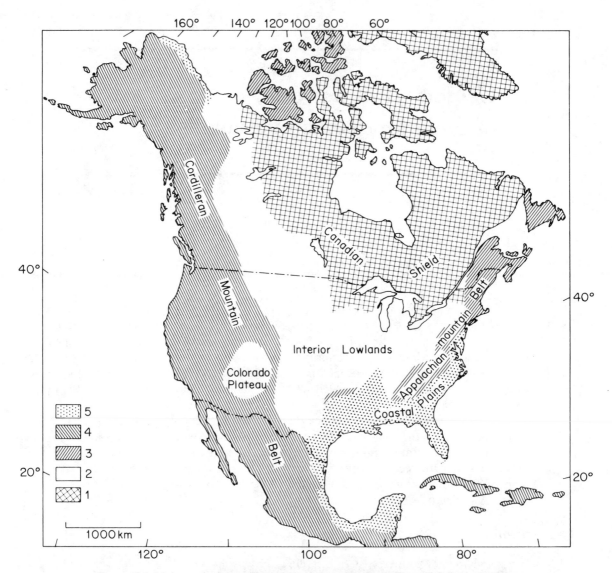

Figure 10.1 Major geologic provinces of North America. Symbols are as follows: (1) Precambrian basement rocks of the Canadian Shield. (2) Flat-lying sediments of the Interior Lowlands, predominantly Paleozoic but with important Mesozoic sequences in the west. Sediment thickness in the Interior Lowlands generally averages 1 kilometer and seldom exceeds 3 kilometers. (3) Older Paleozoic sediments of the Ouachita-Appalachian mountain belt. Stratigraphic thickness of the sediments commonly ranges from 10 to 20 kilometers. (4) Paleozoic through Cenozoic sediments of the Cordilleran mountain belt. Major tectonism is late Mesozoic to early Cenozoic. (5) Late Mesozoic to Cenozoic sediments of the modern continental margin. Sediment accumulations are 5 to 15 kilometers thick and have not undergone tectonic deformation.

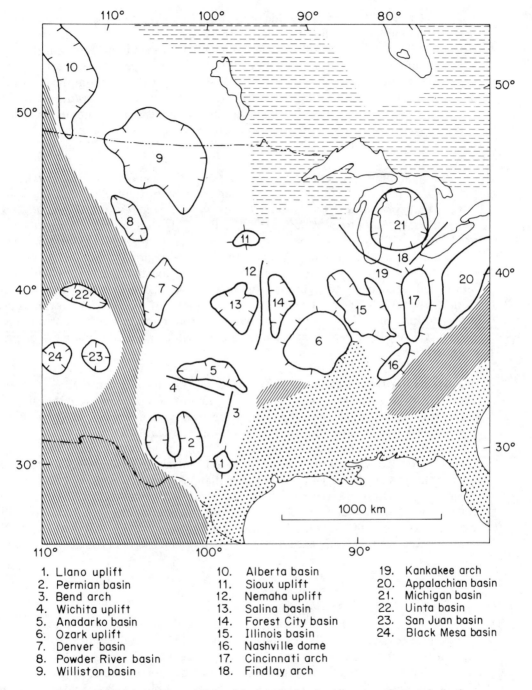

Figure 10.2 Major basins and highs of the Interior Lowlands province. Patterns indicating surrounding geologic provinces are the same as in Figure 10.1.

1. Llano uplift
2. Permian basin
3. Bend arch
4. Wichita uplift
5. Anadarko basin
6. Ozark uplift
7. Denver basin
8. Powder River basin
9. Williston basin

10. Alberta basin
11. Sioux uplift
12. Nemaha uplift
13. Salina basin
14. Forest City basin
15. Illinois basin
16. Nashville dome
17. Cincinnati arch
18. Findlay arch

19. Kankakee arch
20. Appalachian basin
21. Michigan basin
22. Uinta basin
23. San Juan basin
24. Black Mesa basin

posits and average 1 kilometer in thickness. Lower Paleozoic clastic sediment supply was predominantly from the Precambrian rocks that form the core of the continent. Upper Paleozoic and Mesozoic clastic sediment supply came in part from the Appalachian and Cordilleran mountain belts. Low-lying topography and tectonic stability has resulted in the development of classical layer-cake stratigraphy. In this province, geological events happened over very large areas and accordingly left sediments that can be physically correlated for literally thousands of kilometers. Limestones, black shales, coal deposits, and orthoquartzite sandstones are common lithologies of large lateral extent.

Intracratonic Highs and Lows. Within the sediments of the Interior Lowlands, there are general trends in the thickening and thinning of correlative stratigraphic units. The thickened areas are referred to as *intracratonic basins,* and the areas of thinning are referred to as *domes, arches,* and *uplifts.* Note well that some of these are broad, gentle features. On the scale of an outcrop, we would think that these rocks are essentially flat-lying. Yet, on the scale of regional mapping, low-angle dips may persist over large areas, and thickness relationships show systematic variation. Sediments over the highs may thin to zero; sediments in the lows may thicken to more than 3 kilometers. The major highs and lows of the Interior Lowlands are indicated in Figure 10.2. Features far removed from mountain belts probably owe their origin to lithosphere dynamical processes such as aborted pull-aparts or passage across hot spots.

Transition from Craton to Mountain Belts. As indicated in Figure 10.2, the transition from Interior Lowlands geologic province to adjacent mountain belts is commonly marked by the development of foreland basins. The Appalachian basin, the Anadarko basin, and the Alberta basin are such features.

Foreland basins have a unique history. During their early stages, they receive their clastic sediment supply from the craton and accumulate a sedimentary sequence that is correlative with, but thicker than, the cratonic sequence. Later, the basins accumulate thick clastic sequences derived from newly emerging mountain ranges. In addition, structural deformation of the mountain belt commonly spills over into the foreland basin. Thus, the structural complexity within these stratigraphic sequences generally exceeds that of their counterparts in the interior of continents.

The Passive Continental Margins

The coastal regions from Mexico northeastward to Cape Cod (Figure 10.1) are underlain by accumulation of essentially undeformed sedimentary sequences of Jurassic and younger age. For the purposes of this discussion, we shall consider the Texas-Louisiana Gulf Coast province to be an example of clastic infilling of an ocean basin adjacent to a highly focused continental clastic source. We shall consider the East Coast carbonate province to be an example of bank-margin carbonate sedimentation on a post-rifting passive margin.

The Texas-Louisiana Gulf Coast Clastic Province. Note in Figure 10.3 that sediments derived from the Appalachian Mountains to the east and from the Cor-

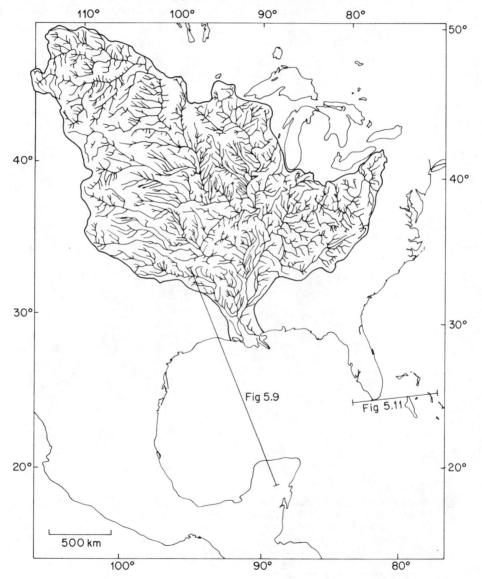

Figure 10.3 The Mississippi River drainage basin. Uplift of the Appalachians to the east and the Rocky Mountains to the west funnels rivers, and their clastic sediments, to the north end of the Gulf of Mexico.

dilleran Mountains to the west find their way down the Mississippi River drainage system to be deposited in the north end of the Gulf of Mexico. This configuration of clastic sediment source to depositional province has been operational for approximately the last 100 million years.

Figures 10.4 and 10.5 summarize the general features of the Gulf Coast clastic province. During Jurassic and Lower Cretaceous time, a broad shelf was developed that was characterized by carbonate bank-margin sedimentation along its seaward margin from Mexico to Florida. With the major uplift of the eastern Cordillera (Rocky Mountains) during late Cretaceous–Eocene time, clastic sedimentation began to dominate the Texas-Louisiana region. The position of major clastic sediment

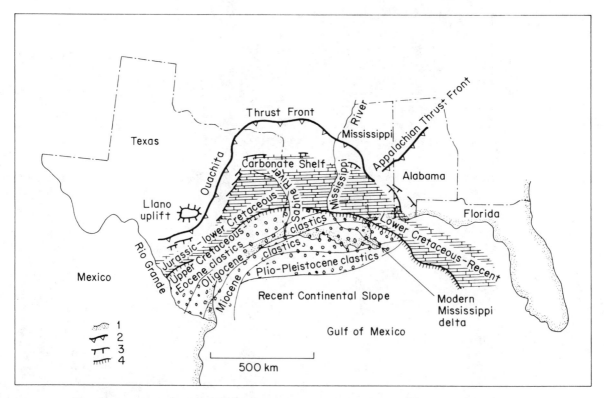

Figure 10.4 Sketch map showing major sediment accumulations in the northern Gulf of Mexico. North and west of the modern Mississippi delta, carbonate sediments accumulated on a broad shallow shelf during Jurassic and Lower Cretaceous time. By Upper Cretaceous time, carbonate sedimentation had given way to predominantly clastic sediments supplied by an ancestral Mississippi River. The position of major accumulation of clastic sediments has prograded seaward 200 kilometers since that time. In sharp contrast, carbonate sedimentation continued to prevail in the area east of the Mississippi River. (After Lehner, 1969, with modification and simplification.)

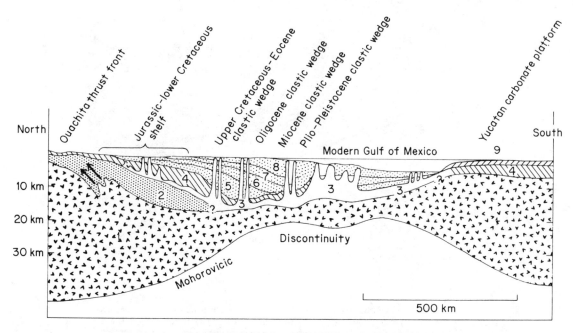

Figure 10.5 Generalized cross section of the Gulf of Mexico geological province. See Figure 10.3 for location. Rock and sediment types are as follows: (1) crystalline basement rocks; (2) Paleozoic sediments of the Ouachitan-Appalachian mobile belt province; (3) salt, presumably Jurassic in age; (4) Jurassic and Lower Cretaceous sediments, predominantly shelf carbonates; (5)–(8) clastic sediments of Upper Cretaceous–Pleistocene age; and (9) Upper Cretaceous–Recent shelf carbonates. (After Lehner, 1969, with modification.)

accumulation during the various epochs of the Cenozoic is indicated in Figures 10.4 and 10.5.

The Gulf Coast clastic sequence records the complex interaction between sediment supply, sea-level fluctuations, and regional isostatic subsidence in response to the weight of new sedimentary load. Note in Figure 10.5 that a Cenozoic sediment thickness of 15 kilometers is indicated by seismic refraction data. Oil wells in southeastern Louisiana have been drilled to a depth of 8 kilometers and have bottomed in middle Miocene sediment. The thickness of Cenozoic sediments is presumed to result from synchronous sedimentation and isostatic subsidence. The major tectonic features of the area are normal faults, presumably related to sediment loading and associated subsidence. Salt domes rise upward through the Cenozoic strata, probably from a source layer of Jurassic salt, and provide local structural deformation of the Cenozoic sequence. These salt domes are extremely important to the localization of large petroleum reserves.

The fundamental sedimentation cycle of the Gulf Coast clastic province is indicated in Figure 10.6. With eustatic sea-level rise or with rapid subsidence, relative

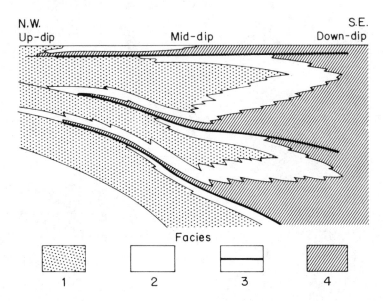

Figure 10.6 Schematic stratigraphic cross section illustrating the primary sedimentary cycle of the Gulf Coast clastic province. Sediment types are as follows: (1) alluvial sediments; (2) marginal-marine sands and shales; (3) marine shale or carbonate facies marking rapid transgression; and (4) deep-water marine shale and pelagic marl. The first-order sedimentary cycle of the Gulf Coast clastic province consists of rapid submergence followed by gradual progradation. (After Lowman, 1949, with modification.)

deep water moves landward. Thus, much of the clastic sediment supply is commonly temporarily trapped in alluvial and coastal environments, allowing pelagic shale or carbonate facies to penetrate far into a sedimentary sequence otherwise dominated by alluvial and marginal-marine sands and shales. With the continued passage of time, marginal-marine environments again advance seaward, followed by alluvial facies. Continued progradation and subsidence accommodates a thick clastic wedge that is finally terminated by yet another episode of rapid transgression of the sea.

The Florida-Bahama Carbonate Province. In the absence of large clastic input, continental margin sedimentation in tropical regions may be dominated by shallow-water carbonate and evaporite deposition. The Florida-Bahama platform provides a good example. The eastern portion of Figure 10.4 indicates the paleogeographic continuity between the Cretaceous carbonate provinces of Texas and Florida. Whereas the Gulf Coast clastic province received a large clastic sediment supply from the Mississippi River during late Cretaceous through Cenozoic time, the Florida-Bahama province continued to accumulate shallow-water carbonate and evaporite sediments throughout the gradual subsidence of the region. Figure 10.7 is a cross section through the Florida-Bahama carbonate province. Note that Cenozoic carbonate accumulation in this region is a scant 1.5 to 3 kilometers as com-

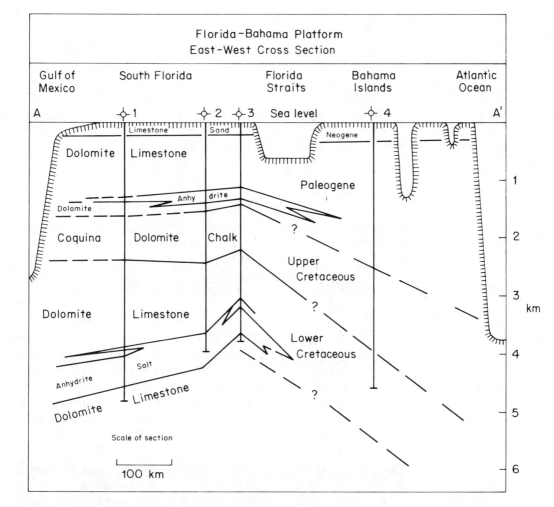

Figure 10.7 Generalized cross section of the Florida-Bahama carbonate province. See Figure 10.3 for location. (After Ginsburg, 1964, with modification.)

pared to the 15 kilometers of clastic sedimentation within the Texas-Louisiana Gulf Coast clastic province.

The East Coast Outer Continental Shelf Province. Beginning in the Triassic, North America became separated from Africa by the continual formation of new oceanic lithosphere at the Mid-Atlantic Ridge. The spreading continues today. Since Jurassic time, the entire East Coast of North America has been the site of gradual subsidence along a lithospheric cooling curve. Driving subsidence at the time of the initial opening of the Atlantic Ocean was probably on the order of 10 centimeters per 1000 years and followed an exponential decay such as depicted in Figure 8.7.

Shallow-water bank-margin carbonates dominated the subsiding passive margin throughout the Jurassic and early Cretaceous. The shelf-slope break during this time existed far to the east of its present location. A mid-Cretaceous transgression overran the old bank margin and established a new continental margin several hundred kilometers landward, leaving the old Jurassic–Lower Cretaceous continental margin as the bathymetric feature known as the Blake Escarpment. These configurations are depicted in a map and cross section in Figures 10.8 and 10.9.

The Active Continental Margin

Mountain belts have long captured the attention of the geologist. Understanding these complex and enormous features has always been taken as a first-order challenge to the science. With the advent of modern plate tectonics in the 1960's, the science has taken great strides toward a genuine understanding of mountain belts. Mountain belts are the surficial expression of tectonic activity at the boundary between two lithosphere plates.

Sediments Associated with Mountain Belts. Sedimentological and stratigraphic terminology for mountain belts predates plate tectonics. For many years, geologists have recognized that mountain ranges exhibit at least three distinct stages of sedimentation. Kay (1951) popularized the use of the terms "miogeosyncline," "eugeosyncline," and "exogeosyncline." The words "flysch" and "molasse" from the European literature are also frequently used. To these, we might add the classical sandstone compositional names "orthoquartzite," "graywacke," and "arkose." Let us place these words within the context of plate tectonics.

Miogeosynclinal sediments are substantially continental passive-margin sediments that predate plate consumption and resultant tectonism. These deposits are commonly the thickened seaward extension of stable craton sedimentation. The miogeosynclinal sequence is characterized as shallow-water carbonates on the continental shelf. Orthoquartzite sandstones are also to be anticipated.

In its broadest context, the term eugeosyncline refers to those sediments that are basinward from, and at least part contemporaneous with, miogeosynclinal deposits. These deposits would include the fine-grained sediments of the continental rise, turbidites and pelagic sediments of the abyssal plain, and commonly significant amounts of volcanic sediment.

The precise origin of the volcanic input to eugeosynclinal deposits is not always clear. The eugeosynclinal sediments commonly undergo such extreme tectonic deformation and metamorphism that stratigraphers are unable to unravel the precise paleogeography. At least two very different paleogeographies are easy to envision as alternative hypotheses. On the one hand, association of volcanic arcs with zones of plate consumption is an outstanding generality of today's world; we may assume

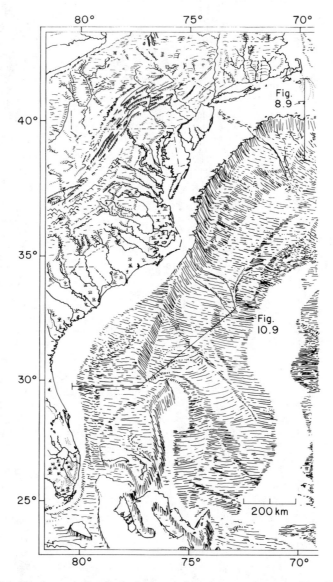

Figure 10.8 Sketch map indicating bathymetry of continent-to-ocean transition, East Coast of the United States. Unshaded area represents continental shelf, generally less than 200 meters water depth at its outermost edge. The Blake Plateau, approximately 27°N to 32°N, is a huge area of continental margin presently submerged to about 2000 meters. The transition from continental margin to ocean basin is indicated by steep hasher marks throughout the length of the diagram from the Bahamas in the south to Georges Bank in the north. (After Heezen and Tharp, 1968.)

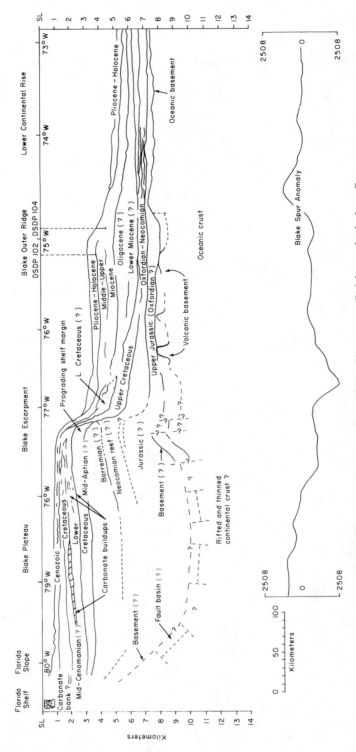

Figure 10.9 Schematic cross section of Blake Plateau compiled from seismic data. Compare with Figure 8.9, which is based on similar data off the New England region. Note the relatively thick accumulation of Jurassic bank-margin sediments during rapid subsidence following rifting of North America from Africa. Following several unconformities in Cretaceous time, bank-margin carbonate sedimentation reestablished itself along the Blake Escarpment. It is interesting to speculate as to why the Cenozoic shelf margin reestablished itself so far to the west. (After Shipley *et al.*, 1978.)

158

that it was a valid generality in the past. Thus, eugeosynclinal volcanics may be the record of the previous existence of arc-trench systems. Alternatively, it may be that the eugeosynclinal sediments represent the tectonic accumulation of abyssal plain sediment along the zone of subduction. For example, Figure 15.5 maps the occurrence of numerous volcanic seamounts in a portion of the North Atlantic. Consider what would happen to these seamounts if a North Atlantic plate were to be subducted beneath a continental plate. We can easily imagine the accumulation of a linear belt of extremely thick volcanic sediments if these materials tended to detach themselves from the descending slab of oceanic lithosphere.

Exogeosynclinal deposits typically postdate the time of major tectonic uplift and typically accumulate continentward of the emerging mountain belt. Arkosic sandstones derived from newly emplaced intrusive rocks to the interior of the mountain range are the hallmark sediment type of the exogeosyncline.

The terms *flysch* and *molasse* refer to the same sediments as the words eugeosyncline and exogeosyncline, but they are rooted in a different conceptual framework. Whereas eugeosyncline and exogeosyncline are defined in terms of *position* relative to tectonic activity, flysch and molasse are defined with relation to the *timing* of major tectonism. Flysch sedimentation takes place *during* the major tectonic deformation of the region. In theory, flysch sediments not only are the product of newly emerged tectonic lands, but also become caught up in the continuation of the tectonic deformation. Flysch sediments, therefore, roughly correspond to eugeosynclinal sediments and to those exogeosynclinal sediments that themselves become highly deformed during continuing tectonism. In contrast, molasse sedimentation refers to those exogeosynclinal sediments that accumulate *after* major tectonic deformation of the region.

The Western United States Mesozoic-Cenozoic Arc-Trench System. For many years, nobody "understood" the geology of California. Geologists "mapped it," said "That's incredible," and went on mapping. Today, we understand at least the source of the confusion. Basically, an arc-trench system has tried to swallow a mid-ocean ridge (see Figure 10.10). The resulting transform continental margin has a rather complex history (Blake *et al.*, 1978, for example). If we wish to understand this history, we must first go elsewhere and view the architecture of arc-trench systems in a simpler context.

Our knowledge of the general geology of arc-trench systems comes from the study of modern circum-Pacific oceanic subduction zones. Figure 10.11 presents a generalized cross section of such a region and provides the basic terminology that we shall carry forward to the western United States.

In Figure 10.11, note especially that the arc-trench system and related tectonic features occupy linear belts that measure several hundred kilometers. The downgoing slab typically plunges at approximately 45° to a depth of perhaps 700 kilometers (the boundary between upper and lower mantle). Lithospheric reheating and

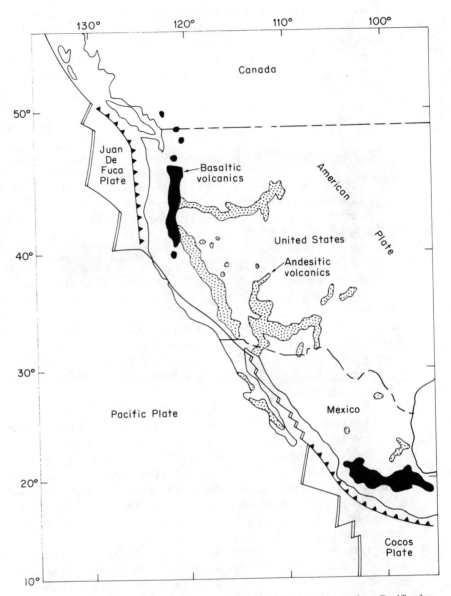

Figure 10.10 Sketch map indicating modern configuration of American plate, Pacific plate, and related subduction zones and ridges. The Juan de Fuca plate and the Cocos plate are remnants of the Farallon plate, now substantially swallowed by the subduction zone. [After Peter W. Lipman *et al.*, "Evolving Subduction Zones in the Western United States as Interpreted from Igneous Rocks," *Science,* **174**, 821–825 (19 November 1971) with modification.] Copyright 1971 by the American Association for the Advancement of Science.

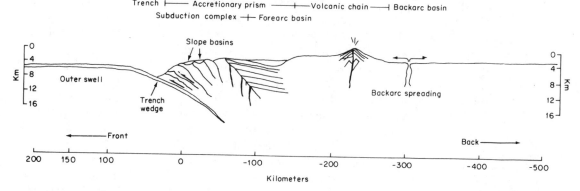

Figure 10.11 Schematic cross section indicating terminology for the arc-trench system. (After Karig and Sharman, 1975 with simplification.)

magmatic activity associated with frictional heating of the downgoing slab account for volcanism and/or backarc spreading to distances of several hundred kilometers away from the deep-sea trench.

Traversing from the deep-sea trench across the upthrown lithospheric slab, we encounter several zones of sedimentary geology that are worthy of specific note. The *subduction complex* can be thought of as the sediment accumulation scraped off of the downgoing slab. This region will also include occasional slabs of oceanic crust up to kilometers in thickness that have likewise become detached from the downgoing slab.

Interior to the subduction complex, there is commonly a *forearc basin*. The forearc basin is substantially a flat-lying sedimentary fill in a topographic basin created by uplift in the subduction complex. Taken together, the subduction complex and the forearc basin are referred to as the *accretionary prism*.

Next in our traverse is the volcanic arc. Frictional heating and partial melting of the downgoing slab produces magmatic activity that injects this region of the upper lithospheric plate and produces volcanism. Sediments derived from the volcanic arc may be important contributors to the accretionary prism and to the backarc basin.

The volcanic arc seldom occurs closer than 100 kilometers of the trench. The separation between volcanic arc and trench can be extended to several hundred kilometers by accumulation of a large subduction complex during the 35-million-year life of a simple arc-trench system.

Traversing beyond the volcanic arc, we find the *backarc basin*. In the most classic case, the backarc basin may be the site of generation of new oceanic lithosphere at localized backarc spreading centers. As with the volcanic arc itself, this activity results from frictional heating within the region of the downgoing slab. Additionally, the backarc basin may be the site of considerable isostatic subsidence under a new load delivered to the region from the volcanic arc region and/or subsequent uplift of granitic terrain postdating the volcanic arc.

Figures 10.12 and 10.13 offer conceptual summary of Mesozoic and Cenozoic geology of the western United States in terms of evolution of an arc-trench system. Note that the coast ranges of northern California are regarded as subduction complex, the San Joaquin and Sacramento valleys as forearc basins, and the Sierra Nevada batholith as the root zone of the former volcanic arc. The situation here is fairly straightforward from the late Jurassic through Paleogene. With Neogene time, there comes the additional complication of the arc-trench system "swallowing" the Pacific spreading center (northward extension of the East Pacific Rise, see Figure 8.4). Attendant with this complication are the addition of an even larger heat source beneath the continental lithosphere and initiation of right-lateral transform faults, shown in Figure 10.12 as the San Andreas Fault delivering igneous rock of the Salinian block to juxtaposition with the Franciscan subduction complex rocks of the northern coast range.

The Eastern United States Paleozoic Arc-Trench System. The sedimentary terrains of the Appalachian Mountains are one of the classic field areas of the world. Description of sequences of rock types and the mapping of thickness relationships among various stratigraphic units far preceded modern understanding of the origin of this mountain chain. It is instructive to review this classical region from early description toward modern understanding.

The terms miogeosyncline and exogeosyncline have their origins in the study of Appalachian Paleozoic stratigraphy of the northeastern United States. Figure 10.14 presents a schematic stratigraphic cross section through the Devonian rocks of New York State. A Devonian sequence that begins everywhere with Helderberg or Onondaga shallow-water limestones gives way upward to clastic rocks that become more and more nonmarine in character up the section. Note also that many of the units from Hamilton group upward are much thicker in the east than they are in the west. These relations have long been recognized as indicative of the emergence of a continental source of clastic materials to the east, in the region now occupied by the crystal-line Appalachians.

Note in Figure 10.15 that a sequence of lithology similar to that depicted for New York State in Figure 10.14 can be observed throughout the sedimentary Appalachians. However, the exact time at which the marine carbonate section becomes more clastic and gives way to nonmarine coarse clastics varies greatly from one end of the mountain belt to the other. As indicated in Figure 10.16, it has long been recognized that these stratigraphic variations could be simplified conceptually by calling upon point sources of clastic sediment at various places at various times throughout the length of the crystalline Appalachians.

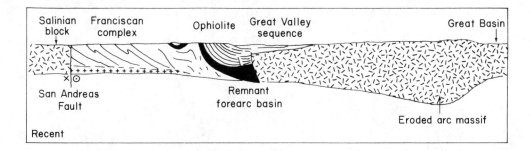

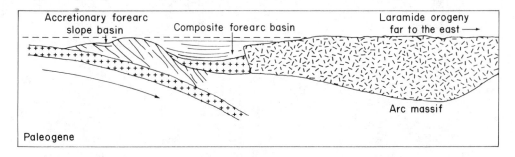

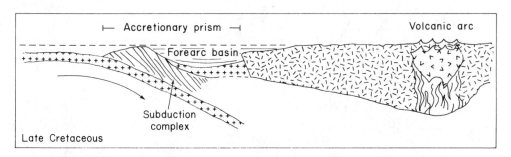

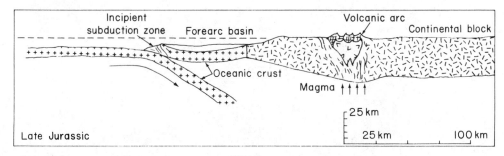

Figure 10.12 Schematic cross sections indicating evolution of northern California sedimentary geology. Note evolution of the accretionary prism from the late Jurassic through Paleogene. During the Neogene, the subduction zone "swallowed" a spreading center (the northward extension of the East Pacific Rise), thus producing thermal uplift of the entire region. (After Dickenson and Seely, 1979, with simplification.)

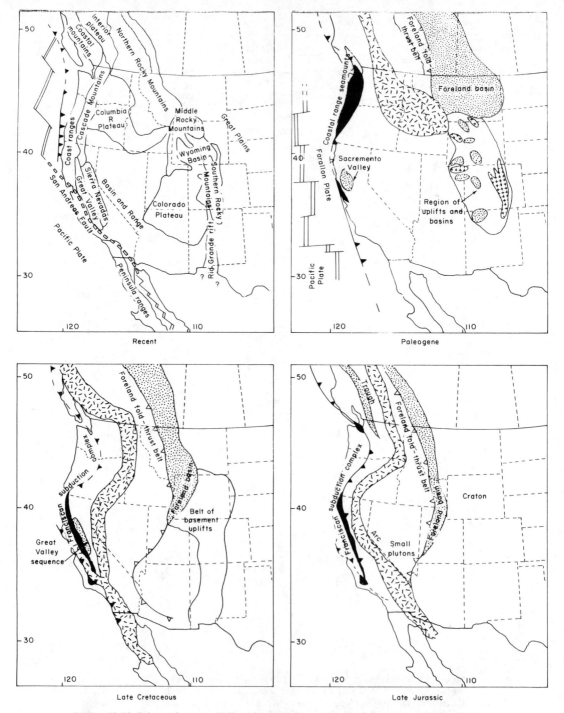

Figure 10.13 Schematic maps indicating tectonic evolution of the western United States from late Jurassic time. Note persistence of arc-trench system throughout the late Mesozoic. Note the Neogene emphasis on vertical tectonic motions resulting from lithospheric reheating and lithospheric thinning associated with placement of continental lithosphere above a mantle heat source. (Compiled from various sources including Dickenson, 1976, with modification and simplification.)

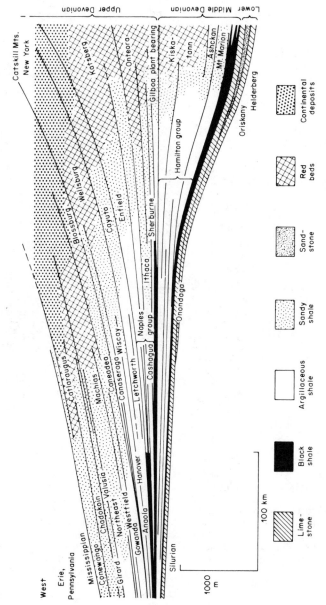

Figure 10.14 Stratigraphic cross section through New York State. The westward advance of the exogeosynclinal clastic wedge during Upper Devonian time is shown. [After Philip B. King, *The Evolution of North America*, (copyright © 1959 by Princeton University Press), Figure 32, p. 58.] Reprinted by permission of Princeton University Press.

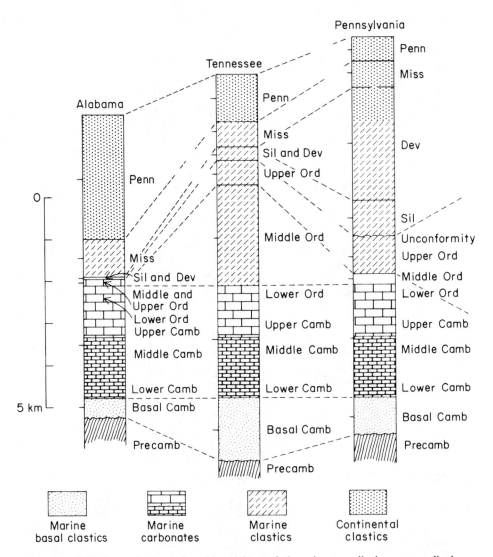

Figure 10.15 Generalized stratigraphic sections of the miogeosynclinal-exogeosynclinal sediment belts of the Appalachians. Note that all sections record essentially the same sequence of events. Basal Cambrian sands give way to Lower Paleozoic miogeosynclinal carbonate sediments, which in turn give way to thick exogeosynclinal clastic sequences, culminating in continental deposits. Note further, however, that the time of transition from miogeosyncline to exogeosyncline sedimentation varies greatly. Similar information is presented in map form in Figure 10.16. [After Philip B. King, *The Evolution of North America* (copyright © 1959 by Princeton University Press), Figure 3.4, pp. 61.] Reprinted by permission of Princeton University Press.

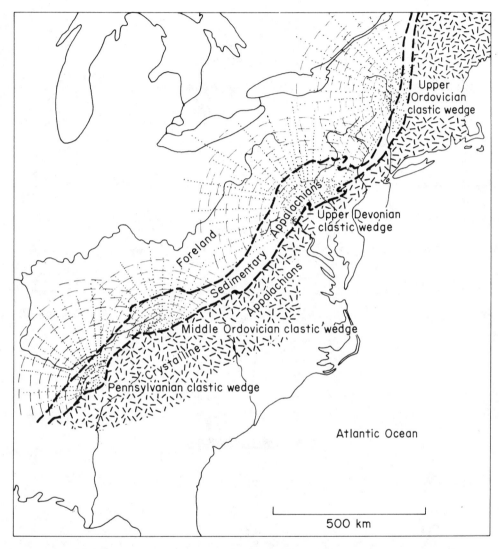

Figure 10.16 Schematic map indicating the location and age of major exogeosynclinal wedges of the Appalachians. [After Philip B. King, *The Evolution of North America* (copyright © 1959 by Princeton University Press), Figure 33, p. 59.] Reprinted by permission of Princeton University Press.

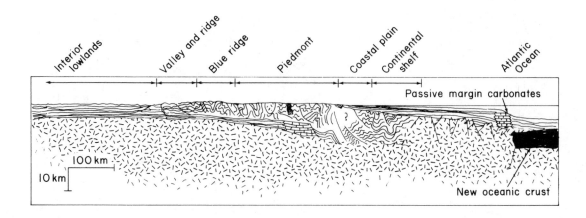

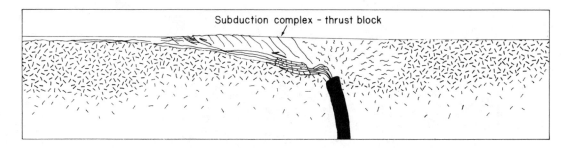

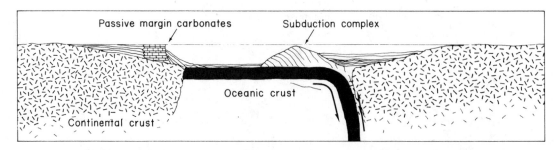

Figure 10.17 Schematic cross sections depicting the evolution of the southern Appalachians as the interaction between a North American Cambro-Ordovician passive margin and a subduction complex. Note that what earlier geologists perceived to be the "crystalline Appalachians" is now taken to represent a former subduction complex that has overridden the Cambro-Ordovician passive margin of the paleo–North American continent. Subsequent to this episode of mountain building, a new North Atlantic Ocean opened throughout Mesozoic time giving rise to the modern coastal plain and continental shelf sequences depicted in the upper diagram. (After Dickenson, 1977, and Cook *et al.*, 1979; with modification and generalization.)

For years, description of the sedimentary geology of the Appalachians stood at about the level depicted in the three foregoing figures. In recent years, major progress has been made toward understanding these field relations in terms of modern concepts of passive-margin and active-margin sedimentation. The Cambro-Ordovician carbonate sequences are taken to be passive-margin sediments of the old, tectonically quiet North American continent. The formation of the clastic source regions in the areas now occupied by the crystalline Appalachians is taken to represent the closure of the North American continent into a subduction zone.

While innumerable cartoons have been put forward to explain this or that region of the Appalachians in terms of plate tectonics, by far the simplest model with the most supporting evidence is that depicted in Figure 10.17. In this model, one simply takes a preexisting passive continental margin and tries to stuff it down an arc-trench system. As the passive margin closes on the arc-trench system, the passive margin tends to subside and the subduction complex tends to rise out of the ocean. In the more advanced stages, the continent fails to be swallowed by the subduction zone, thus bringing the closure of the ocean basin to a grinding halt. The resulting configuration is that of a clastic sequence in the foreland basin overlying the carbonate sequence of the continental region. In the southern Appalachians, COCORP data provides spectacular example of Cambro-Ordovician continental margin now buried 15 kilometers beneath the Piedmont, which is taken to be the subduction complex under which the North American continent refused to be swallowed.

It is common to presume that the geology of the Appalachian Mountains reflects the closure of the North Atlantic basin concurrent with Africa crashing into North America. While such a configuration of continental plates may be reasonable, continent-continent collisions are not necessary to produce the geological features described above; it is totally sufficient that a continent such as North America simply has buoyancy to counterbalance subduction forces as the continent approaches the arc-trench system.

The Strata of North America

Embedded within all of that "big-picture geology" discussed above, there are the sedimentary rocks of the North American continent. Most of them have formation names that preceded the development of the "big picture." The names of numerous formations in common use over large areas of North America are indicated in Tables 10.1, 10.2, and 10.3. While these formations are physical stratigraphic units by strict definition, in many cases their names have been elevated to formal local and regional chronostratigraphic usage. A more detailed complilation of lithostratigraphic information should be available through the American Association of Petroleum Geologists COSUNA project (Correlation of Stratigraphic Units in North America).

Table 10.1 North American Cenozoic Formation Names

		California			Western Interior	
Pleistocene		Terrace Deposits			Glacial and Alluvial Dep.	
Pliocene		Purisima Ss	Pico Ss			Ogallala Gr Ss
			Repetto Ss			
Miocene	U	Maricopa Sh			Brown's PK Ss	
	M	Temblor Ss	"Monterey Sh"	Sisquol Sh	Hemingford Gr	
	L			S. Magarita Ss		
				Rincon Sh	Arikaree Gr Clastics	
		Vaqueros Ss				
Oligocene	U	Sespe Congl				Brule Cl St
	L				White R. Gr Clastics	
Eocene	U	Kreyenhagen Sh			Duchesne R. Ss	
					Uinta Ss	
					Bridger Sh and Ss	
	M	Domengine Fm			Green R. Sh	
	L	Lodo Sh			Wasatch Gr Volcanic	
Paleocene	U	"Martinez Ss"	Silverado Ss		Fort Union Ss and Sh	
	L				Cannonball Ss	
					Ludlow Ss and Sh	

Gulf Coast	Florida	South Carolina	Maryland
Houston Ss	Anastasia to Caloosahatchee Ls	Pamlico Ss	Kempsville Ss
		Waccamaw and Croatan Ss	Norfolk Ss
Citronelle Fm		Bear Bluff Fm	
	Jackson Bluff Sh	Duplin Sh	Yorktown Ss
Pascagoula Cl	Chocktahatchee Ss	Hawthorn Ls	St. Marys Sh
Hattiesburg Cl			Choptank Ss
	Alum Bluff Ss		Calvert Ss
Fleming Gr — Lagarto Fm	Tampa Ls		
Fleming Gr — Oakville Fm			
Catahoula Gr — Anahuac Ss and Sh		U. Cooper Marl	Trent Fm	
Catahoula Gr — Frio Ss and Sh			
Vicksburg Ss and Sh	Suwanee Ls		
	Byram Ls		
Jackson Gr — Whitsett Ss and Ls	Ocala Ls	L. Cooper Marl	
Jackson Gr — McElroy Ls		Santee Ls	
Jackson Gr — Caddell Ls			
Claiborne Gr	Avon PK Ls	McBean Fm	
	Lake City Ls		Nanjemoy Cs and Ls
Wilcox Gr	Oldsmar Ls	Black Mingo Ss	
			Marlboro Fm
Midway Gr	Cedar Keys Ls	Clayton Marl	Aquia Cr. Fm
			Brightseat Fm

171

Table 10.2 North American Mesozoic Formation Names

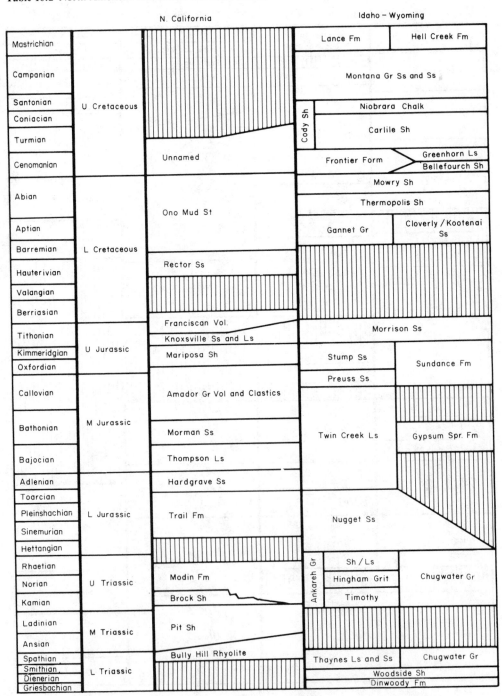

		N. California	Idaho – Wyoming		
Mastrichian			Lance Fm	Hell Creek Fm	
Campanian	U Cretaceous		Montana Gr Ss and Ss		
Santonian			Niobrara Chalk		
Coniacian			Carlile Sh (Cody Sh)		
Turmian					
Cenomanian		Unnamed	Frontier Form	Greenhorn Ls / Bellefourch Sh	
Abian	L Cretaceous	Ono Mud St	Mowry Sh		
			Thermopolis Sh		
Aptian			Gannet Gr	Cloverly / Kootenai Ss	
Barremian		Rector Ss			
Hauterivian					
Valangian					
Berriasian		Franciscan Vol.			
Tithonian	U Jurassic	Knoxsville Ss and Ls	Morrison Ss		
Kimmeridgian		Mariposa Sh	Stump Ss	Sundance Fm	
Oxfordian			Preuss Ss		
Callovian	M Jurassic	Amador Gr Vol and Clastics	Twin Creek Ls	Gypsum Spr. Fm	
Bathonian		Morman Ss			
Bajocian		Thompson Ls			
Adlenian		Hardgrave Ss			
Toarcian	L Jurassic	Trail Fm	Nugget Ss		
Pleinshachian					
Sinemurian					
Hettangian					
Rhaetian	U Triassic	Modin Fm	Sh / Ls (Ankareh Gr)	Chugwater Gr	
Norian			Hingham Grit		
Kamian		Brock Sh	Timothy		
Ladinian	M Triassic	Pit Sh			
Ansian					
Spathian	L Triassic	Bully Hill Rhyolite	Thaynes Ls and Ss	Chugwater Gr	
Smithian			Woodside Sh		
Dienerian			Dinwoody Fm		
Griesbachian					

Stratigraphic correlation chart.

Arizona – New Mexico	Gulf Coast	South Carolina	New Jersey
McDermott Fm	Navarro Gr	Peedee Fm	Monmouth Gr
Lewis Sh		Black Creek Fm	Matawan Gr
La Ventana Ss	Taylor Marl		
Menefee Fm	Austin Chalk	Middendorf Fm	Magothy Ss and Cl
Pt. Lookout Ss			
Mancos	Eagle Ford Sh	Cape Fear Fm	Raritan Ss and Cl
Gallop Ss	Woodbine Ss		
Greenhorn Ls	Washita Gr Carb		Patapsco Ss and
Dakota Ss	Edwards Ls / Stuart City Ls	Unnamed Unexposed Carbonates	
Burro Canyon Fm	Trinity Gr: Glenrose Ls, Pearsall Fm, Sligo Ss, Hosston Clastics		Potomac Gr
?			Arundel Gr
Morrison Ss	Cotton Valley Gr		Patuxent Ss
Bluff Ss	Buckner Sh	?	?
Summerville Marl	Smackover Ls and Dol		
Entrada Ss	Norphet Redbeds		
Carmel Ls	Louann Salt		
	Werner Redbeds Anhy		
Glen Canyon Gr: Navajo Ss, Kayenta Ss, Wingate Ss	Eagle Mills Salt (?)		
Chinle Sh	?	Newark Gr – Ss Volc and Sh	
Shinarump Congl			
Moenkopi FM (LS)			

Table 10.3 North American Paleozoic Formation Names

Period		Utah	Texas	
Permian	L	Woodside Sh	Rustler Ls	
		Park City Fm	Castille Fm	
			San Andres Carb and Evap	
			Leonard Fm	
	E	Weber Qtzite	Wolfcamp Clastics	
Pennsylvanian	L	Weber Qtzite	Cisco Fm	Gaptank
	M		Canyon Fm	Haymond Ss and Sh
			Strawn Fm	
	E	Morgan Ss	Dimple Ls	
Mississippian	L	Brazer Ls and Ss	Tesnus Ss and Sh	
	E	Madison Ls		
Devonian	L	Mowitza Sh	Woodford Fm	Caballos Novaculite
		Guilmette Ls		
	M	Simonson Fm		
	E	Sevy Fm	Slator Fm	
Silurian	L			
	M	Laketown Dol	Fusselman Dol	
	E			
Ordovician	L	Fish Haven Dol	Montoya Ls	Ft. Pena Ls and Ch
	M	Swan Peak Ss / Eureka Qtzite	Simpson Ls	Woods Hollow Ss
				Maravillas Chert
				Alstate Slate
	E	Garden City Ls	Ellenburger Dol	Marathon Ls
Cambrian	L	Notch Peak Ls	Wilberns Fm	
		Orr Ls	Cap Mtn. Ss	
	M	Marjun Ls		
		Pioche Sh		
	E	Prospect Mtn Ss		

174

Oklahoma /- Mississippi Valley   Pennsylvania   New York

Oklahoma	Mississippi Valley	Pennsylvania	New York
		Greene Fm	
		Dunkard Fm / Washington Fm	
	Shawnee Gr Ls	Monongahela Fm	
	K. C. Gr Ls	Conemaugh Fm	
Atoka Fm	Cherokee Gr	Allegheny Coal	
John's Valley			
Jackfork Ss		Pottsville Ss and Sh	Potsville Ss and Sh
Stanley Fm	? ? ?	Mauchunk Fm	
		Pocono Gr	Pocono Ss and Sh
Arkansas Nov.	Chattanooga Sh	Catskill Fm	Blossberg Red Beds
	Grand Tower Ls	Hamilton Gr	Hamilton Gr
		Needmore Fm	Onandago Ls
Frisco Ls	Clear Cr Ls	Old Port Fm	Helderberg Gr
Haragan Sh	Bailey Ls	Keyser Ls	Salina Gr
Henry House	Lafferty Ls	Bloomsberg Red Beds	Lockport Dol
Missouri Mtn. Fm	St. Clair Ls	Clinton Sh and Ss	Clinton Sh and Ss
Blaylock Ss	Cason Sh	Tuscagora Ss	Medina Ss
Viola Ls /	Galena Ls	Juanita Ss	
Big Fork Ch	Decorah Sh	Oswego Ss	Oswego Ss
	Platteville Fm	Reedsville Sh	
Womble Sh	St. Peter Ss	Trenton Ls	Pawlet Sh
		Loysberg Ls	Normanskill Sh
Arbuckle Ls	Prarie Du Chien	Bellefonte Dol	Poultney Sl
Crystal Mtn. Fm	Eminence Dol.	Gatesburg Ls	Hatch Hill Sh / Potsdam Ss
Collier Fm	Derby Dol.		
Reagan / Lukfata Ss	Lukfata Ss	Warrior Ls	
		Plesant Hill Ls	Westcastleton Sh
		Waynesboro Ss	Bull Sh
			Rensselaer Grywk

175

Citations and Selected References

ATWATER, T. 1970. Implications of plate tectonics for the Cenozoic tectonic evolution of western North America. Bull. Geol. Soc. Amer. 81: 3513–3536.

AUSTIN, JAMES, A., JR., ELAZAR UCHUPI, D. R. SHAUGHNESSY, III, and R. D. BALLARD. 1980. Geology of New England passive margin. Bull. Amer. Assoc. Petrol. Geol. 64 (4): 501–526.

BECK, MYRL, ALLAN COX, and DAVID L. JONES. 1980. Penrose Conference Report: Mesozoic and Cenozoic microplate tectonics of western North America. Geology 8 (9): 454–456.

BIRD, J. M., and J. F. DEWEY. 1970. Lithosphere plate-continental margin tectonics and the evolution of Appalachian orogene. Bull. Geol. Soc. Amer. 81: 1031–1060.

BLAKE, M. C., JR., R. H. CAMPBELL, T. W. DIBBLEE, JR., D. G. HOWELL, T. H. NILSEN, W. R. NORMARK, J. C. VEDDER, and E. A. SILVER. 1978. Neogene basin formation in relation to plate tectonic evolution of San Andreas fault system. Bull. Amer. Assoc. Petrol. Geol. 62: 344–372.
Development of the West Coast North American "transform margin."

COOK, FREDERICK. A., DENNIS S. ALBAUGH, LARRY D. BROWN, SIDNEY KAUFMAN, JACK E. OLIVER, and ROBERT D. HATCHER, JR. 1979. Thin-skinned tectonics in the crystalline southern Appalachians: COCORP seismic-reflection profiling of the Blue Ridge and Piedmont. Geology 7: 563–567.

DEWEY, J. F., and J. M. BIRD. 1970. Mountain belts and the new global tectonics. J. Geophys. Res. 75: 2625–2647.
A very helpful review of the mountain chains of the world in the light of ocean-floor spreading.

DICKINSON, WILLIAM R. 1976. Sedimentary basins in western North America. Can. J. Earth Sci. 13: 1268–1287.

———. 1977. Tectono-stratigraphic evolution of subduction assemblages, p. 33–40. *In* M. Talwani and W. C. Pitman III (eds.), Island arcs, deep sea trenches, and back-arc basins. Amer. Geophys. Union, Maurice Ewing Series 1.

DICKINSON, WILLIAM R., and D. R. SEELY. 1979. Structure and stratigraphy of forearc regions. Bull. Amer. Assoc. Petrol. Geol. 63: 2–31.

DICKINSON, WILLIAM R., and WALTER S. SNYDER. 1979. Geometry of triple junctions related to San Andreas transform. J. Geophys. Res. 84 (B2): 561–572.

DIETZ, R. S. 1972. Geosynclines, mountains, and continent-building. Science 226 (3): 30–38. Easy reading and well illustrated.

FISHER. G. W., F. J. PETTIJOHN, J. C. REED, and K. N. WHELLER. 1970. Studies of Appalachian geology, central and southern. Wiley-Interscience, New York, 460 p.

GINSBURG, R. N. 1964. South Florida carbonate sediments: guide-book for field trip no. 1. Geol. Soc. Amer. Annual Convention, Miami Beach, Florida. 72 p.

HALES, A. L., C. E. HELLSLEY, and J. B. NATION. 1970. Crustal structure study on Gulf Coast of Texas. Bull. Amer. Assoc. Petrol. Geol. 54: 2040–2057.

HEEZEN, B. C., and M. THARP. 1968. Physiographic diagram of the North Atlantic Ocean. Geol. Soc. Amer. Spec. Pap. 65.

HILL, M. L. 1971. A test of new global tectonics: comparison of northeast Pacific and California structures. Bull. Amer. Assoc. Petrol. Geol. 55: 3–9.
One of the "grand old men" of southern California geology suggests that things are not quite so simple as proponents of "the new global tectonics" might have us believe.

KAY, M. 1951. North America geosynclines. Geol. Soc. Amer. Mem. 48. 143 p.
A landmark in the literature of mountain belt geology.

KING, P. B. 1959. The evolution of North America. Princeton University Press, Princeton, N.J. 189 p.
A delightful synthesis of the geology of North America. Easy reading.

———. 1965. Tectonics of Quaternary time in middle North America. p. 831–870. *In* H. E. Wright, Jr., and D. G. Frey (eds.), The Quaternary of the United States. Princeton Univ. Press, Princeton, N.J.

LEHNER, P. 1969. Salt tectonics and Pleistocene stratigraphy on continental slope of northern Gulf of Mexico. Bull. Amer. Assoc. Petrol. Geol. 53: 2431–2479.

LIPMAN, PETER W., HAROLD J. PROSTKA, and ROBERT L. CHRISTIANSEN. 1971. Evolving subduction zones in the western United States, as interpreted from igneous rocks. Science 174: 821–825.

LIVACCARI, RICHARD F. 1979. Late Cenozoic tectonic evolution of the western United States. Geology 7 (2): 72–75.

LOWMAN, S. W. 1949. Sedimentary facies in the Gulf Coast. Bull. Amer. Assoc. Petrol. Geol. 33: 1939–1997.

NILSEN, TOR H., and JOHN H. STEWART. 1980. Penrose Conference Report: the Antler orogeny—mid-Paleozoic tectonism in western North America. Geology 8 (6): 298–302.

PRICE, R. A. 1973. Large-scale gravitational flow of supracrustal rocks, southern Canadian Rockies, p. 491–502. *In* K. A. Dejong and Robert Scholten (eds.), Gravity and tectonics. John Wiley & Sons, New York.

RODGERS, J. 1970. The tectonics of the Appalachians. Wiley-Interscience, New York. 288 p.

SCHLEE J., J. C. BEHRENDT, J. A. GROW, J. M. ROBB, R. E. MATTICK, P. T. TAYLOR, and B. J. LAWSON. 1976. Regional geologic framework off northeastern United States. Bull. Amer. Assoc. Petrol. Geol. 60: 926–951.

SEELY, D. R. 1979. The evolution of structural highs bordering major forearc basins, p. 245–260. *In* J. S. Watkins, L. Montadert, and P. W. Dickerson, Geological and geophysical investigations of continental margins. Amer. Assoc. Petrol. Geol. Mem. 29.

SHERIDAN, R. E. 1974. Conceptual model for the block fault origin of the North American continental margin geosyncline. Geology 2: 465–468.

SHIPLEY, THOMAS H., RICHARD T. BUFFLER, and JOEL S. WATKINS. 1978. Seismic stratigraphy and geologic history of Blake Plateau and adjacent western Atlantic continental margin. Bull. Amer. Assoc. Petrol. Geol. 62 (5): 792–812.

STANLEY, K. O., W. M. JORDON, and R. H. DOTT, JR. 1971. New hypothesis of early Jurassic paleogeography and sediment dispersal for western United States. Bull. Amer. Assoc. Petrol. Geol. 55: 10–19.
Island arc and shallow seas of the western one-third of the United States as the Atlantic began to open.

WICKHAM, JOHN, DEITRICH ROEDER, and GARRETT BRIGGS. 1976. Plate tectonics models for the Ouachita foldbelt: Geology 4: 173–176.

ZEN, E., W. S. WHITE, J. B. HADLEY, and J. B. THOMPSON, JR. (eds.). 1968. Studies of Appalachian geology, northern and Maritimes. Wiley-Interscience, New York. 475 p.

Dynamics
in the Context
of Environment

IV

In the previous sections, we looked at the stratigraphic record in a very general fashion. Now we shall learn to recognize specific sedimentary environments within the stratigraphic record in the hope that we may develop a more accurate understanding of how stratigraphic sequences originated.

Each of the following chapters deals with a major suite of sedimentary environments. Each chapter begins with a review of facts based on the observation of Recent sediments accumulating in these environments. Then these facts are formed into a general model of (1) what the sediments of these environments would look like if we saw them in the stratigraphic record and (2) how the environment would react to dynamic events such as submergence and emergence. Finally, with a model based on Recent sedimentation firmly in mind, we shall examine selected Ancient sequences in order to test the model and in order to get some feeling for the adequacy of this approach to the stratigraphic record.

Our discussion of Recent environments is oriented toward the development of principles that will have general application. Specifically, our discussion of Recent environments is *not* oriented toward the detailed analysis of every little facet of sedimentation in this or that area. Rather, we shall pick up specific points that will be useful to us in the formulation of a general model.

The following treatment of sedimentary environments and the stratigraphic record, therefore, focuses on the development of general models. A model with constant sea level is first constructed from Recent sedimentation data. Then it is expanded to a theoretical dynamic model showing how the constant sea-level model should react to conditions of emergence or submergence. The models are of value as a *tool* for solving problems, not as an *answer* to our problems. We must organize our thoughts, recognize our misconceptions and inconsistencies, improve our models, and so work forward in an iterative process that leads to better understanding of the stratigraphic record through the use of ever-improving models.

Clastic Sedimentation in Stream Environments

11

Our look at the details of sedimentary environments begins with the action of running water on terrigenous clastic debris derived from mountainous terrains. Physical and chemical weathering processes yield boulders, sand, silt, and clay from the mountainsides. These materials immediately begin to accumulate in sedimentary environments that we hope to recognize in the geologic record.

Recent Clastic Sedimentation in Stream Environments

In the following pages, we shall primarily study four examples of stream sedimentation. The Donjek River, a tributary of the Yukon, provides an example of the braided-river environment in mountainous terrain (Williams and Rust, 1969). The Platte River of Colorado and Nebraska is a braided river that extends some 800 kilometers outward from the mountain region (Smith, 1970; Blodgett and Stanley, 1980). The Brazos River of southeastern Texas is an exceedingly well-studied meandering stream (Bernard et al., 1970). Finally, the Red River of northwestern Louisiana is an interesting example of a complicated meandering-stream pattern (Harms et al., 1963).

The Donjek River. Figure 11.1 indicates the location of the Donjek River. Note that the study area is only a few tens of kilometers downstream from the glaciers that form the headwaters of the river. The modern braided river is 2 to 4 kilometers in width and occupies a valley cut into glacial tills, outwash sands and gravels, and loess.

Figure 11.2 presents the generalized physiography of the study area, and Figure 11.3 illustrates the complex system of channels and bars that exists within the var-

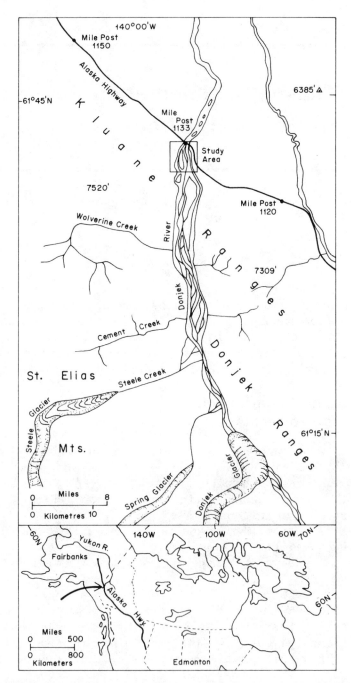

Figure 11.1 Locality map of the Donjek River and study area. The Donjek River is an example of a braided stream in mountainous terrain very near the source area. (After Williams and Rust, 1969.)

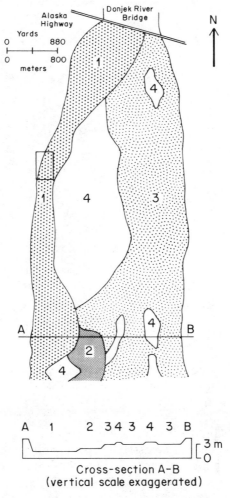

Figure 11.2 Areal distribution and diagrammatic profile section of stream channels and levels within a portion of the Donjek River. Note location of Figure 11.3. (After Williams and Rust, 1969.)

ious levels. Most of the water discharged today is carried by channel level (1). Figure 11.3 shows the major elements of bar and channel geometry in a portion of level (1). The first-order topographic highs are longitudinal bars, separated from one another by the large channels that carry the majority of the water at intermediate- and low-river stage. Deposition and migration of first-order longitudinal bars occur primarily during flood stage. Because they are deposited at flood stage, longitudinal bars are the most poorly sorted sediment within the braided-stream environment. They commonly consist of pebbles and cobbles mixed with sand. As flood waters recede, first-order bars are usually dissected by smaller channels.

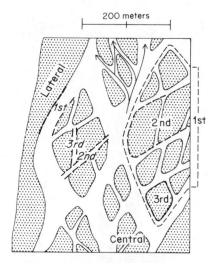

1st, 2nd, 3rd, = Channel (Orders)
1st, 2nd, 3rd, = Bar (Orders)

Figure 11.3 Map demonstrating the hierarchal organization of channels and bars within a portion of level (1) of the Donjek River. First-order longitudinal bars, deposited during flood stage, are later dissected by second-order and third-order channels as the river stage subsides. (After Williams and Rust, 1969.)

During low stage of the river, sedimentation on the longitudinal bars more or less ceases. Continued transport and redeposition of sand-sized material continue to occur within the channels. Transverse sandbars commonly have steep avalanche faces in the downstream direction and small-scale asymmetrical ripples on the top of the sandbar. Both features tend to suggest an almost continuous transportation of sand-sized material across the top of the transverse bar, with sand deposition occurring along the avalanche slope at the front of the bar. Migration of such bars produces large-scale planar cross strata.

Figure 11.4 shows size distribution within a number of samples from the Donjek River. Note the incredible variation in both size and size distribution within the confines of this braided-stream environment. This variation results from the wide range of sedimentation conditions that predominate in the area at varying times throughout the year. During flood stage, sediment supply is large and the accumulation beneath the main flow of water is coarse-grained and extremely poorly sorted. This is also the time of occasional flooding of the higher terrace levels of the river valley. Flood waters reaching these higher levels are relatively slow-moving and deposit only silt and clay. At low stage of the river, sediment supply is also low. Stream power is insufficient to transport the gravels of the longitudinal bar deposits. Whereas the flood waters were extremely muddy, low-stage water is likely to be crystal clear. At this time, only sand-sized material is being transported and resedimented. As indicated in Figure 11.4, some of the medium to fine sands are the best-sorted sediments to be found in the environment.

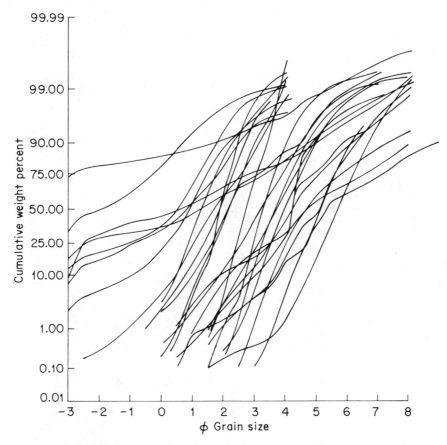

Figure 11.4 Cumulative size-frequency curves of sediment samples from the Donjek River. Note the extreme variation in grain size and sorting that occurs in samples from a relatively small area within a braided stream. (After Williams and Rust, 1969.)

Figure 11.5 attempts to portray lateral facies relationships within any single bar-channel complex. Observe that transitions from poorly sorted bar gravels to well-sorted channel sand can occur within a matter of a few meters. Where channels lie dormant, they will eventually be filled with silt and clay. Then the entire area will become vegetated unless fluvial processes reoccupy that area within a year or so.

A good indication of bar mobility within the various levels is the colonization of the bar crests by trees. In level (1), bar mobility associated with annual flooding has not allowed trees to establish themselves.

Level (2) is the site of widespread fluvial activity just during flood stage. At other times, only the major first-order channels contain a slight flow of water. Bar mobility during flood stage is not so pronounced in level (2) as it is in level (1). Frequently, willow trees have established themselves on the crests of bars; some are as much as 12 years old (as estimated by tree-ring analysis).

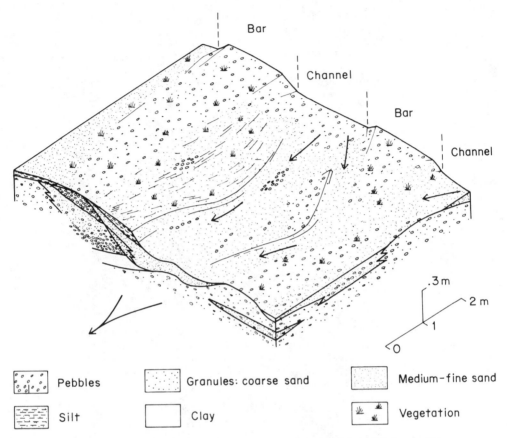

Figure 11.5 Three-dimensional model illustrating rapid variation in lithology within an area of dissected bars and small channels. (After Williams and Rust, 1969.)

Although levels (3) and (4) retain geomorphic features that are clearly channels and bars, these levels have not been sites of active bar migration for many years. Today, level (3) is covered with water only during maximum stage; at that time, sedimentation is predominated by slack-water settling of suspended muds. Willow trees on level (3) are as old as 27 years, and spruce development on level (4) indicates that the topographic surface there is at least 250 years old.

The ages of vegetation on the various topographic levels of the braided-river valley provide a good demonstration of a fundamental characteristic of the braided-stream environment: From time to time the site of major fluvial activity shifts from one area of the river valley to the other in an unsystematic fashion. Level (3) provides the most graphic example of this shift. At some time in the not too distant past, fluvial activity was sufficiently intense in this area of the river valley to develop the system of bars and channels quite similar to those now developing in level (1). Today, these bars and channels are largely inactive, and they are being covered over with a drape of mud deposited only during maximum flood stage. If we may judge from other braided-rivers studies (such as Doeglas, 1962), the eastern side of the river valley will likely become the site of energetic braided-stream activity again at some time in the near future. The present active channel system need only become glutted with sediment or blocked by a rock slide or a logjam; then the site of active transportation and recent sedimentation would rapidly shift back to the east.

Figure 11.6 gives a composite stratigraphic view of the entire study area. Consider whether or not you could untangle this stratigraphic complexity if it faced you in an outcrop of Ancient rocks. We have already noted that lithologic variations occur on a very local scale. The various levels of the river valley that are now so clear in maps and cross sections would appear only as zones of cut and fill in an outcrop of Ancient rocks. The relationship between level (3) and level (2) is well illustrated in Figure 11.6.

The Platte River. Whereas the Donjek River exemplifies braided-stream sedimentation within mountainous regions, the Platte River shows braided-stream sedimentation extending as far as 900 kilometers away from the source area. As indicated

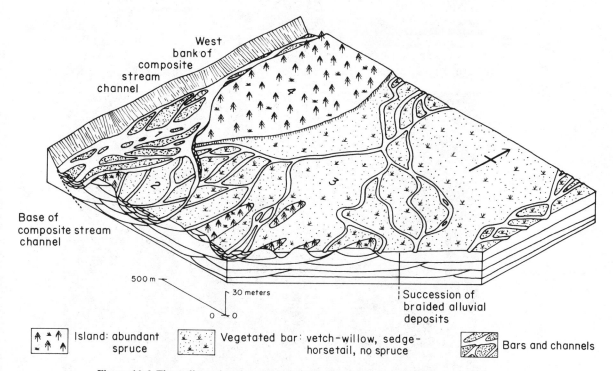

Figure 11.6 Three-dimensional model indicating the complex relationships among the various stream-channel systems and levels within the Donjek study area. Note especially the extreme lack of lateral continuity, both on the scale of the relationship of one level to the other and even within a single level. Cut and fill and cut again is the ever-recurring story of braided-stream sedimentation. (After Williams and Rust, 1969.)

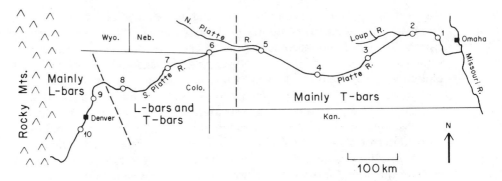

Figure 11.7 Map of the South Platte and Platte rivers showing sample localities in the generalized distribution of longitudinal bars and transverse bars. The Platte River provides an example of braided-stream sedimentation out to considerable distances beyond the mountainous source area of the sediments. (After Smith, 1970.)

in Figure 11.7, Smith (1970, 1971) has studied portions of the Platte River system from the front ranges of the Rocky Mountains to the junction of the Platte River and the Missouri River in eastern Nebraska.

As with the Donjek River, the story of Platte River sedimentation is written in terms of *longitudinal bars* and *transverse bars*. Yet this study incorporates a sufficiently long stretch of river so that important differences are noted from west to east. Figure 11.7 indicates that longitudinal bars predominate near the source area,

Figure 11.8 Longitudinal bar near Fort Morgan, Colorado, South Platte River. Bar is about 8 meters wide. (Photo by N. D. Smith.)

Figure 11.9 Transverse bars near Valley, Nebraska, lower Platte River. Note especially dissection of the point bar in the left foreground. (Photo by N. D. Smith.)

whereas transverse bars predominate in the downstream area. Figures 11.8 and 11.9 depict typical Platte River longitudinal bars and transverse bars.

As in the previous example, the deposition of longitudinal bars begins as a flood-stage phenomenon. While the general shape of the bar is dictated by flood-stage deposition, subsequent slackening of river discharge dissects the bar and results in sediment redistribution in a number of microenvironments around the overall geomorphic feature of the longitudinal bar (Blodgett and Stanley, 1980). In their simplest form, the internal characteristics of these bars are (1) a relatively large mean grain size, (2) relatively poor sorting, (3) crude development of horizontal stratification, and (4) trough cross-stratification near the top and around the perimeter of the bar. However, repeated flood and ebb cycles may complicate this generality.

In contrast, transverse-bar migration is a relatively more simple process occurring within channels at moderate- to low-river stage. The internal characteristics of transverse bars include: (1) relatively smaller mean grain size, (2) relatively better sorting, and (3) a preponderance of high-angle, planar cross-stratification (Figure 11.10) topped by small-scale, tough cross-stratification. During low stage of the river, the predominant process is the dissection of transverse bars (see Figure 11.9). Such dissection results in extreme lateral discontinuity of individual sand bodies, a characteristic that will aid our recognition of braided streams in the stratigraphic record.

Figure 11.11 provides a graphic summary of mean grain size and standard deviation of sands from longitudinal bars and transverse bars. Because the relative

Figure 11.10 Planar cross strata in a dissected transverse bar, Platte River. Accretion on the avalanche slope causes transverse-bar migration and leaves this sedimentary record. (Photo by N. D. Smith.)

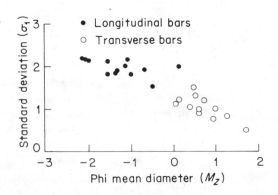

Figure 11.11 Plot of grain size versus sorting for 25 samples of Platte River bar sand. Observe the sharp distinction between coarse-grained, poorly-sorted, longitudinal bars and finer-grained, better-sorted, transverse bars. (After Smith, 1970.)

proportion of longitudinal bars and transverse bars varies from west to east, these individual parameters likewise show systematic trends downstream. Figure 11.12 plots mean grain size and standard deviation as a function of distance downstream. Size decreases and sorting becomes better downstream. Likewise, stratification types can be recast as a function of distance downstream (Figure 11.13). Trough cross-

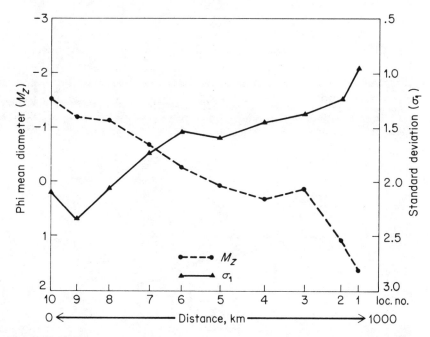

Figure 11.12 Downstream changes in mean grain size and sorting in the South Platte–Platte River. Grain size decreases and the sediment becomes better sorted with increasing distance from the source area. (After Smith, 1970.)

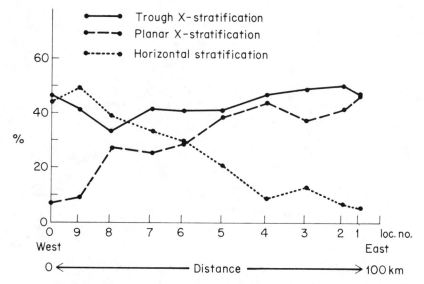

Figure 11.13 Downstream variation in relative amounts of the principal stratification types, South Platte–Platte River. Whereas trough cross-stratification is uniformly present throughout the study area, planar cross-stratification (associated with transverse bars) is observed to become more abundant downstream, and horizontal cross-stratification (associated with transitional flow regime and longitudinal bars) is most prevalent near the source area. (After Smith, 1970.)

stratification occurs typically as the upper few tens of centimeters on top of both longitudinal bars and transverse bars. The relative importance of trough cross-stratification shows little variation downstream. On the other hand, high-angle, planar cross-stratification occurs predominantly in transverse bars and thus increases systematically downstream. Conversely, horizontal stratification is more typical of longitudinal bars and shows a systematic decrease in importance downstream.

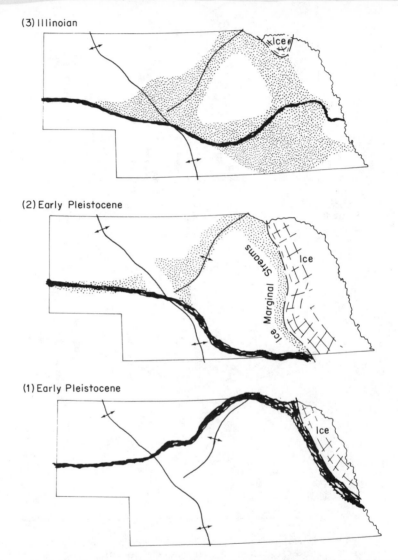

Figure 11.14 Sketch maps of Nebraska showing the shifting pattern of the Pleistocene "paleo-Platte River." The northwest-southeast trending anticlinal feature is the Chadron-Cambridge Arch. Epeirogenic uplift along this arch influences stream erosion-deposition (see Figure 11.15). (After Stanley and Wayne, 1972, with simplification.)

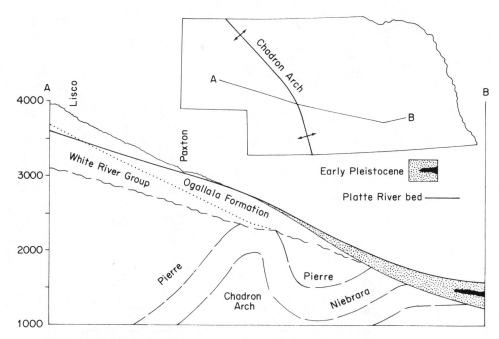

Figure 11.15 Schematic cross section of Nebraska showing the relation between the profile of the Platte River and the Chadron-Cambridge Arch. Whereas pre-Pleistocene stream deposition deposited the Ogallala formation in western Nebraska, subsequent epeirogenic uplift constrains Quarternary stream sedimentation to the eastern half of the state.

Stanley and Wayne (1972) have shown that the position of the Platte River has varied greatly during the Quaternary. As indicated in Figure 11.14, virtually the entire eastern half of Nebraska has experienced sediment deposition from the "paleo-Platte River" braided-stream system over the course of the last million years or so. Figure 11.15 provides a schematic cross section indicating incisement (erosion) of the Platte River in western Nebraska and aggradation of stream sediments in eastern Nebraska is substantially controlled by tectonic activity along the Chadron-Cambridge Arch. These observations are important to us for two reasons. First, we see that braided streams in relatively flat-lying terrain do not tend to maintain a single preferred position, but they migrate widely along the general strike. Second, we see here a striking example of interaction of river profile with local epeirogenic movement. Areas which are rising become incised, relatively low areas are infilled by aggradating stream deposits.

The Brazos River. The Brazos River, southwest of Houston, is an exceedingly well-studied example of a recent meandering stream (Figure 11.16). Whereas the braided stream is characterized by numerous small, shallow channels dissecting a river bed composed of an irregular arrangement of longitudinal and transverse bars, the meandering stream is typified by a single, relatively deep channel in which erosion and sedimentation occur very systematically.

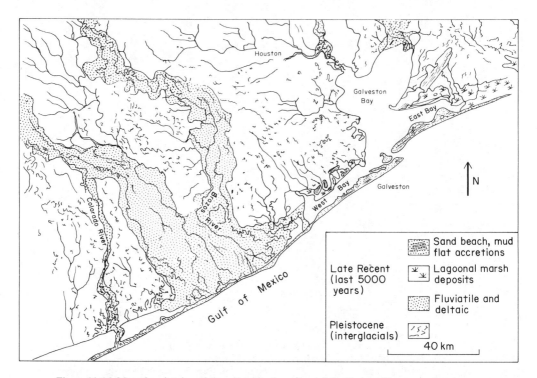

Figure 11.16 Map showing location and geologic setting of the Brazos River, southeastern Texas. The Brazos is an extremely well-studied example of a meandering stream. Note the numerous abandoned meander loops throughout the coastal plain of southeast Texas. (After Bernard *et al.,* 1970.)

The major elements of the meandering-stream system are indicated in Figure 11.17. The basic subsystem is the individual *meander loop.* The deepest portion of the river channel lies very close to the outer bank of the loop. The outer bank of the loop is the site of active erosion and is commonly referred to as the *cutbank*. The inside bank of the meander loop is the site of active deposition. Sandbars deposited in this position are referred to as *point bars*.

Figure 11.18 is a cross section through the river channel and point-bar deposit. A well-developed oxidized soil zone provides a convenient lower boundary for studying Recent meandering-stream deposits. Above the soil are approximately 12 meters of river deposits not related to the deposition of the modern point bar under investigation. Consider first the surficial sediments of the river channel and point bar. Note that the deepest portion of the channel lies at the base of the cutbank. Sediment that accumulates in the deep channel consists of poorly bedded gravel and coarse sand. Grain size decreases systematically, with the coarsest material in the deep channel and the finest sand on top of the point bar. Furthermore, shallow pits dug in the surface of the point bar reveal systematic variations in primary structures. Near the low-water level, there is large-scale, high-angle cross-stratification. Somewhat higher on the bar, stratification is nearly horizontal. The top of the bar is

Figure 11.17 Oblique aerial photograph exhibiting the major elements of meandering-stream erosion and sedimentation. Erosion occurs on the outside of the bend (cutbank) and deposition occurs on the inside of the bend (point bar). (After Bernard *et al.,* 1970.)

characterized by small-scale, trough cross-stratification. Observe that cores taken through the point bar confirm a vertical sequence of variation in grain size and primary structures. This observation is consistent with observations in the modern river channel and surface of the point bar. Channel migration and concomitant point-bar sedimentation develop a laterally extensive sand body that shows a strong tendency to systematic variation in grain size and primary structures from bottom to top.

Clearly, most of the point-bar deposition process occurs during high stages of the river. Variations in water depth and velocity produce systematic sorting of sediment that is eroded from the cutbank as well as brought down the river from upstream. Coarse sands and gravels remain as a channel-lag deposit in the floor of the deepest portion of the river. In sharp contrast, the upper portions of the point bar record deposition from shallow, relatively slow-moving water. Fine sand accumulates here. The middle portions of the point bar are intermediate between these two extremes of grain size. Similarly, variations in velocity, water depth, and grain size lead to systematic variations in primary structures. In deep, fast-moving water,

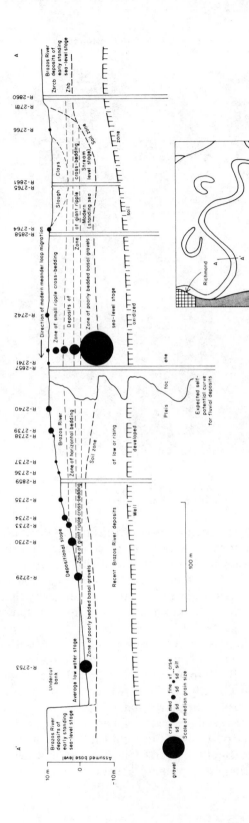

Figure 11.18 Cross section of a Brazos River point-bar deposit. Traversing from the deepest portion of the river channel (left) up onto the point bar (center), note the systematic decrease in grain size as well as a sequence of distinctly different primary structures. This sequence of variation in grain size and primary structures is repeated in vertical sequence in the borehole near the center of the diagram. The point-bar deposit has accumulated by a lateral migration from right to left as the river eroded the cutbank to the outside of the meander loop and deposited sediment on the point bar to the inside of the meander loop. (After Bernard *et al.*, 1970.)

coarse-grained sediments typically form large-scale, high-angle cross strata. In somewhat shallower water, sand-sized sediments form near horizontal stratification indicative of transitional low regime. Still higher on the bar, fine-grained sand accumulating in relatively slow-moving shallow water forms small-scale, ripple cross-stratifications. The point-bar deposit will be a major portion of the fining-upward unit that we shall come to associate with meandering-stream deposition.

As indicated in Figure 11.19, the present meander belt of the low stage Brazos River occupies but a small portion of the area dominated by the river during flood stage. During flood stage, large amounts of fine-grained sediment may be deposited throughout the flood plain. Such deposits are commonly referred to as *backswamp* or *overbank* deposits. On the Brazos flood plain, these deposits are generally brown silty clay. Inasmuch as the flood plain is dry most of the year, evidences of soil development and mottling by burrowing animals and plant roots are common.

Note in Figure 11.16 the numerous abandoned stream channels (indicated by wiggly lines) in the Pleistocene terrain which start and end abruptly. Whereas the last 5000 years have seen the Brazos and the Colorado Rivers restricted to modern flood plains such as depicted in Figure 11.19, it would appear that Pleistocene flood plains of these same rivers have varied widely in their location. As with our Platte River example to the interior of the continent (Figures 11.14 and 11.15), streams appear to seek out the low spots and fill them in with aggradating stream deposits.

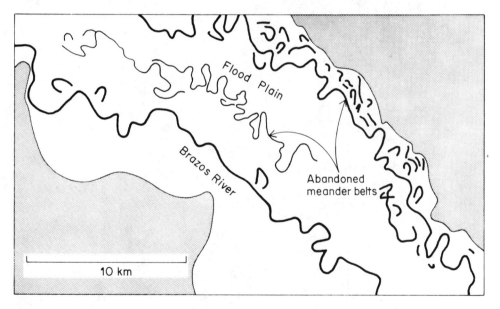

Figure 11.19 Map of the Brazos alluvial valley, illustrating the modern meander belt, abandoned meander belts, and flood plains. During flood stage, water occupies the entire alluvial valley. At such time, overbank deposits of silty clay are deposited throughout the flood plains. (After Bernard *et al.,* 1970.)

The Red River. Whereas the Brazos River provides a simple and pleasing picture of the systematics of meandering-stream sedimentation, the study of point-bar sedimentation in the Red River of northwestern Louisiana (Harms *et al.,* 1963) provides some insight into the complexities that can arise from sedimentological processes operating during various stages of the river. Figure 11.20 indicates the location of the study area. Note once again that the main channel of the present meandering stream occupies just a small portion of the total flood plain. Two topographic surfaces are apparent in the construction of the Beene point bar (Figure 11.21). Both levels are constructed by point-bar deposition of silt, sand, and gravel in a fashion quite similar to that described for the Brazos River. As indicated in Figure 11.22, the upper-level point bar appears to represent sediment accumulation while the river was at or near flood stage, whereas the lower level represents point-bar sediment

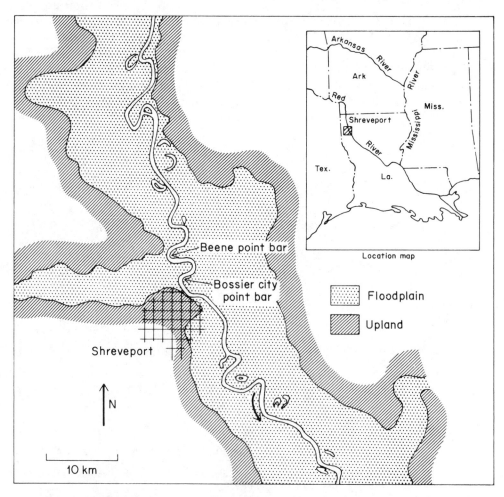

Figure 11.20 Map showing location of the study area, Red River, northwestern Louisiana. (After Harms *et al.,* 1963.)

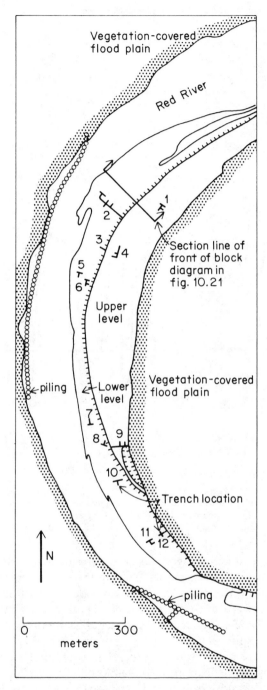

Figure 11.21 Map of Beene point bar, Red River. This point bar consists of an upper level and a lower level. The upper level stands some 7 to 9 meters above the main channel of the river; the lower level stands approximately 4 meters above the main channel. [After J. C. Harms *et al.*, *Jour. Geol.*, **71**, 566–580 (1963).]

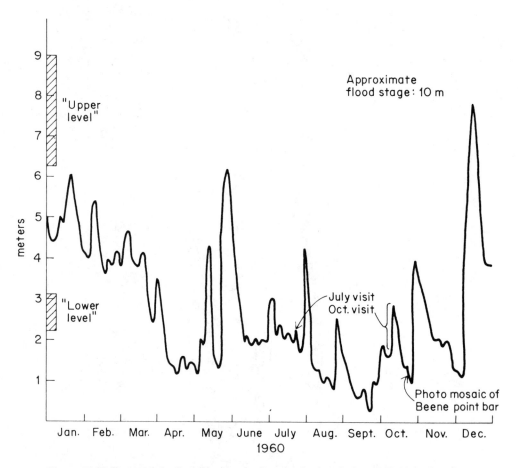

Figure 11.22 Stages of the Red River at the Beene point bar during 1960. Note that the position of the lower level point bar was covered throughout the January–March spring floods and on numerous other brief occasions throughout the remainder of the year. On the other hand, the upper level was only partially flooded for a brief time during December. [After Harms *et al., Jour. Geol.,* **71,** 566–580 (1963).]

accumulation while the river was at an intermediate level, still well confined within the banks of the main channel. During the entire year represented by river gauge data in Figure 11.22, the river did not reach the top of the upper level. Thus, upper-level point-bar sedimentation must be regarded as an extremely sporadic process.

Figure 11.23 summarizes the present distribution of sediment types in a profile of the Beene point bar. According to grain size and type of cross-stratification, the internal construction of the upper level is fairly familiar to us from the Brazos River. Trough cross-stratification in gravelly sands is overlain by horizontal laminations in sand-sized materials, which in turn are overlain by large-scale and small-scale trough cross strata in sand or silty sand. In the lower level, cross-stratification in sand and silty sand predominates. Note that we know little about stratification relationships between the lower level and the upper level.

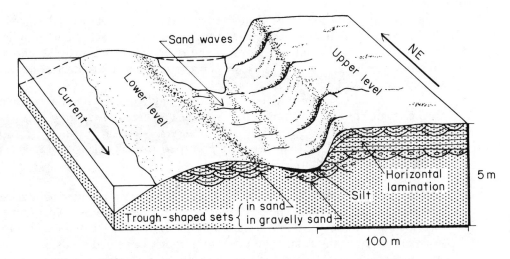

Figure 11.23 Block diagram showing the complex relationships between grain size and primary structures of upper and lower levels of the Beene point bar. [After Harms *et al.*, *Jour. Geol.*, **71**, 566–580 (1963).]

Figure 11.23 illustrates some rather complicated sequences of primary structures that developed during the continued growth and westward migration of the Beene point bar. It would appear that point-bar migration of the lower level is a common process occurring during a net three to four months out of each year (Figure 11.22). Point-bar migration of the upper level seems to occur only during an unusual flood stage. At this time, large-scale, cross-stratified gravel deposits may be expected to migrate across the area formerly occupied by the lower-level point bar, thus forming a complicated vertical sequence of grain sizes and primary structures.

Alternatively, the upper level may be essentially inactive today, a relict surface recording a time in the near recent past when the Red River flood stage usually reached higher levels than are commonly attained now. At any rate, the complexities of this point bar serve to warn us that a certain amount of flexibility must be built into our generalities concerning meandering-stream sedimentation.

A Basic Model for Clastic Sedimentation in Stream Environments

In the preceding pages, we have noted braided-stream sedimentation in mountainous regions (the Donjek River), braided-stream sedimentation on the plains far out in front of mountain ranges (the Platte River), and meandering-stream sedimentation of the coastal plains (the Brazos and Red rivers). We must now attempt to draw conclusions from these various local situations.

Variation in Grain Size and Sorting. In the Donjek River, we saw that longitudinal bars have large grain size and are rather poorly sorted. In contrast, transverse bars generally consist of sand-sized material and are rather well sorted. In the Platte River, we noted the same generalities as well as the fact that longitudinal bars tend to dominate the upper regions of the river, whereas transverse bars tend to dominate the lower reaches. Thus, we can expect grain size to decrease and sorting to improve away from this source area.

The transition from braided stream to meandering stream is accompanied by further regimentation of grain size. The coarsest material tends to accumulate as channel-lag deposits; finer-grained, but relatively well-sorted, sands tend to accumulate on the top of the point bar; and poorly sorted, silty clays tend to accumulate in the overbank position.

Variation in Composition of the Sediment. Transportation in the braided-stream environment during flood stage is an extremely violent process. This fact is demonstrated by the compositional variation found away from the sediment source area. For example, we can see large boulders of a poorly cemented sandstone in braided-stream deposits near the outcrop of sandstone. At some distance downstream, however, the rigors of transportation will have smashed other sandstone boulders into their constituent sand particles. Similarly, less durable rock types, such as shale or schist, will be ground to a finer powder by continued transport. In general, the shorter the transportation history, the more unstable the rock fragments contained

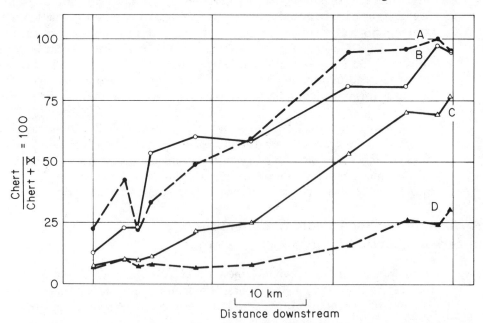

Figure 11.24 Effect of transport on rock constituents in the gravel of Rapid Creek, South Dakota. As the ratio of chert to (chert $+ x$) approaches 1, the component x has been removed from the gravel-size fraction by mechanical destruction during the transportation. Component (A) is a rather poorly cemented sandstone; (B), a limestone; (C), Precambrian metamorphic rock fragments; and (D), quartz and quartzite. [After W. J. Plumley, "Black Hills Terrace Gravels: A Study in Sediment Transport," *Jour. Geol.,* **56,** 526–577 (1948).]

by the sediment. With longer transport, unstable rock fragments are destroyed. These relationships are exemplified particularly well in the data of Plumley (1948). (see Figure 11.24.)

Primary Structures. Structures characteristic of unidirectional transport distinguish stream environments. Large-scale and small-scale, high-angle cross strata predominate. Horizontal laminations of transitional flow regime are noted in high-gradient regions of braided streams and at intermediate levels in point-bar deposits. In these locations, a combination of stream velocity, water depth, and grain size combine to produce transitional conditions. Upper-flow-regime primary structures (antidunes) are not common. If they are found anywhere, they will be in shallow-water environments of high-gradient braided streams.

In the meandering-stream environment, we would commonly expect to find a systematic vertical sequence of grain size and primary structures, the so-called *fining-upward unit* (Figures 11.25 and 11.26). As illustrated in Figure 3.15, the lower

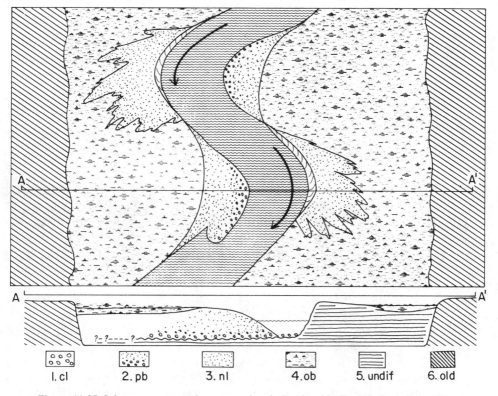

1. cl	2. pb	3. nl	4. ob	5. undif	6. old

Figure 11.25 Schematic map and cross section indicating idealized facies relationships within meandering-stream deposits. Sediment types are as follows: (1) channel-lag deposit, commonly gravel; (2) point-bar deposits; (3) natural levee (may be poorly developed); (4) overbank or backswamp deposits; (5) undifferentiated, previously deposited, meandering-stream sediments, now subjected to erosion and resedimentation downstream; (6) older sediments or rock.

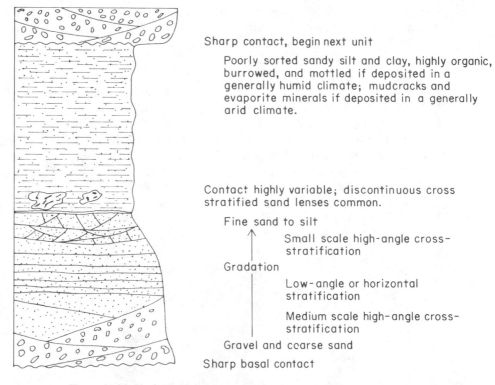

Sharp contact, begin next unit

Poorly sorted sandy silt and clay, highly organic, burrowed, and mottled if deposited in a generally humid climate; mudcracks and evaporite minerals if deposited in a generally arid climate.

Contact highly variable; discontinuous cross stratified sand lenses common.

Fine sand to silt

Small scale high-angle cross-stratification

Gradation

Low-angle or horizontal stratification

Medium scale high-angle cross-stratification

Gravel and coarse sand

Sharp basal contact

Figure 11.26 The basic fining-upward unit of meandering-stream deposition.

portion of the fining-upward unit results from the systematic lateral migration of a point bar. Both erosion of the cutbank and scouring in the deeper portions of the river channel produce the sharp lower boundary of the unit. The erosional surface is overlain by the coarsest material present, typically channel-lag gravel and coarse sand. These sediments usually display medium-scale, high-angle cross strata (Figures 3.12 and 3.13). The sequence becomes finer-grained toward the top of the point bar, with primary structures changing from medium-scale, high-angle cross strata to low-angle or horizontal stratification and finally to small-scale, high-angle cross strata. This vertical sequence through the point bar constitutes the lower portion of the fining-upward unit. The horizontal stratification in the middle of the point-bar sequence is not universally developed, but some form of gradation from coarse-grained, medium-scale, high-angle cross strata to fine-grained, small-scale, high-angle cross strata is the invariable hallmark of the point-bar deposit. Allen (1970) offers a theoretical discussion of vertical successsion of grain size and primary structures in point-bar deposits as a function of various characteristics of the stream. While such calculations are instructive, it must be noted that variation in characteristics of stream from flood stage to low-water stage superimpose a wide variety of conditions upon the sediment. For many purposes, it will be sufficient to look for some indication of coarser grain size, larger-scale cross-stratification low in the point bar and finer grain size, small-scale cross-stratification high in the point bar.

The upper portion of the fining-upward unit consists of overbank deposits. These are the fine-grained sediments that occupy the entire flood plain of the river system. Typically, they are organic-rich shales, recording swamp deposition. Discontinuous sand bodies within overbank deposits are taken to represent natural levees or places where breaks in levees have allowed rapid local flooding by a moderately high stream level that is otherwise contained to its channel (commonly referred to as "crevasse splays").

In contrast to the systematic variation of grain size and primary structures in the vertical sequence deposited by meandering streams, braided streams would be expected to deposit highly variable sequences. Whereas the deposition of a point bar occurs by systematic lateral migration, sediment accumulation in braided-stream environments occurs largely by vertical aggradation during times of flood. Thus, each new deposit in the braided-stream vertical sequence is not necessarily related to depositional conditions of the sediment that it overlies. Miall (1977) provides a detailed breakdown of braided-stream facies and offers some hope that sequences in braided-stream deposits may be subject to a more specific environmental interpretation than is suggested here.

Braided versus Meandering Habit of Streams. For physical reasons beyond the scope of this book, streams naturally should meander. Experiments involving constant gradient of the stream bed and constant discharge rate tend to confirm this theory. Thus, we can consider that a stream showing a braided pattern has not reached the meandering stage.

On the basis of field observations, we can offer several reasons why braided streams do not achieve a meandering pattern. For example, if the banks of a stream are composed of easily eroded material, intermittent collapse of the cutbank may block the main channels of the river and force the stream to follow numerous smaller channels, which may likewise become blocked by slumping sediment. Similarly, it is frequently observed that braided streams tend to have more rapid and more extreme fluctuations in discharge rate than do meandering streams.

Abundant sediment load tends to produce a braided-stream pattern. This situation may be particularly important in those areas where stream gradient changes rapidly from high to low. A high-gradient stream delivers a large volume of sediment, whereas the lower-gradient stream can carry away only a portion of that sediment. The net result is an oversupply of sediment, which tends to glut the channels and causes a braided pattern. Finally, steep stream gradients generally characterize braided streams; low stream gradients tend to characterize meandering streams.

The Concept of a Graded-Stream Profile. So far we have limited our discussion of steam sedimentation to the characteristics of a single bed of sand as it is laid down by the stream. We must now begin to think of stream sedimentation on a more regional scale.

The ability of a stream to transport sediment is dependent upon the gradient of the stream. Other factors being equal, streams flowing down a steep surface will flow faster and therefore transport a larger suspended load and bed load. Thus, if we imagine an initial condition in which a stream of high gradient suddenly becomes a stream of low gradient, we can expect load to be deposited at the change in gradient. Such a situation might occur where a mountain stream comes out onto the prairie. As the net result of this sediment deposition, the transition from high stream gradient to low stream gradient will be smoothed out. Eventually, a stream profile will be obtained in which the capacity of the stream to transport sediment will be uniform throughout the length of the profile. Such a stream is referred to as a *graded* stream.

It is convenient to visualize a graded stream in terms of any two stations, *A* and *B,* where sediment flux may be measured. If the stream is graded, the average sediment flux passing *A* and *B* over some appropriate time interval will be the same. Although it is true that both deposition of sediment and erosion of previously deposited sediment have occurred within the region between *A* and *B,* the amount of new sedimentation is precisely balanced by the amount of concurrent erosion. Although we could go to a graded stream and observe newly deposited sandbars, we would realize that ultimately all deposits in the graded stream are eroded and carried a little further down stream. To a sedimentologist, this situation may be very interesting; but no net stratigraphic record is accumulating in a graded stream.

Response of Graded Streams to Dynamics of the Earth's Surface. The statement that graded streams do not accumulate a stratigraphic record carries with it a corollary: Streams accumulate a stratigraphic record only when the equilibrium profile is undergoing change (Figure 11.27). For example, if the source area is undergoing tectonic uplift, the equilibrium profile will be raised from its source area end. Meandering-stream or braided-stream sedimentary sequences should accumulate to raise the floor of the river valley up to the new equilibrium profile. Alternatively, if sea level rises, the equilibrium profile will be raised from its seaward end. Meandering-stream deposits should accumulate to fill the river valley up to the new equilibrium profile. Similarly, local tectonic subsidence may drop the floor of the river valley below the equilibrium profile. A vertical sequence of alluvial sediments should accumulate to maintain the equilibrium profile throughout such tectonic subsidence.

A second corollary to the concept of a graded-stream profile is the fact that a stratigraphic record will not be accumulated at a time when the equilibrium profile is being lowered. If the equilibrium profile is lowered, the stream will become incised into the surrounding rock types. Net removal of material from that stretch of the river will occur, not net deposition. Similarly, a graded profile will tend to incise into an area that is undergoing gradual tectonic uplift.

All of these relationships between a graded stream and its geographic context are fascinating to stratigraphers. It is their job to unravel the detailed history of tectonism and eustatic sea-level fluctuations that is recorded by the interaction of these events with the graded profile.

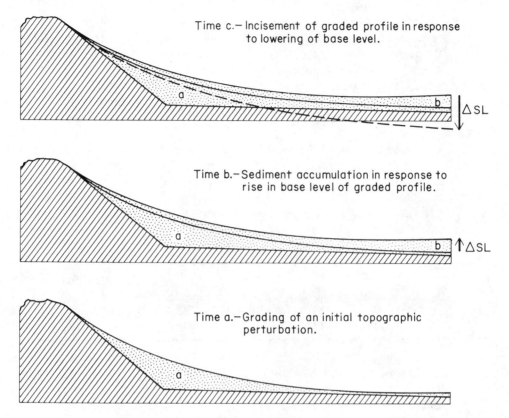

Time c.– Incisement of graded profile in response to lowering of base level.

Time b.– Sediment accumulation in response to rise in base level of graded profile.

Time a.– Grading of an initial topographic perturbation.

Figure 11.27 Schematic example depicting the response of a graded stream to dynamic events. Given an initial topographic perturbation (time a), a stream will seek to fill in relatively low areas and establish a smooth graded profile from source area to base level. Given rising base level (time b), the entire graded profile is raised and sedimentation will occur to maintain equilibrium grade. Similarly, sediment will accumulate in response to subsiding substrate at the distal end of the profile in response to rising source area. Given lowered base level (time c) or lowered source area, the graded stream will erode deeply into its substrate and there will be no net accumulation of sediment throughout the length of the graded profile. Numerous combinations are possible on a regional scale.

While the concept of the graded profile is most easily understood with regards to stream profile, a similar concept probably applies to coastal-plain and continental-shelf sedimentation. Sediments in bays will not accumulate significantly above sea level. Thus, a zero sum for sedimentation and erosion can be envisioned in these environments once an equilibrium has been attained. Similarly, outer continental shelves in terrigenous clastic provinces may be thought of as accumulating to "*wave base.*" Intermittent sedimentation is followed by intermittent erosion and no net deposition results unless something happens to change "wave base." Thus, a vast array of sedimentary environments from braided streams near mountainous regions to outer continental-shelf marine clay muds can be conceptually linked together as responding to perturbations of the equilibrium profile. Recall that this concept is fundamental to the graphic correlation technique (Chaper 5). Consider the perturbation of equilibrium profile to be sea-level rise. If there is variation in magnitude within a time series of sea-level rises, then accumulation of new sediment should vary proportionally regardless of sedimentary environment from the braided streams of the mountain regions to the deep-water clay muds of the outer continental shelf.

Progradation by Meandering-Stream Deposits. In Chapter 9, we noted that regression is a general term meaning simply that the shoreline appears to have moved basinward for some reason. The reason may be eustatic or tectonic emergence of the coastal area relative to the level of water in the basin. Or the basinward movement may result from progradation of coastal sediments outward into the basin and be independent of submergence-emergence considerations. Note especially that progradation may occur even at a time of submergence, provided only that sediment supply is rapid enough to offset the submergence effect (see Figure 12.18).

In the geologic record, we often observe meandering-stream deposits overlying marine deposits in a more or less continuous fashion. Such a situation is portrayed in Figure 10.14 and is commonly referred to as a *regressive relationship*. Can we say more about the precise earth history recorded by these relationships? Must we use the general term "regressive" or can we use more specific terms such as "emergence," submergence," or "progradation"? In view of our discussion of the graded profile, we must emphatically state that this relationship could not have been generated by eustatic or tectonic emergence of the coastal area. In either of these situations, the stream valley would tend to become a site of net *erosion* as the stream incised a new bed in response to lowered sea level (eustatic emergence) or to rising land surface (tectonic emergence). Widespread meandering-stream deposits would not be preserved under these conditions.

Consideration of the dynamic modification of the graded-stream profile (Figure 11.27) suggests three alternatives for the generation of stratigraphic relationships indicated in Figure 10.14. Perhaps we are looking at the effect of the combination of progradation accompanied by tectonic or eustatic submergence of the coastal region. Alternatively, the equilibrium profile may be raised from the other end by uplift of the source area.

We can debate whether there might exist stratigraphic relationships that would allow distinction among the three possibilities just outlined. Let us simply note now that adherence to models based on the modern stream allows us to totally discredit one of the possible interpretations that had been previously hidden in the background behind the general term "regression."

Isostatic Considerations. As stream sediments accumulate into a new equilibrium profile, they are adding new load to that portion of the face of the earth. As we have seen in Chapter 9, isostatic subsidence in response to this new load is to be expected. Thus, the thickness of sediment accumulated will likely be two or three times as great as the initial perturbation of the relationship between the graded-stream profile and the area that it traverses.

An Ancient Example: The Shawangunk Conglomerate and Tuscarora Quartzite, Silurian of Pennsylvania, New Jersey, and New York

In our discussion of geosynclinal sequences in Chapter 10 we noted the common occurrence of thick, clastic wedges marginal to the present position of the crystalline core of the Appalachians. Even on the broad scale of that discussion, it was clear that the thick accumulation of sediment marginal to the crystalline core recorded the uplift and erosion of a large volume of rock in the area now occupied by that core. What we said earlier about clastic wedges sounds very similar to what we are saying now about modern braided streams: Large volumes of debris are derived from the rapid erosion of mountains and deposited relatively nearby. What better place to look for braided-stream deposits than in the clastic wedges of the Appalachians! A modern attempt at paleogeographic reconstruction has been carried out by Smith (1967, 1970). Let us review his work and see if we can agree with his conclusions.

Earlier Stratigraphic Work. The section under consideration is comprised primarily of the Shawangunk conglomerate, the Tuscarora quartzite, and the Clinton formation of Silurian age in Pennsylvania, New Jersey, and New York (Figures 11.28 and 11.29). As with most of the rocks in the Appalachians, early stratigraphers named the formations of the area on the basis of gross lithology, regional unconformities, and a few fossils. When Smith began his studies, the geologists' understanding of the section had been summarized as appears in Figure 11.29.

The Tuscarora quartzite and the Shawangunk conglomerate are very hard rocks; therefore, they form the ridges of the area and are a starting point for stratigraphic correlation throughout the outcrop belt. Below these quartzites and conglomerates, there is usually a shale or shale and fine-grained sandstone (the Hudson River shales and the Martinsburg formation). Fossils of the Martinsburg formation are decidedly Ordovician. However, these fine-grained sandstones and shales contrast sufficiently with conglomeratic quartzites so that a formation distinction can be made on the basis of lithology alone.

In Pennsylvania, the Tuscarora quartzite is overlain by the Clinton formation. The distinction between the two formations is based on gross lithology and fossil content. The Tuscarora is predominantly sandstone. As we proceed up the section, shale beds begin to come in. Along with the alternation of shale and sandstone, we begin to find more and more marine fossils. Thus, the early geologists found it convenient to set apart the shale and sandstone sequence as a new formation, the Clinton formation.

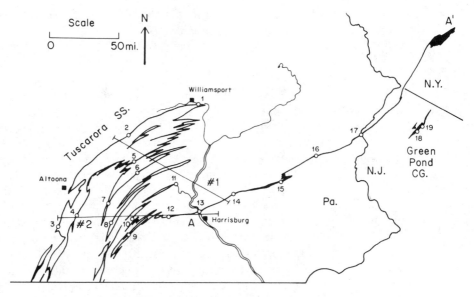

Figure 11.28 Map showing outcrop and distribution of Tuscarora sandstone and Shawangunk conglomerate, northeastern United States. AA′, denotes the location of stratigraphic cross section Figure 11.29. (After Smith, 1970.)

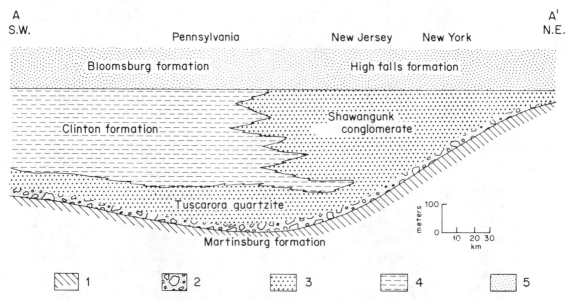

Figure 11.29 Stratigraphic cross section of Middle Silurian formations in the northcentral Appalachians. Major lithologies are as follows: (1) Ordovician shales and sandstones of the Martinsburg formation, unconformably overlain by (2) basal conglomerates; (3) quartzites and conglomeratic sandstones; (4) interbedded shales and sandstones with some *in situ* marine fossils; and (5) red beds. (After Smith, 1967.)

In this discussion, we need not concern ourselves with the section above the Clinton, other than to say that the Bloomsburg and High Falls formations are similarly defined on the basis of lithologic contrast.

Sedimentological Observations within the Shawangunk and Tuscarora. The Shawangunk conglomerate catches the eye of anyone looking for braided-stream environments in the stratigraphic record. Here are 500 meters of conglomeratic sandstone apparently contemporaneous with the marine sands and shales of the Clinton formation. Obviously, the Shawangunk is coarse debris derived from the erosion of a paleogeographic high area to the southeast.

Figures 11.30 through 11.33 offer various sedimentological observations within the Tuscarora and Shawangunk. Figure 11.30 illustrates a rather straightforward example of medium-scale, high-angle, planar cross strata in Tuscarora sandstone. Figure 11.31 shows extremely local irregularities in bedding within Tuscarora sandstone. Some individual beds have irregular thickness, whereas other beds are truncated by scour surfaces. Figure 11.32 presents schematic diagrams of six different types of sedimentation units that recur within these rocks. Variations in grain size and in primary structures suggest extremely local variations in both flow conditions and sediment sorting. Clearly, these variations are reminiscent of the extremely local variations that we discovered in the Donjek River (Figure 11.4). In Figure 11.32,

Figure 11.30 Planar cross-stratification within the Tuscarora sandstone. Compare especially with Figure 11.10, which showed planar cross-stratification in a transverse bar of the modern Platte River. (Photo by N. D. Smith.)

Figure 11.31 Irregular bedding surfaces in the Tuscarora sandstone. Strata are now nearly vertical; depositional up is to the left. Observe the cut-and-fill structures and extreme irregularity of bedding thickness. (Photo by N. D. Smith.)

observe that some sedimentation units record vertical decreases in flow regime, whereas other record vertical increases in flow regime. Furthermore, intraclast of shale and siltstone are common in these rocks. These particles are locally derived by erosion of overbank deposits and other desiccated muds. Their abundance indicates the alternating processes of sedimentation, erosion, and resedimentation that we have seen to be so typical of braided-stream environments. Within each stratigraphic section, sedimentation units are seldom more than 1 or 2 meters thick. There is no discernible repeated sequence in which the various sedimentation units may be expected to occur. The deposition of each successive bed appears practically unrelated to the deposition of beds above or below it.

There is systematic variation in the type of cross-stratification that is perpendicular to the general strike of the paleogeography. Figure 11.33 plots the relative abundance of various types of cross strata into traverses from east (near the source area) to west (basinward). Comparison of these data with similar data for the modern Platte River (Figure 11.33) led Smith (1970) to conclude that there was a Silurian geographic distribution of longitudinal bars and transverse bars similar to that observed in the modern Platte River.

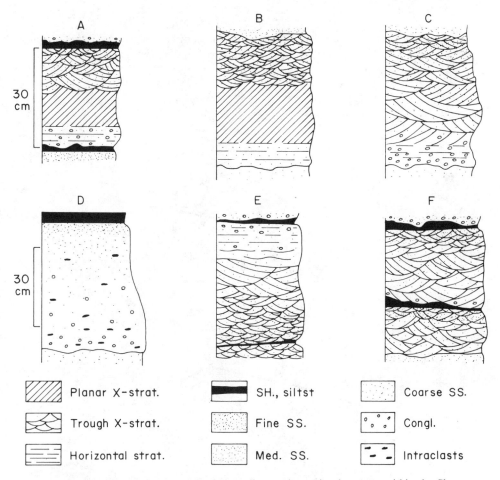

Figure 11.32 Schematic diagrams of six sedimentation units that recur within the Sha-wangunk conglomerate. Note that numerous combinations of variation in grain size and in primary structures are observed. (After Smith, 1970.)

Interpretation of Earth History Recorded by the Shawangunk, Tuscarora, and Clinton Formation. Smith (1967) provides a paleogeographic and paleoenvironmental interpretation for each of the formations displayed in Figure 11.29. The Tuscarora and the Shawangunk are braided-stream deposits; the High Falls formation, meandering-stream deposits; and the Clinton and Bloomsburg formations, marginal-marine to subtidal, open-shelf, marine clastics. In this chapter, we have emphasized Smith's evidence for the recognition of braided-stream deposits. Let us assume for the sake of discussion that his interpretations concerning the other formations are correct. Likewise, the Martinsburg formation appears to consist largely of turbidites that presumably accumulated in marine waters of significant depth.

The Tuscarora and Shawangunk formations record activation of major uplift to the southeast. The intertonguing of marine and marginal-marine Clinton formation rocks with Shawangunk conglomerates offers us an additional puzzle con-

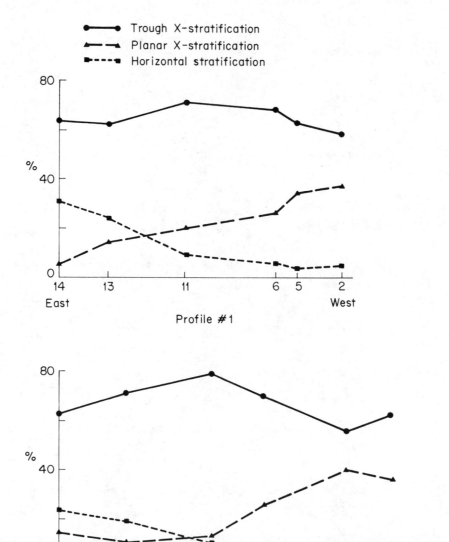

Figure 11.33 Downslope changes in principal stratification types along two east-to-west profiles in the western portion of the outcrop area depicted in Figure 11.28. As noted in Figure 11.13, the relative importance of planar cross strata increases away from the source area and the importance of horizontal stratification decreases away from the source area. (After Smith, 1970.)

cerning the earth's history as recorded by these sequences. Specifically, does the Clinton formation record a major submergence of the North American continent, or does it record local submergence in response to local tectonic conditions, or does it record isostatic submergence in response to deposition of new gravitational load onto this portion of the earth's crust? Recall that calculation of isostatic for similar situations was treated in Chapter 9. A quick calculation of the isostatic adjustment attendant to Shawangunk and Clinton deposition (use data in Figure 11.29, for example) demonstrates that the local isostatic component of the Clinton transgression is probably quite large relative to any eustatic or tectonic component that may be present.

An Ancient Example: Interaction between Meandering-Stream Cycles and Regional Submergence, the Catskill "Delta," Devonian of New York State

Utilizing the concept of the equilibrium profile of meandering streams, McCave (1969) has put together an interesting proposal for event correlation between marine and nonmarine units. He has taken the unifying concept of a model depicting what *should* happen sedimentologically and made seemingly unrelated lithologies the basis for time correlation between marine and nonmarine rocks.

Regional Setting and Earlier Stratigraphic Work. McCave's work is primarily in the Middle Devonian relics of the Catskill Mountains. He seeks to correlate these nonmarine rocks with marine rocks to the west.

As indicated in Figure 10.14, the basic history of this area has been understood in general terms for a long time. You can consult numerous papers in Shepps (1963) and Klein (1968) for additional regional detail.

During the Lower Devonian period, shallow carbonate seas covered the area. By the Middle Devonian period, thick accumulations of clastic sediments began to pour into the area from a source to the east. By Upper Devonian time, continental clastic sedimentation extended well into western New York. Marker beds of black shale and limestone allow physical stratigraphic correlation within the marine units. But how can these correlations be extended into a nonmarine unit of the Catskill front?

Field Observations. While measuring sections in the Catskill Mountains, McCave came to recognize the basic fining-upward unit that is familiar to us from earlier discussions. Figure 11.34 presents his version of the basic fining-upward unit of the Middle Devonian of the Catskills. Measured sections through these sequences can easily be divided into a lower channel sand and an upper overbank shale. Figure

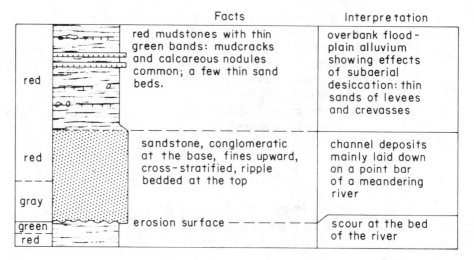

Figure 11.34 Principal sedimentological features of the Catskill fining-upward unit. (After McCave, 1969.)

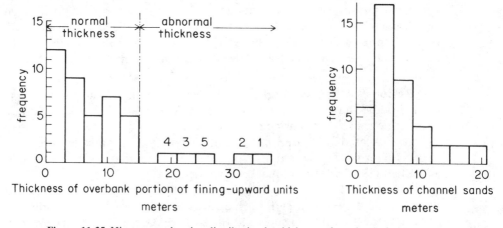

Figure 11.35 Histograms showing distribution by thickness of overbank shale and channel sandstone in the Catskill fining-upward units. Whereas the thickness of channel sands is more or less normally distributed, overbank deposits of 20 to 40 meters appear to be anomalously thick. (After McCave, 1969.)

11.35 is a histogram of the thickness of channel sands and overbank deposits in the sequences measured by McCave. The thickness of channel sands is more or less normally distributed. The thickness of overbank deposits, however, show marked deviations from normal distribution. The high abundance of very thin overbank deposits is easily explained by erosional removal of this material as the next meander belt crossed the area. The occurrence of five units having overbank deposits 18 to 36 meters thick cannot be so easily explained.

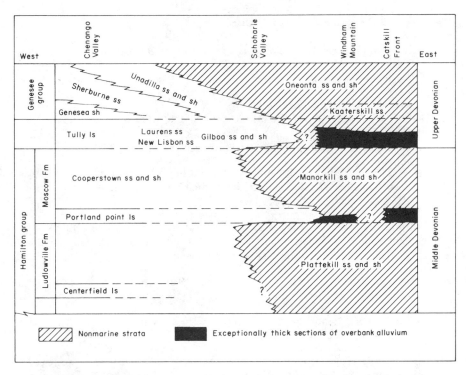

Figure 11.36 Correlation chart for the upper part of Middle Devonian and the lower part of Upper Devonian in New York State, indicating the event correlation between limestone units within the marine section and abnormally thick overbank deposits in the alluvial section. (After McCave, 1969.)

Sedimentological Interpretation. McCave proposes that the accumulation and preservation of abnormally thick overbank deposits reflect a rise in equilibrium profile resulting from a rise in sea level. With this concept in mind, he proposes the correlations outlined in Figure 11.36.

Specifically, an abrupt rise in sea level is a single event that produces different consequences in the marine environment and in the meandering-stream environment. In the marine environment, rising sea level momentarily traps clastic sedimentation in the heads of drowned estuaries. This situation causes brief episodes of clear-water sedimentation (Portland Point limestone and Tully limestone) in areas previously dominated by deposition of marine sandstones and shales.

In the meandering-stream environment, rising sea level has lifted the equilibrium profile of the meandering stream, allowing for the accumulation and preservation of abnormally thick overbank deposits. Thus, the conceptual model dictates correlation of Portland Point limestone and Tully limestone with zones of thickened overbank deposits in the meandering-stream facies. Seemingly unrelated rock types, limestones in the marine environment and thickened overbank deposits in the meandering-stream environment, are both the logical consequences of a single *event:* an abrupt rise in sea level.

Citations and Selected References

ALLEN, J. R. L. 1964. Six cyclothems from the lower Old Red sandstone, Anglo-Welch Basin. Sedimentology 3: 163–198.
Cyclic repetition of sand and shale units is shown to record meandering-stream deposition.

———. 1965. A review of the origin and characteristics of Recent alluvial sediments. Sedimentology 5: 89–191.
Extensive compilation of the literature into the early 1960's.

———. 1970. Studies in fluviatile sedimentation: a comparison of fining-upwards cyclothems, with special reference to coarse-member composition and interpretation. J. Sed. Petrology 40: (1): 298–323.

BERNARD, H. A., C. F. MAJOR, JR., B. S. PARROTT, and J. R. LeBLANK, SR. 1970. Recent sediments of southeast Texas. Univ. of Texas at Austin, Bureau of Economic Geology, Guidebook 11.16 p. (plus figures and appendices).

BLODGETT, ROBERT H., and K. O. STANLEY. 1980. Stratification, bed forms, and discharge relations of the Platte braided river system, Nebraska. J. Sed. Petrology 50 (1): 139–148.

BRIDGE, JOHN S., and LEEDER, MICHAEL R. 1979. A simulation model of alluvial stratigraphy. Sedimentology 26 (5): 617–644.

BULL, W. B. 1964. Geomorphology of segmented alluvial fans in western Fresno County, California, p. 79–129. U. S. Geol. Survey Prof. Paper 352-E.

CAMPBELL, C. V. 1976. Reservoir geometry of a fluvial sheet sandstone. Bull. Amer. Assoc. Petrol. Geol. 60 (7): 1009–1020.
A good introduction to some really outstanding outcrops in the Jurassic of New Mexico.

CANT, D. J., and R. G. WALKER. 1976. Development of a braided-fluvial facies model for the Devonian Battery Point Sandstone, Quebec. Can. J. Earth Sci. 13: 102–119.
Well illustrated, somewhat quantitative.

DOEGLAS, D. J. 1962. The structure of sedimentary deposits of braided rivers. Sedimentology 1: 167–190.

ECKIS, R. 1928. Alluvial fans of the Cucamonga District, Southern California. J. Geology 36: 224–247.

FAHNESTOCK, R. K. 1963. Morphology and hydrology of a glacial stream. p. 1–70. U. S. Geol. Survey Prof. Paper 422-A.

GIGNOUX, M. 1955. Stratigraphic geology. W. H. Freeman and Co., San Francisco. 682 p.

HARMS, J. C., D. B. MacKENZIE, and D. G. McCUBBIN. 1963. Stratification in modern sands of the Red River, Louisiana. J. Geology 71: 566–580.

HEIN, FRANCES J., and ROGER G. WALKER. 1977. Bar evolution and development of stratification in the gravelly, braided, Kicking Horse River, British Columbia. Can. J. Earth Sci. 14: 562–570.

HOOKE, R. L. 1967. Processes on arid region alluvial fans. J. Geology 75: 438–460.

JOHNSON, K. G., and G. M. FRIEDMAN. 1969. The Tully clastic correlatives (Upper Devonian) of New York State: a model for recognition of alluvial, dune (?), tidal nearshore (bar and lagoon), and offshore sedimentary environments in a tectonic delta complex. J. Sed. Petrology 39 (2): 451–485.

KLEIN, G. DEV. (ed.). 1980. Late Paleozoic and Mesozoic continental sedimentation, northeastern North America. Geol. Soc. Amer. Spec. Paper 106. 309 p.
Numerous papers concerning Paleozoic stream sedimentation in the Appalachians.

LEOPOLD, L. B., and M. G. WOLMAN. 1957. River channel patterns: braided, meandering, and straight. U. S. Geol. Survey Prof. Paper 282-B. 85 p.

LEOPOLD, L. B., M. G. WOLMAN, and J. P. MILLER. 1964. Fluvial processes in geomorphology. W. H. Freeman and Co., San Francisco, 522 p.

MCCAVE, I. N. 1969. Correlation of marine and non-marine strata with example from Devonian of New York State. Bull. Amer. Assoc. Petrol. Geol. 53: 155–162.
Event-correlation between marine limestone and thickened overbank deposits in meandering-stream facies.

MIALL, ANDREW D. 1977. A review of the braided-river depositional environment. Earth-Science Rev. 13: 1–62.

NANSON, GERALD C. 1980. Point bar and floodplain formation of the meandering Beatton River, northeastern British Columbia, Canada. Sedimentology 27 (1): 3–30.
Very detailed study; well illustrated.

PLUMLEY, W. J. 1948. Black Hills terrace gravels: a study in sediment transport. J. Geology 56: 526–577.

RIGBY, J. K., and W. K. HAMBLIN. 1972. Recognition of Ancient sedimentary environments. Soc. Econ. Paleont. Mineral. Spec. Pub. 16. 340 p.
Papers from a 1969 symposium dealing with a wide variety of sedimentary environments. Each paper discusses the recognition of a specific environment or cluster of environments. A good starting point for many topics. Generally well-referenced.

SHEPPS, V. C. (ed.). 1963. Symposium on Middle and Upper Devonian stratigraphy of Pennsylvania and adjacent states. Pa. Topog. Geol. Surv. Gen. Geol. Rep. G-39. 301 p.

SMITH, DERALD G., and NORMAN D. SMITH. 1980. Sedimentation in anastomosed river systems: examples from alluvial valleys near Banff, Alberta. J. Sed. Petrology 50: (1): 157–164.

SMITH, N. D. 1967. A stratigraphic and sedimentologic analysis of some Lower and Middle Silurian clastic rocks of the north-central Appalachians. Ph.D. dissertation, Brown University. 195 p.

———. 1970. The braided stream depositional environment: comparison of the Platte River with some Silurian clastic rocks, North-central Appalachians. Bull. Geol. Soc. Amer. 81: 2993–3014.
Excellent discussion of cross-stratification and grain-size variation. Includes recognition of the proximal and distal portions of the braided-stream environment.

————. 1971. Transverse bars and braiding in the lower Platte River, Nebraska. Bull. Geol. Soc. Amer. 82: 3407–3420.

STANLEY, K. O., and W. J. WAYNE. 1972. Epeirogenic and climate controls of early Pleistocene fluvial sediment dispersal in Nebraska. Bull. Geol. Soc. Amer. 83 (12): 3675–3690.

VAN HOUTEN, FRANKLYN B. 1973. Origin of red beds: a review—1961–1972. Ann. Rev. Earth Planet. Sci. 1: 39–61.

WALKER, R. G. 1976. Facies models—3. Sandy fluvial systems. Geoscience Canada 3: 101–109.

WARD, W. C., and K. C. McDONALD. 1979. Nubia formation of central eastern desert, Egypt—major subdivisions and depositional setting. Bull. Amer. Assoc. Petrol Geol. 63 (6): 975–983.
Makes extremely interesting use of current direction information. See also Klitzsch *et al.*, same journal issue.

WILLIAMS, P. F., and B. R. RUST. 1969. The sedimentology of a braided river. J. Sed. Petrology 39: 649–679.

YEAKEL, L. S. 1962. Tuscarora, Juniata, and Bald Eagle paleocurrents and paleogeography in the central Appalachians. Bull. Geol. Soc. Amer. 73: 1515–1540.

Clastic Sedimentation
in Coastal Environments

12

In stream sedimentation, the driving force is gravity acting upon the water. As gravity causes the water to run downhill, the movement of the water causes sediment transportation and deposition. At sea level, a whole new set of processes begins to act on the sediment supplied by streams. Gravity no longer provides the major energy input to sedimentation. Instead, tidal motions and wind-driven currents are the primary transporting agents in coastal environments.

In the fluvial environments, the key word for water motion is "flow." As we begin considering clastic sedimentation in coastal environments, the key word for water movement becomes "swash." The flow of a fluvial stream is more or less constant and unidirectional. In contrast, marine water is continually moving onto and off of clastic coastal sediments. There is the rise and fall of each tidal cycle, and there are the endless waves that roll up onto a beach, only to roll off again. The oscillatory motion of waves usually generates persistent longshore currents, but the oscillatory components of sediment transport also make major imprints on the sediments accumulated in these environments.

In addition, as we study the coastal clastic environments, we see that fossils begin to play an important role in our recognition of depositional environments.

All in all, there are more and more factors to consider in coastal areas. Changes in the relationship of the sea to the land are among the most profound changes recorded in the strata. Coastal clastic environments offer an excellent opportunity to come to grips with the dynamics of earth history.

Recent Clastic Sedimentation
in Coastal Environments

A *delta* forms as the basic sediment accumulation where the river meets the coastal area. It is here that stream processes end and marine processes (tides and currents) begin to operate.

It is convenient to think of the delta as a loading dock. Just as the loading dock separates the manufacturing activity of a factory from the distribution and sale of the product, the delta separates the sediment factory (the continental areas) from the distribution processes of the shallow-marine environment. Continuing the analogy, we can write a budget for sediment inventory on the delta in terms of the capability of the streams to deliver sediment to the delta and the capability of the shallow-marine processes to transport sediment away from the delta. If the stream can deliver a large volume of sediment and the marine processes can remove only a small volume of sediment, a large amount of sediment will accumulate in the delta area. On the other hand, if the stream can deliver only a small amount of sediment and the marine processes are capable of moving a large amount of sediment, the stream may build no delta at all. It may be that all of the sediment delivered by the stream is transported away from the mouth of the stream as soon as it is delivered.

In the following pages, we shall look at coastal environments that represent a wide spectrum in the balance between the sediment supply capability of the streams and the sediment reworking and transport capability of the marine environment. Several of our examples will be from the northwest Gulf of Mexico. In the Mississippi delta, the sediment supply far exceeds the reworking capabilities of the marine environment. A very large and unique type of delta is accumulating there today. On the other hand, in the Niger delta, sediment supply and reworking capabilities of the marine environment are in much closer balance. Although a large delta is accumulating, the sediment is being significantly reworked and locally transported. The result is a delta that is quite different from the Mississippi. As an example of longshore sedimentation, we shall return to southwest Louisiana and the Texas Gulf Coast. Here the sediment has escaped the delta area and is being transported and deposited by processes acting in the shallow-marine environment. Finally, to round out our model of clastic sedimentation in coastal environments, we shall look at tidal-flat sedimentation.

It is often convenient to discuss clastic sedimentation in coastal environments in terms of topset deposits, foreset deposits, and bottomset deposits. The delineation of these terms in this chapter involves the *strand line,* the general area of interaction between open water and the shallow-water sediments associated with the coastal area. The beach, of course, most clearly exhibits the strand line. This line is less well defined in the rough outline of a delta like the Mississippi. *Topset sediments* are those sediments that accumulate landward from the strand line. Sediment accumulation here is essentially vertical.

Bottomset sediments are the offshore, deeper-water marine sediments that are not deposited by the wave and tidal processes dominating the coastal and shallow-marine areas. These sediments are typically clay muds with a normal marine benthonic fauna and perhaps a sizable representation of pelagic organisms. Accumulation of bottomset sediments is essentially vertical.

Foreset sediments occupy that position between the strand line and bottomset sediments. It is a position of almost continual activity. The top of the foreset deposits are the river-mouth bars and the barrier beaches. Seaward from these fea-

tures, bathymetric gradients are pronounced. Because of this slope to the foreset bathymetry, sediment accumulation is essentially lateral or progradational.

The major task facing the stratigrapher is recognizing and understanding relationships within foreset deposits in clastic sedimentation in coastal environments. Topset and bottomset beds are useful primarily as stratigraphic markers, valuable for setting apart the discrete sedimentary packages.

The Mississippi Delta. A brief glance at Figure 12.1 confirms that the modern Mississippi delta is a major physiographic feature of the northwest Gulf of Mexico. From the Rio Grande to western Louisiana, modern coastal sedimentation more or less parallels the strike of older sedimentary units and extends 15 to 30 kilometers seaward from the Pleistocene outcrop surface. In southeastern Louisiana, modern Mississippi delta sediments have built outward onto the continental shelf as much as 100 kilometers seaward from the Pleistocene outcrop surface. This area is indeed a site of major deltaic accumulation of clastic sediments.

Sedimentation in the Mississippi delta particularly interests us because of the economically important oil and gas accumulations in similar deposits throughout Southern Louisiana. The Mississippi River has been dumping large volumes of sediment into this region from Oligocene time to the Recent epoch. Sedimentation and internal stratigraphy of the modern Mississippi delta is undoubtedly a valuable clue in prospecting for oil and gas in older deposits within this sedimentary basin.

Although the entire Mississippi delta has accumulated since the post-Wisconsin sea-level rise, active delta sedimentation is occurring today only in the modern Birdfoot delta (Figure 12.2). This area has been studied in detail (Gould, 1970, for example) and constitutes in large part the basic model for our understanding of the Mississippi delta as a whole.

Sediments of the modern Birdfoot delta can be divided into three general categories. The mouth of the distributary is the main site of sedimentation. Here the coarsest material brought down the river becomes deposited in a river-mouth bar. Outward from this bar in all directions, the sediment becomes more fine-grained (Figure 12.3). The fine-grained sediment accumulates both in front of the river-mouth bar and also between river-mouth bars. Finally, wherever fine-grained material fills in near sea level, marsh deposits develop. These marsh deposits prograde over shallow, subtidal, fine-grained sediments and form the laterally extensive, and most visible, (upper) topset unit of the delta sequence.

Note that the coarsest sediment accumulates in the shallow water of the river-mouth bar. Grain size becomes progressively finer offshore and between river-mouth bars. Primary structures also reflect the decreasing carrying capacity from river mouth outward into the Gulf. High-angle cross-stratification is common in the shallow-water sands of the river-mouth bar but is uncommon seaward.

The transition from river mouth to open marine is also apparent in the fauna contained in the sediments. Sediments accumulating in front of the river-mouth bar will commonly contain a burrowing marine fauna. The shells of these organisms not only become an important attribute of the sediment, but their burrowing activity

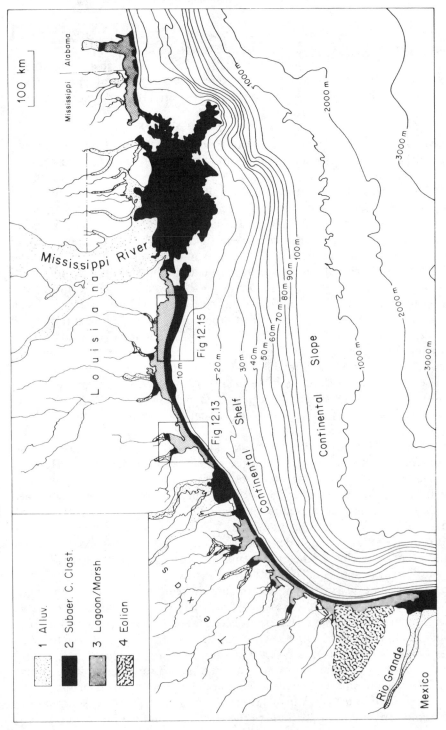

Figure 12.1 Generalized features of Recent coastal clastic sedimentation, northwest Gulf of Mexico. In this chapter, we shall look at Recent sedimentation in the Mississippi delta (Figure 12.2), the Galveston barrier-bar–lagoon complex (Figure 12.13), and the Chenier Plain of southwestern Louisiana (Fig 12.14). Generalized sediment types are as follows: (1) alluvial plain deposits, predominantly from meandering streams; (2) subaerial coastal clastic sediments, deltaic complexes, and progradational barrier bars and cheniers; (3) lagoons and marshes behind progradational features; and (4) eolian sand transported inland from the barrier-beach environment.

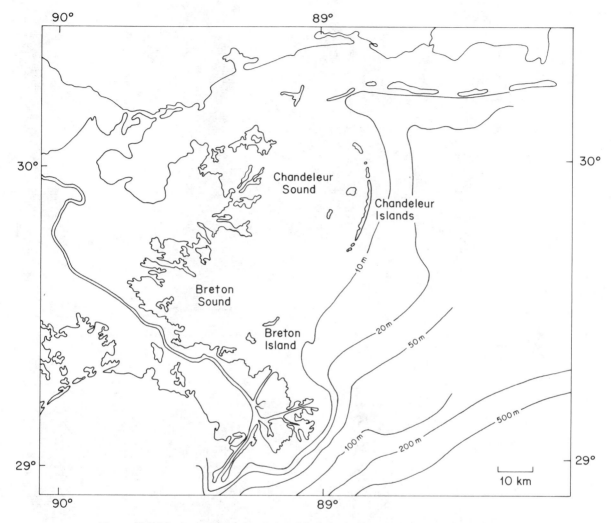

Figure 12.2 Physiography of the modern Mississippi River Birdfoot delta and environs.

also reworks and mottles the sediment. If primary structures were formed in the subtidal environments in front of the delta, they would likely be destroyed by the burrowing activity of these organisms.

Historical data, combined with coring of the modern Birdfoot delta, indicate that the mouths of the river distributaries have been building seaward at a very rapid rate. The net result of this seaward progradation of the distributary channels is the construction of *bar-finger sands* (Figure 12.4). The characteristics of present-day sediments, previously discussed in relation to areal distribution from the distributary mouth, are repeated in the vertical profile of any single bar-finger sand [Figure 12.3 (B)]. The coarsest material in shallow water and the finest material in deeper water

(A)

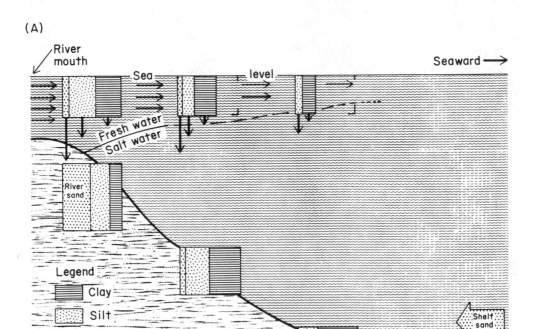

(B)

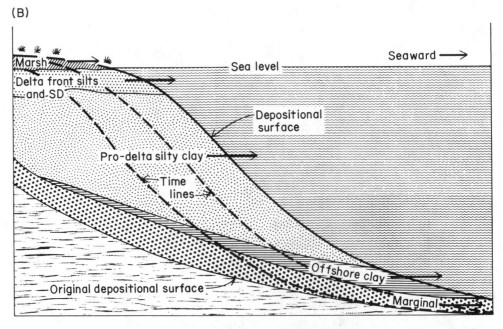

Figure 12.3 Sedimentation seaward from the mouth of the delta distributary. Figure (A) traces the composition of suspended sediment and deposited sediment from the river mouth out to sea. Figure (B) schematically summarizes progradational relationships resulting from this deposition. (After Scruton, 1960.)

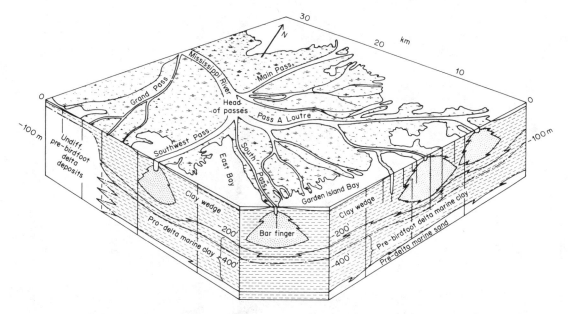

Figure 12.4 Block diagram indicating geometric arrangement of bar-finger sand, fine-grained sediments, and topset marsh deposits of the modern Birdfoot delta of the Mississippi River. (After Fisk *et al.*, 1954.)

in front is translated into the coarsest material at top of the bar-finger sand and the finest material at its base. This process continues for the various characteristics of these sediments.

As indicated in Figure 12.4, coarse-grained material in the Mississippi delta accumulates only in the bar-finger sands, resulting in progradation of major distributaries. Areas between major bar-finger sands are filled in with fine-grained sediment, not unlike the muds accumulating in front of the mouths of passes. The basic geometry of modern Birdfoot delta sedimentation is, then, the radial arrangement of bar-finger sands surrounded by finer-grained sediments. As the area between bar-finger sands becomes filled in with muddy marine sediment, marsh deposits prograde seaward to fill in between the "spokes" of the radially arranged bar-finger sands. Although these marsh deposits are really extensive, they are generally thin. They may prove useful for correlation in Ancient sequences, but they do not make up the bulk of the typical deltaic stratigraphic sequence.

The development of radially arranged bar-finger sands of the modern Birdfoot delta can be considered an example of sediment supply grossly exceeding the reworking and transportation capabilities of the local marine environment. As we move on to look at other areas, we shall see that the sediment delivered to the mouth of a distributary is very often reworked and transported to build other deposits. In the Mississippi delta, the sediment supply is so great that the effective reworking of these deposits does not occur so long as the sediment is being supplied to the area. The sediment simply piles up at or near the distributary mouth.

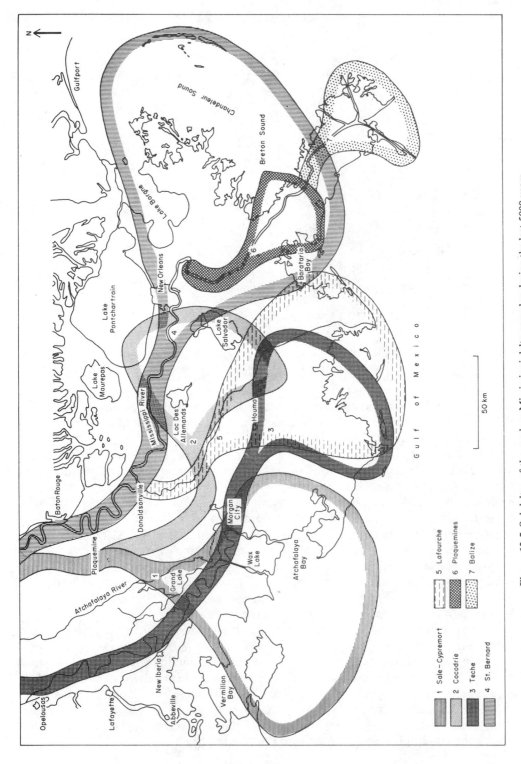

Figure 12.5 Subdeltas of the modern Mississippi delta complex during the last 5000 years. The site of major deltaic sedimentation has shifted numerous times. Chronology of these shifts are given in Figure 12.6. (After Kolb and Van Lopik, 1966.) Courtesy Houston Geological Society.

1 Sale–Cypremort
2 Cocodrie
3 Teche
4 St. Bernard
5 Lafourche
6 Plaquemines
7 Balize

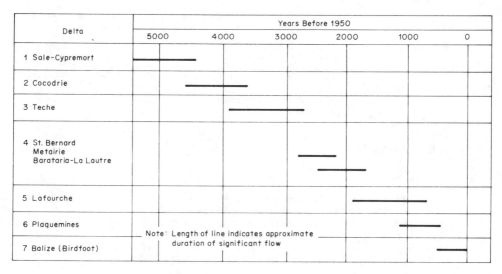

Figure 12.6 Chronology of the major delta lobes of the Mississippi during the past 5000 years. Position of each delta is indicated in Figure 12.5. (After Kolb and Van Lopik, 1966.) Courtesy Houston Geological Society.

Another outstanding attribute of Mississippi delta sedimentation is the varying position of delta sedimentation. As indicated in Figures 12.5 and 12.6, the focal point of delta sedimentation has changed radically at least seven times during the past 5000 years. Presumably, each of these subdeltas was built in the regular fashion, as was the modern Birdfoot delta, until the river was abruptly diverted from its channel.

Thus, there is a strongly discontinuous and nonsystematic aspect to Mississippi delta sedimentation. The internal stratigraphy of any one delta lobe has no relationship to the internal stratigraphy of another delta lobe, no matter how close they are. Each delta lobe is capped with a marsh deposit, and each delta lobe contains bar-finger sands that have the coarsest material at the top and the fine downward. At that point, the systematics of sedimentation and stratigraphy end. As an entity, each delta lobe is unique.

The abrupt discontinuation of sediment supply to a delta lobe provides an opportunity for marine processes to rework the delta deposits. As long as sediment is being supplied to an active delta front, new sediment is added so rapidly that redistribution by marine processes does little to change the basic radial arrangement of bar-finger sands. When the sediment supply is shifted to another site, a destructive phase of delta history may occur (Figure 12.7 and Scruton, 1960). For example, the Chandeleur Islands (Figure 12.2) mark the former seaward position of the St. Bernard delta. When the river shifted its site of major deposition, the St. Bernard delta was left exposed to marine processes without the protective cover of rapid addition of new sediments. Wave action winnowed the sediment, leaving relatively clean sand and silt and carrying away the fine-grained sediment. The net result is twofold. First, deflation of the old St. Bernard delta has produced a subtidal area

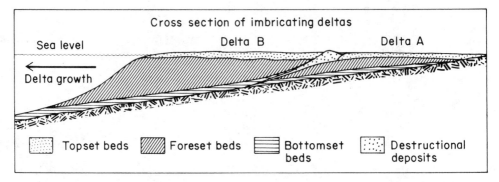

Figure 12.7 Schematic cross section showing the vertical relationships between constructional and destructional deposits in imbricating deltas. Stratigraphic units of successive deltas merge and may appear correlative over wide areas if destructional deposits are not recognized. (After Scruton, 1960.)

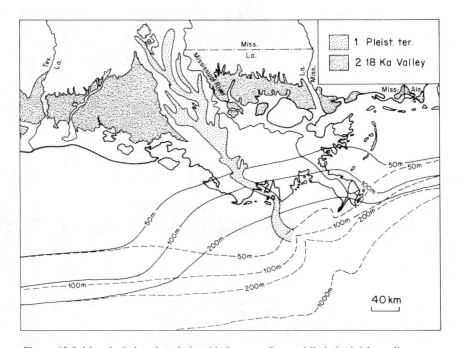

Figure 12.8 Map depicting the relationship between Recent Mississippi delta sediment accumulation and the Pleistocene surface. Dashed contours indicate present-day bathymetry of the Recent sediment surface. Solid contours indicate the present-day topography on the Pleistocene erosional surface that underlies Recent delta sediment. Pattern (2) indicates the present-day subaerial outcrop of Pleistocene erosional surface. Pattern (1) indicates the deeply incised Mississippi trench, through which the Mississippi River flowed at the time of the Wisconsin lowstand of the sea. Radiometric dating of materials obtained from boreholes demonstrates that the former Pleistocene erosional surface has been deformed into its present configuration by the loading of Recent Mississippi delta sediment onto the area. (After Fisk and McFarlan, 1955, with modification.)

where there was once topset marsh deposits. Secondly, the destructive phase has resulted in a relatively clean sand deposit of large areal extent. The geometry of these sand deposits associated with the destructive phase of delta sedimentation is in sharp contrast to the bar-finger sands of the constructive phase of Mississippi delta sedimentation. Thus, Mississippi delta sedimentation results in two distinctive types of sand-body geometry: (1) the constructive radial bar-finger sands and (2) the laterally extensive blanket sands of the destructive phase.

Extensive coring in southern Louisiana has provided unparalleled information concerning the relation of modern sediments to the underlying uncomformity surface. Of particular interst is the isostatic and compactional deformation of unconformity surface under the load of delta sediments. Figure 12.8 is a map of the present configuration of the Pleistocene subaerial surface beneath the modern Mississippi delta, and Figure 12.9 presents two generalized cross sections through the Recent delta. In cross section A-A', note that the Pleistocene subaerial surface is lowest beneath the thickest accumulation of Recent delta material. To the west, as delta material thins gradually, the Pleistocene subaerial surface rises gradually. To the east, as delta thickness decreases abruptly, the Pleistocene subaerial surface rises abruptly. The pronounced embayment of contours in Figure 12.8 conveys the same information.

Radiometric dating of deposits that trangress this surface indicates that the present configuration of the subaerial surface is not paleogeographic but rather is

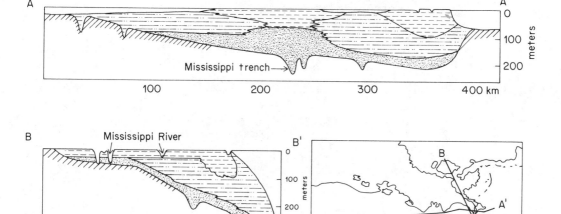

Figure 12.9 General facies relationships within the modern Mississippi delta and the relationship of these sediments to the former Pleistocene subaerial erosion surface. Sediment types are as follows: (1) deltaic plain facies, bar-finger sand, and marsh deposits; (2) prodelta facies, predominantly muds with marine fauna; (3) alluvial facies associated with Mississippi River sedimentation during the late Wisconsin lowstand; (4) former Pleistocene subaerial surface.

the result of local subsidence brought about by the loading of sediment onto the area. Before the loading of sediment onto this area, the 100-meter contour, for example, was simply that line connecting 100-meter contours to the west and to the east of the modern delta. With the loading of sediment into this area, the paleo-geographic 100-meter contour has been depressed by as much as 100 to 150 meters.

Thus, as predicted from general considerations in Chapter 9, a sea-level rise allows for the accumulation of sediment thicknesses at least twice as great as the amount of the sea-level rise. Compaction and isostatic adjustment associated with the loading of the unconformity surface allow for the accomodation of a thickness of marine sediments significantly in access of the actual amount of the sea-level rise.

The Niger Delta. The Niger delta (Figure 12.10) provides an example in which marine processes of transportation and redeposition are in closer balance with sediment supply. In the Mississippi delta, the constructional phase involved such rapid

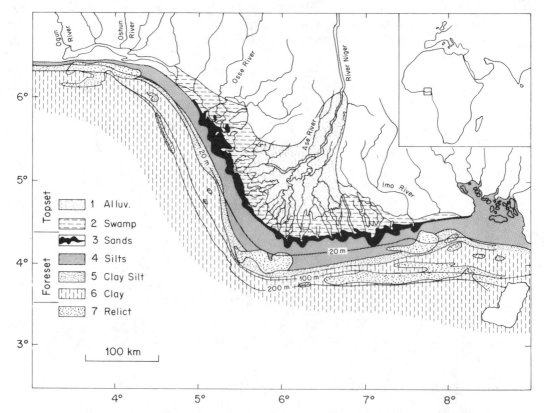

Figure 12.10 Major sedimentary facies of the modern Niger delta. Contemporaneous delta sedimentation is occurring around the entire perimeter of the Niger delta. Principal sedimentary facies are as follows: (1) alluvial flood plain, (2) mangrove swamp, (3) beaches and river-mouth bars, (4) very fine sand and coarse silt, (5) clayey silt, (6) silty clay, (7) nondepositional (older sediments). (After Allen, 1965, with modification.)

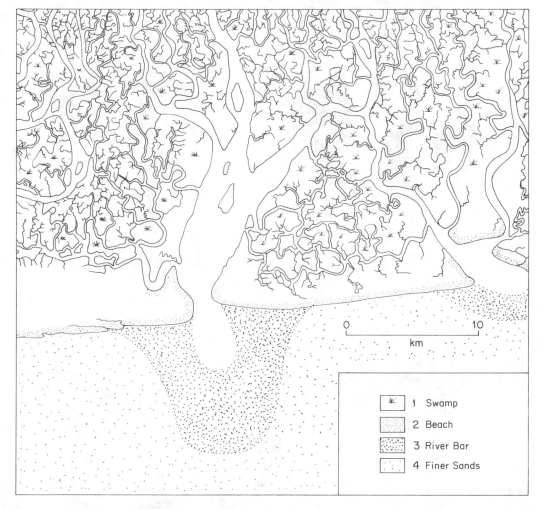

Figure 12.11 Map showing the physiography and sediment types near the mouth of the Brass River, Niger delta distributary. Note the intricate pattern of tidal distributaries within the mangrove swamp topset deposits. Sediment types are as follows: (1) mangrove swamp, (2) barrier beaches, (3) subtidal river-mouth bar, (4) offshore fine sand. Water depth throughout the offshore area is less than 10 meters. (After Allen, 1965, with modification.)

sedimentation that bar-finger sands prograded rapidly into open-marine waters. Redistribution of delta sediments occurred only after the subdelta entered the destructive phase following a major shift in river course. In contrast, in the Niger delta, constructional and destructional phases are essentially superimposed upon each other contemporaneously. Marine processes are capable of reworking newly deposited river-mouth bars almost as fast as the bars are being deposited. The net result, then, is concentric growth along the entire front of the Niger delta. Whereas the area between river-mouth bars in the Mississippi delta was filled in by fine-grained sediments, the area between the river-mouth bars in the Niger delta is filled in by the extensive development of beaches composed of sand derived from those bars (Figures 12.11 and 12.12). Consequently, there is an almost continuous body of sandbars and beaches at or near sea level around the entire perimeter of the Niger delta.

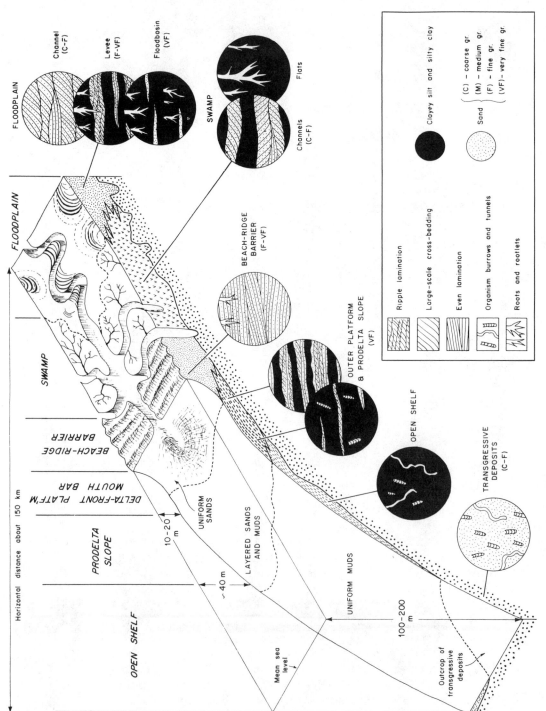

Figure 12.12 Schematic block diagram of facies and facies geometry of the modern Niger delta. (After Allen, 1970.)

Instead of rapid progradation of single bar-finger sands, as in the Mississippi delta, the Niger delta barrier-bar complex is prograding simultaneously over its entire front.

As in the Mississippi delta, grain size of the marine sediment decreases away from the delta front. Behind the barrier-bar complex, extensive mangrove swamps occupy the topset position.

The Barrier Island, Galveston, Texas. In the Mississippi delta, river-supplied sediments are deposited so fast that river-mouth bars prograde to form bar-finger sands. In the Niger delta, river-supplied sediments are locally reworked to form an extensive barrier-beach complex around the whole perimeter of the delta. Now let us look at an example of coastal sedimentation in which the sand supply is from the marine environment rather than from intimately associated river discharge. The marine environment may supply sediment for coastal deposition either by reworking material in front of the coastline or by longshore transport from deltas located elsewhere along the coast.

Galveston Island lies some 500 kilometers west of the modern Birdfoot region of the Mississippi delta (see Figure 12.1). The predominant longshore current throughout the northwestern Gulf of Mexico is westward. Because Galveston Island is far removed from the sediment source to the east, sedimentation here is a rather simple example of transportation and sedimentation by longshore currents. Therefore, we shall examine Galveston Island as a simple example of longshore sedimentation. Then we shall proceed eastward into the Chenier Plain of southwestern Louisiana to look at additional complications brought about by greater sediment supply.

In the Galveston area (see Figure 12.13), the post-Wisconsin transgression has trapped local sediment supplies at the head of estuaries. Both the San Jacinto River and the Trinity River supply sediment to Galveston Bay, but the sediment supply has not been sufficient to fill in the bay during Recent time. As a result, the sand supply that has accumulated to form Galveston Island and Bolivar peninsula was brought to the area from the marine environment rather than from local river discharge.

Cross section A-A' in Figure 12.13 depicts the relationship of Galveston Island to the preexisting topography of the Pleistocene subaerial surface. The island consists of a series of accretionary beach ridges that have prograded seaward about 5 kilometers since the post-Wisconsin sea-level rise. The first accumulation of beach conditions appears to be associated with the rapid convergence of Pleistocene surface with sea level. Wave action delivers the coarsest sediment to the beach environment and winnows out silt and mud that are deposited in the marine environment in front of the beach. With continued accumulation of sand on the beaches and mud in the offshore area, the beach has prograded seaward by some 5 kilometers.

Behind the barrier island, West Bay is now a relatively protected body of shallow water. Tidal marshes prograde toward the center of the bay from both the mainland and the barrier island. West Bay retains fairly good communication with

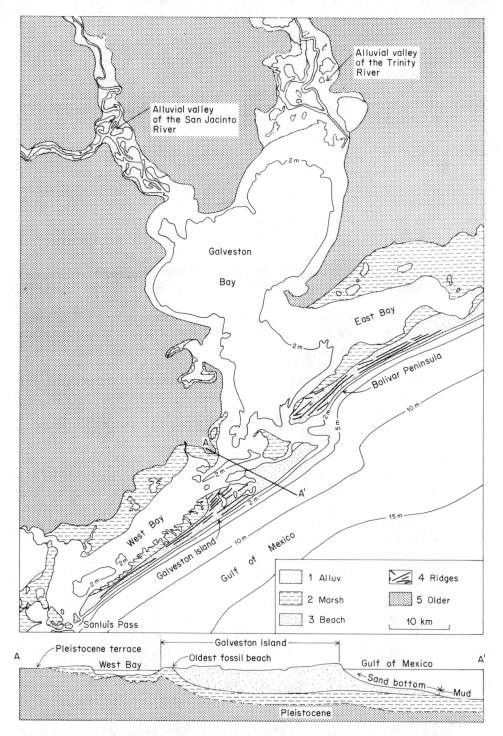

Figure 12.13 Physiography and sediment types of Galveston Island, and surrounding area, Texas. Sediment types are as follows: (1) alluvial sediments, mostly backswamp; (2) marine to freshwater marsh deposits; (3) beach sand; (4) predominant lineations within the beach-bar complex; (5) older sediments, mostly Pleistocene. (After LeBlanc and Hodgson, 1959, with modification.)

the open Gulf of Mexico through the two channels, Bolivar Roads and San Luis Pass.

Topset deposits are essentially lacking on the barrier island complex. Sand accretion ridges form the very low-lying topography of the island. In the swales between ridges, relatively thin marsh deposits are accumulating. On the ridges themselves, there is only a thin cover of grass.

From the barrier island seaward, once again we see a basic pattern of accumulation of the coarsest material at or near sea level and a general fining of the sediment outward into the marine area.

The Chenier Plain of Southwestern Louisiana. The Chenier Plain (see Figures 12.14 and 12.15) has a history quite similar to Galveston Island, but the internal stratigraphy is somewhat more complicated because of its closer proximity to the sediment supply, the Mississippi delta. Extensive radiometric dating of these deposits by oil company researchers makes them especially interesting to the sedimentologist-stratigrapher who is interested in the dynamics of how stratigraphic units develop.

In comparison to Galveston Island, the effects of increased sediment supply to the Chenier Plain are several. Note, for example, that progradation has moved the shoreline seaward by some 15 kilometers in the area south of Grand Lake. This distance contrasts to the progradation of about 5 kilometers in the Galveston area. In addition, Calcasieu, Grand, and White lakes are now freshwater lakes, because coastal sedimentation has effectively cut off large-scale communication of these water bodies with the open Gulf. The area between Calcasieu Lake and White Lake was once a shallow bay behind a barrier island, not unlike West Bay behind Galveston Island today. In the Chenier Plain, these bays have become filled in and are now supratidal marshes.

By far the most striking contrast between Galveston Island and the Chenier Plain is the wide spacing between sets of accretion beach ridges. Although accretion beach ridges in the Galveston area are each separated by no more that a 100-meter swale, sets of accretion beach ridges in the Chenier Plain are commonly separated by as much as 3 to 10 kilometers. On a map this separation appears as marsh deposits in between the sandy ridges rising above the marsh surface. In cross section (Figure 12.15), we see that the marsh is simply a thin veneer over tidal-flat and shallow, gulf-bottom, silty clay deposits. The development of Galveston Island represents a rather continuous process of sediment supply, reworking, and deposition; whereas the Chenier Plain represents a discontinuous process of rapid sedimentation followed by a destructional phase. The scheme is the same as the one that we saw in the Mississippi delta. Occasional superabundant sediment supply from the Mississippi delta results in rapid deposition of poorly sorted tidal-flat and shallow-marine coastal environment does not have time to produce well-sorted beach sediments. As sediment supply slackens, winnowing processes can begin to develop well-sorted beach sands on the seaward margin of the newly deposited silty clay coastal sediments. As winnowing continues, the well-sorted chenier sand bodies become a protective armor, preventing the further erosion of the silty clay tidal-flat deposits from which they themselves were formed.

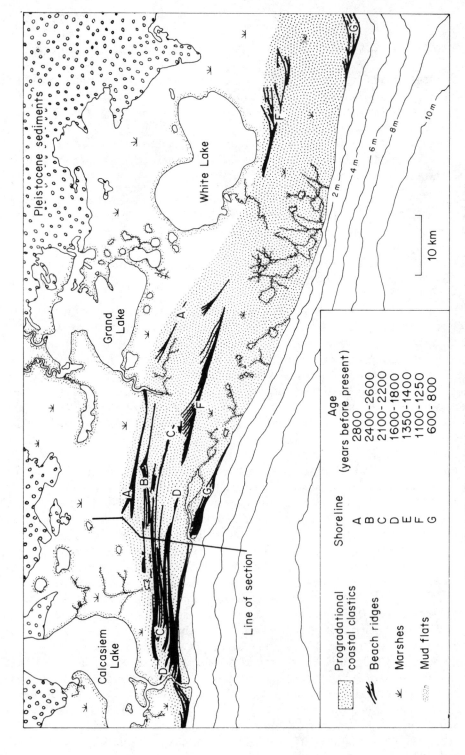

Figure 12.14 The Chenier Plain of southwest Louisiana. View in conjunction with Figure 12.15. Westerly longshore drift occasionally causes rapid progradation of tidal flats southward into the gulf of Mexico. As the sediment supply slackens, reworking of the tidal flats by wave activity and longshore currents produce beach-sand ridges. (After Gould and McFarlan, 1959, with modification and simplification.)

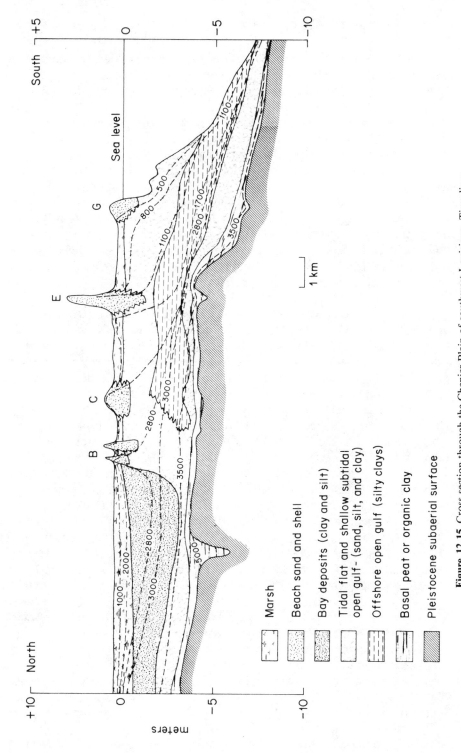

Figure 12.15 Cross section through the Chenier Plain of southwest Louisiana. Time lines are provided by numerous radiocarbon dates on peat and mollusc shells. (After Gould and McFarlan, 1959, with modification and simplification.)

Marsh

Beach sand and shell

Bay deposits (clay and silt)

Tidal flat and shallow subtidal open gulf- (sand, silt, and clay)

Offshore open gulf (silty clays)

Basal peat or organic clay

Pleistocene subaerial surface

The net result of this two-phase process—sedimentation followed by winnowing—is the development of a far more complicated internal stratigraphy within the Chenier Plain as compared to the stratigraphy of Galveston Island. For example, the general rule of the coarsening-upward unit begins to break down in the Chenier Plain. If we have a core or other stratigraphic section that culminates in a chenier sand ridge, we shall observe the expected unit. If, however, we have a core through an area in between chenier ridges, our general rule will fail to apply.

Nevertheless, sedimentation models based on Galveston Island or on the Chenier Plain have certain similarities. In both cases, major accumulations of bay deposits occur only on the updip end of the transgression. With seaward progradation of foreset facies (beach and offshore marine), topset sedimentation is restricted to a very thin veneer of marsh deposits. In both cases, seaward progradation into the Gulf of Mexico results in a clastic wedge that systematically thickens seaward, its upper surface being governed by sea level and its lower surface being governed by topography on the Pleistocene subaerial erosion surface.

The Continental Shelf of the Northwest Gulf of Mexico, Offshore East Texas and Western Louisiana. A discussion of sedimentation on Galveston Island and the Chenier Plain would be incomplete if we did not mention Recent sedimentation on the continental shelf offshore in these areas. In short, there is very little truly modern sedimentation occurring on that shelf. This fact makes the story of the post-Wisconsin transgression even that much more interesting.

The continental shelf south of Galveston and the Chenier Plain is a remarkably flat piece of subtidal real estate. The average slope of the bottom is about 0.5 meter per kilometer. The surficial sediment over all of this continental shelf is relict sediment. When geologists set out to study the Recent sedimentation in this area, they very rapidly realized that they were studying a thin veneer of abandoned beach, lagoonal, and marsh deposits that bear no relationship to the present-day bathymetry. Beneath this layer of shallow-water deposits is the Pleistocene subaerial surface. It is evident that the post-Wisconsin transgression passed over this area so fast that it left only a very thin and scattered record of its passing. Today, the bulk of sedimentation activity is taking place in the foreset position, from the beaches out to the mud deposits several kilometers offshore. For example, note in Figure 12.15 that Recent sediment above the Pleistocene unconformity surface is approximately 8 meters thick beneath the position of the modern beach; whereas Recent sediment is only about 1 meter thick, a scant 5 kilometers seaward from the modern beach in 8 meters of water. Shallow water over the continental shelf extends more than 150 kilometers seaward from this position. Sedimentation there today is essentially nil. The fact that a transgression such as the post-Wisconsin transgression can pass over a vast area and leave such an exceedingly scanty record will be an important part of any model concerning the interaction of clastic coastal sedimentation with events of submergence.

Tidal-Flat Sedimentation. Where longshore transport provides sand to build barrier-beach islands, extensive lagoons may be cut off between the barrier island

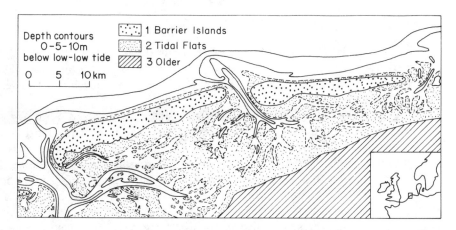

Figure 12.16 Tidal flats of the Wadden Sea, the Netherlands. Where barrier islands take the brunt of wave activity, extensive coastal environments may be influenced primarily by tidal action. Sediment types are as follows: (1) barrier islands; (2) tidal flats, which are sand-deposits that are submerged at high tide, exposed at low tide; (3) older sediments. [After L. M. J. U. Van Straaten, "Sedimentation in Tidal Flat Areas," *Jour. Alberta Soc. Petrol. Geologists,* **9,** 203–226, (1961), with modification.]

and the mainland (see Figure 12.16). As the beach environments accept the brunt of wave energy from the open-marine environment, the lagoons behind the beach may be predominantly influenced by tidal currents. Tidal currents in semirestricted lagoons are generally asymmetrical; flood tides come in faster and ebb tides go out slower. As the net result, sediment is pumped from the open-marine environment into the lagoon and is accumulated under the influence of flowing tidal currents. Klein (1970) provides an excellent example of primary structures resulting from tidal action.

During each half of the tidal cycle, sediments may be transported and deposited by the flowing water. Thus, sedimentation in tidal channels and on tidal flats is to a large extent quite analogous to meandering-stream sedimentation: The channels are like the streams; the tidal flats, like the backswamps. The coarsest material accumulates as high-angle cross strata in the bottom of tidal channels; the finest material accumulates on the tidal flat. With time, progradation occurs, producing a fining-upward unit (see Klein, 1971, for example). However, the presence of marine or brackish-water fauna and the presence of multidirectional cross-stratification distinguishes tidal-flat deposits from meandering-stream deposits.

A Basic Model for Clastic Sedimentation in Coastal Environments

In accord with our systematic approach to the development of models, let us first look at the general attributes of the sediment themselves, then examine the geometric relationships among sediment types, and finally consider the complications that can develop within the model under conditions of emergence or submergence.

The Coarsening-Upward Cycle. Clearly, the most striking generality to come out of our review of coastal clastic sedimentation is the coarsening-upward cycle (see Figure 12.17). Whether we are studying the bar-finger sands of the Mississippi Birdfoot delta, the beach deposits of Galveston Island, or any complex in between, somewhere within the strata we expect to see a well-developed coarsening-upward cycle.

The *coarsening-upward cycle* is the cycle generated by foreset deposition. Foreset deposition may begin in sharp contact with an underlying erosion surface or in gradational contact with fine-grained, open-marine, bottomset beds. The finest-grained sediment present in the cycle occurs at the bottom of the section. It must be emphasized that we cannot say precisely what this grain size will be. Grain size will be a function of size distribution supplied by the river and also a function of the currents in the marine environment at the site of deposition. Nevertheless, the finest gain-size present will occur at the bottom of the cycle. These fine-grained sediments will typically have a good marine fauna and burrow mottling will be the predominant primary structure. Toward the top of the section, grain size gradually increases, burrow mottling becomes less predominant, fossils generally become less abundant and more typically disarticulated and abraded. In addition, low-angle cross-stratification (on a beach) or high-angle cross-stratification (in a river-mouth bar or tidal channel) become the predominant sedimentary structures. The foreset

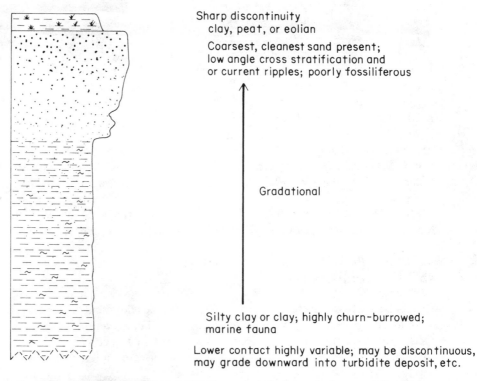

Sharp discontinuity
clay, peat, or eolian

Coarsest, cleanest sand present;
low angle cross stratification and
or current ripples; poorly fossiliferous

Gradational

Silty clay or clay; highly churn-burrowed;
marine fauna

Lower contact highly variable; may be discontinuous,
may grade downward into turbidite deposit, etc.

Figure 12.17 General model of a coastal clastic coarsening-upward cycle.

coarsening-upward unit may be abruptly overlain by topset beds consisting of natural levee, marsh, or eolian deposits.

Tidal-Flat Deposits. Because of the rapid rise in post-Wisconsin sea level, clastic tidal-flat deposits are not aerially extensive today. Stratigraphic evidence suggests that such deposits were much more abundant at various times in the geologic record. It is therefore appropriate that we study their identification.

Tidal-flat deposition does not obey the general rule of the coarsening-upward unit that applies to most other coastal clastic sediments. Transportation and deposition within the tidal flat are much more akin to those of a meandering stream. The generalized cycle is a fining-upward one, but it has important attributes to set it apart from the fluvial meandering streams.

Marine or brackish-water fossils are likely to be present in tidal-flat deposits. Furthermore, cross-stratification tends to indicate multiple directions of sediment transport in the tidal-flat environment, whereas similar features are more or less unidirectional in the fluvial meandering-stream facies.

Systematic Seaward-Thickening of Coastal Clastic Sedimentation Units. The thickness of a coastal clastic sedimentation unit is governed by the plane of sea level above and the preexisting topographic (bathymetric) surface below. For a great many paleogeographies that will interest us, preexisting topographic surfaces slope generally seaward. Thus, as progradation occurs, by whatever style it may be occurring, we would expect the development of a gross relationship in which overall thickness of the sedimentation unit increases in the seaward direction, (for example, see Figures 12.9, 12.12, 12.13 and 12.15). Henceforth, we may find it convenient to summarize this relationship by referring to the *coastal clastic wedge.*

Geometric Arrangement of Coarsening-Upward Units within the Coastal Clastic Wedge. Sedimentation in the modern Birdfoot delta of the Mississippi River and longshore transport and deposition of sand in the Galveston Island area provide us with two end-members in a spectrum depicting geometric configurations of coarsening-upward units. The Niger delta and the Chenier Plain of southwestern Louisiana illustrate two situations intermediate between the end-members. All four examples have already been described, so they require only a brief summary in this discussion of a general model.

At one end of the spectrum, the radial arrangement of bar-finger sands generates an extremely complicated stratigraphy of sand bodies. We cannot predict either the number or the direction in which the bar-finger sands will develop in any one subdelta. Moreover, relationships within one subdelta are totally independent of relationships within any other subdelta. Deflationary sheet-sand deposits of the destructive phase may provide good lithologic correlation over large areas within any one subdelta. Bear in mind, however, that deflationary sheet sands of any one subdelta are almost certainly unsystematically diachronous with the deflationary sheet sands of other subdeltas. But note that a number of deflationary sheet sands were associated with essentially the same period of high sea level. Thus, although they are diachronous in any strict sense, they may represent the closest approach to a time line that is available within the section. The search for a good time line is open-ended. We take the best we have and keep on looking for something better.

At the other end of the spectrum, a more or less continuous supply of sand to a barrier beach results in an extremely simple, widespread development of the coarsening-upward cycle (Galveston Island, for example). Virtually, the entire coastal clastic wedge is deposited as a coarsening-upward unit. Wherever we take a core or wherever we see an outcrop, this phenomenon will be apparent. In this situation, the top of the coarsening-upward unit will probably be a very good surface for regional lithologic correlation. We must keep in mind, however, that this surface is systematically diachronous; the oldest is toward the land and youngest toward the sea.

The Niger delta provides an example of something in between the two preceding extremes. River-mouth bars do, in part, prograde seaward; but they also supply sand for the development of well-sorted beaches in between river-mouth bars. Therefore, the coarsening-upward unit is being developed throughout the broad concentric front of the delta. In some places, the upper portion of the cycle is river-mouth bar sedimentation; in other places, the upper portion of the cycle is beach sedimentation. The whole complex is prograding seaward and is leaving behind the coarsening-upward cycle around the entire perimeter of the delta.

Discontinuous sedimentation and beach deflation in the Chenier Plain produce a stratigraphy of coarsening-upward unit arranged subparallel to the general strike of the coastal clastic wedge. Thus, individual beach sands are excellent time-stratigraphic markers along the strike, but lithostratigraphic correlations of these markers downdip is systematically diachronous.

Associated Lagoons and Topset Deposits. Having recognized coarsening-upward units and the general geometry of the coastal clastic wedge, what do we expect to find overlying the top of the foreset unit (on a beach or a river-mouth bar)? First and foremost, we have noted topset deposits in all of our examples from the Recent epoch. These are typically marsh, swamp, or eolian sand. Secondly, our examples of longshore transportation and deposition commonly have a subtidal lagoon updip from the coarsening-upward unit. Even though we did not see it in the Recent, perhaps that lagoon could migrate seaward as the coarsening-upward unit progrades into slowly rising sea level. Finally, in the delta areas, there are always fluvial environments upstream that could presumably prograde out onto the delta as the delta itself progrades out into the sea. Let us examine each of these situations in a bit more detail.

Wherever we have examined marsh and swamp deposits in the topset position (Figures 12.4, 12.13, and 12.15), we have found them to be rather thin. Limiting their thickness from underneath is the fact that progradation has built a surface at or near sea level before the marsh or swamp takes over. Limiting their thickness above is the fact that these environments must be very wet to keep going. If they begin to accumulate too thick a deposit, the top of the deposit becomes too high above sea level and conditions become too dry for the marsh or swamp to continue to thrive. The process, therefore, is self-limiting and closely related to a water table controlled by sea level.

At first glance, eolian sand dunes appear to offer a considerable opportunity for the development of thick topset deposits over the progradational foreset deposits. However, here again we find that what we see in the environment today is not necessarily what we would expect to see incorporated into the stratigraphic record. Whereas stabilization of eolian sand is the first fundamental step in its incorporation into the geologic record, the large dunes that we see today in south Texas, for example (see Figure 12.1), are transient, migrating features.

An effective way to stabilize eolian sand is to permanently wet it near the water table. Dry sand at the top of the dune continues to blow and migrate, but wet sand near the water table tends to remain in place so long as it remains slightly wet. Thus, it is not the eolian dunes that we would expect to see incorporated into the stratigraphic record as topset deposits. Instead, the planar deflation surfaces associated with the water table offer good opportunity for the lower portion of migrating eolian dunes to become incorporated into the geologic record. Once again, therefore, we see that the thickness of topset deposits over progradational foreset deposits is effectively limited by a water table directly related to sea level. So long as the sea level remains constant, we would not expect to see development of thick topset marsh, swamp, or eolian deposits, no matter how far the foreset beds may prograde.

Similar arguments apply to the possible development of a subtidal lagoon over foreset deposits during a time of progradation into constant sea level. There is simply no room between sediment and sea level. If there were room, swamp and marsh deposits would likely expand quickly to fill that space.

Thus, the basic model based on the Recent epoch would suggest that the best way to get lagoonal deposits progradational over foreset coarsening-upward deposits would be to prograde into rising sea level. We shall explore this fascinating possibility shortly.

Finally, we might expect to find fluvial meandering-stream deposits prograding into the topset position over a delta as the foreset of the delta prograde seaward into constant sea level. Such a model is in keeping with our concept of the equilibrium profile in meandering-stream deposition. As the ultimate seaward end of the equilibrium profile (the river-mouth bar) prograde seaward, the equilibrium profile will be raised somewhat, thereby accommodating meandering-stream deposition over earlier deltaic deposits. However, it is axiomatic that this relationship can develop only were rivers join the coastal clastic wedge. Progradational foreset deposits resulting from longshore transportation and deposition need not necessarily have fluvial deposits updip and therefore may not be prograded by meandering-stream facies.

These considerations have two important consequences for us when we look at the stratigraphic record. First, classical stratigraphers often attached formation significance to such descriptions as "alternating sand and shale." Alternating sand and shale describes both the coastal clastic wedge and the fluvial meandering-stream facies. Thus, puzzling thickness relationships and puzzling lithostratigraphic correlations may result from the fact that one section of "alternating sand and shale" involves both coastal clastic wedge and meandering-stream deposits whereas nearby sections involve only coastal clastic wedge. Second, progradation of meandering-stream facies onto the delta generates greater topographic relief than will be present in adjacent coastal clastic environments resulting from longshore transportation. Differences in geometry of the depositional surface may grossly affect the development of facies with the next transgression.

Foreset Progradation. As we talk of progradation in Recent sedimentary environments, we commonly speak of changes in map view: "the beach has prograded 3 kilometers seaward," and so forth. We must bear in mind, however, that progradation is not a function of area or distance but rather a function of sediment volume. For a beach to prograde seaward, the area in front of the beach must first be filled in. Thus, the rate of progradation of foreset deposits is a function of sediment supply and preexisting foreshore bathymetry. However, the situation can become more complicated.

We must consider the question of sediment *deposition in* an area as opposed to sediment *transportation through* an area. Why do some sediments come to their final resting place on that particular stretch of beach, whereas other sediments are transported past that spot and deposited further down the beach? Clearly, there exists some balance between sediment supply and the energy conditions of that particular stretch of coast. The more sediment brought to the area, the better chance there is that some of it will come to its final rest in that area. The more wave and current action on the beach, the more sediment may be carried on past that particular area.

Size distribution of the sediment supply and the concept of effective wave base may further complicate foreset progradation. When ocean waves approach a coast line, they begin to exert a significant force on the bottom as water depth shallows to 50 or 100 meters. Whether or not this force will resuspend a sediment is a function of the grain size of the bottom sediment. If abundant fine-grained sediment is present, then it may be mobilized by incoming waves and redistributed to form an equilibrium profile in which wave energy is dissipated evenly from deep water into shallow water. Thus, an incoming ocean wave will first expend some energy in mobilizing and transporting fine-grained sediment in 100 meters of water, utilize still more energy mobilizing and transporting silt and fine sand in less than 50 meters of water, and finally utilize its last energy mobilizing and transporting coarse sand in the beach environment. With continuing sediment supply, all sediment types, from fine-grained mud in deep water to coarse-grained sand on the beach, will be transported and deposited. With continued sorting, transportation, and deposition, the whole coastal clastic wedge progrades seaward. Note, however, that if fine grained material is lacking in the offshore area, inconsistencies may enter this scheme.

Consider a coastal area to which a given mix of sand, silt, and clay are transported by longshore currents. At first, this mix of particle sizes is adequate for initiating progradation. The sand accumulates, let us say, from the beach out to 10 meters of water; the silt, in water depth from 10 to 30 meters; and clay-sized material, in water depths of 30 meters and greater. As progradation continues, the amount of sand and silt required to produce one areal unit of seaward progradation (that is, to prograde the beach 1 kilometer seaward, for example) remains constant. However, the amount of clay-sized material required for fill in water depth of 30 meters or greater in front of the prograding beach becomes greater and greater as the clastic wedge progrades further and further out onto the continental shelf. Sand and silt may not be suitable substitutes for the clay. The weak wave agitation and low-velocity currents acting in water depths of 30 meters or greater may not be capable of sand or silt transportation.

So what happens? Beach sand and offshore silt go on accumulating for a while as though they were continuing to prograde. This process results in a steepening of the forebeach depositional topography, which puts the beach in closer proximity to incoming ocean waves and allows less distance for ocean wave energy to be dissipated before reaching the beach. The net result is increased agitation and stronger currents in the beach and offshore silt environments. Increased agitation and stronger currents result in greater transportation of sediment *through* the environment and less accumulation of sediment *in* the environment. Progradation, therefore, may become self-limiting in the absence of sufficient fine-grained sediment to produce equilibrium profiles in deeper foreshore waters.

Shelfslope Break as Foreset Progradation. Where energy conditions on the continental shelf are sufficiently vigorous that fine-grained sediment does not accumulate, spectacular "foreset" progradation may occur on the continental slope. It is useful to think of upper slope sedimentation as "foreset" because sediment accumulation is on an inclined surface and outbuilding of the shelfslope geographic boundary is precisely analogous to the outbuilding of a delta front or prograding beach. Continuing the analogy, the "topset beds of the shelfslope break would presumably be lag deposits of coarse-grained sediment such as shelf material left behind during the continued winnowing of fine-grained sediment off the shelf and onto the slope.

Basin Filling by Vertical Accumulation of Bottomset Beds. It is the natural fate of geologic basins to become filled with sediment. Progradation by coastal clastic wedges is a common method of basin filling. It is also the method involved in the origin of a great many sedimentary units that are of economic importance for their petroleum accumulations. For these reasons, we have spent a lot of time on foreset progradation and related phenomena of coastal clastic wedges. Let us note, at least in passing, that essentially vertical sedimentation of clays and pelagic organisms can also contribute to basin filling.

Isostatic Considerations. As a final point in the consideration of our constant sea-level model for coastal clastic wedges, we must note the profound importance of isostatic subsidence during the loading of basin margins. Consider, for example, a basin such as the Gulf of Mexico. For the sake of discussion, assume that we are going to replace 4000 meters of water with sediment of the coastal clastic wedge environment. Four thousand meters of water with a density of 1.0 is being replaced by sediment with a density of approximately 2.5. Isostatic considerations suggest that the floor beneath the basin margin subsides from 3000 to 6000 meters to accommodate the new load in isostatic equilibrium. Indeed, seismic investigations indicate that some 15,000 meters of Tertiary sediment have accumulated in southern Louisiana.

Similarly, each eustatic transgression will accommodate sediment accumulation to at least twice the amount of the eustatic sea-level change (see Figures 12.8 and 12.9, for example).

Response of Coastal Clastic Wedges to Emergence. As we begin to consider the response of coastal clastic wedges to conditions of changing relative sea level, we must keep in mind two general questions. First and foremost, what happens to sediment supply? Secondly, how are existing deposits affected by the change?

With a condition of relative sea-level lowering (emergence), the answer to both of these questions is rather simple. Meandering streams continue to supply sediment from continental areas. In fact, probably a little more sediment may come down to the coastal area as the meandering-stream facies incises into older deposits to approach the equilibrium profile dictated by the new lower sea level.

Old marshes, bays, and lakes sit high and dry, ceasing to accumulate sediment and perhaps undergoing considerable surficial modification and erosion as terrestrial soil zones develop. Actively prograding river-mouth bars and beaches go right on prograding; their rate of seaward progradation is somewhat speeded up by the fact that decreasing water depth allows a given volume of sediment to prograde farther seaward.

The systematic seaward thickening of coastal clastic wedges accumulated under constant sea-level conditions clearly does not apply to coastal clastics accumulating under conditions of emergence.

Progradation under conditions of constant sea level produces an extremely flat upper surface to the coastal clastic wedge. The subsequent transgression will flood a large area of essentially uniform bathymetry. But coastal clastic sediments deposited under conditions of emergence result in an upper surface that may have considerable seaward slope to the top of the regressive unit. The subsequent transgression will therefore have a more or less continually rising topography on which to transgress.

Response of Coastal Clastic Wedges to Submergence. Rising sea level may seriously diminish sediment supply to coastal clastic environments. Initially, rising sea level raises the lower end of the equilibrium profile of the meandering stream that is supplying sediment to the coastal area. The meandering-stream environment may temporarily become the deposition site of clastic sediments that otherwise would have been carried downstream to the coastal area (recall Figures 11.34, 11.35, and 11.36, for example). In addition, rivers commonly occupy valleys that are incised into preexisting topography. With rising water, the site of new deltaic sedimentation may be pushed well up into the head of the newly formed estuary as the river valley becomes flooded. Given this situation, all clastic sediments supplied by the river must go into building a new delta that will fill the estuary before any sediment can again be supplied to longshore currents. In Galveston Bay, for example (see Figure 12.13), sediment supplied by the Trinity River has barely begun to fill the bay from the northeast end. It will be many years before the Trinity River supplies a significant amount of sediment to the barrier beaches of the Texas Gulf Coast.

Thus, we have two generalities concerning sediment supply to coastal clastic wedges during a time of submergence: (1) The supply is decreased by rise of equilibrium profile in the meandering stream; and (2) those clastics that are supplied to the coastal environment may be required to build new deltas out of the estuaries before longshore transport of sediment can assume its former importance. While variation in grain size and transport mechanism of the sedimentary load carried by the rivers renders precise calculation rather complicated, it is instructive to consider some simple arithmetic with regard to this situation. Let us continue to use the Mississippi River drainage basin as our example.

In round numbers, the drainage basin of the Mississippi River is approximately 3 million square kilometers and the annual discharge of suspended sediment load is approximately 300 billion kilograms (Holeman, 1968, for example). Let us presume that this suspended load becomes deposited somewhere and that the deposited sediment has a density of approximately 1.5 grams per cubic centimeter. According to our simple model for the equilibrium profile of streams, a rise in sea level will cause a rise in the equilibrium profile throughout the drainage basin. Rise in sea level creates new room for sediment accumulation throughout the system. As one end-member calculation, assume that all of the suspended sediment load becomes deposited as new alluvial deposits throughout the drainage basin. The amount of sediment divided by the area of the drainage basin yields an average sedimentation rate of 6.7 centimeters per 1000 years. Note well that this is a rather small number compared to late Pleistocene rate of glacio-eustatic sea-level rise (8 meters per 1000 years; see Chapter 9). Thus, our simple end-member calculation indicates that the modern Mississippi River must surely be operating well below our theoretical equilibrium profile.

Given a Holocene sea-level rise of a mere 10 meters, the total sediment supply coming down the river would require 150,000 years to fill in the drainage basin to equilibrium profile if all of the sediment were accumulated in the river valley. Similarly, given 100 meters of post-Wisconsin sea-level rise (a more realistic number), it would require over 1 million years for the drainage basin to attain a new equilibrium profile. Thus, the effects of rising sea level on equilibrium profile in drainage basins is quite significant. However, surely it is unrealistic to expect all of the sediment to be trapped in the alluvial valley; today's "muddy Mississippi" is a very simple empirical fact. Thus, we simply note the tendency indicated by this calculation and turn to empirical data as to how rising sea level really affected the post-Wisconsin Mississippi coastal clastic wedge.

Carbon-14 data concerning the post-Wisconsin transgression suggests that sea-level rises of 8 meters per 1000 years (the approximate average for post-Wisconsin sea-level rise) are sufficient to generate grossly discontinuous responses of coastal environments to the rising water. Figure 12.6 shows that the first subdelta of the recent Mississippi delta formed approximately 5000 years ago. However, as indicated in Figures 12.14 and 12.15, development of the Chenier Plain did not get into full swing until approximately 3000 years ago, some 2000 years after the sea had effectively flooded this area. This lag time between the Mississippi delta and Chenier Plain is presumably related to lack of sediment supply from the infant Recent Mississippi delta. At any rate, a sea-level rise of 8 meters per 1000 years is sufficient to wreck the continuity of sediment accumulation in coastal clastic environments.

Clearly, there is some slow rate of sea-level rise that is sufficiently slow that deltaic sedimentation and longshore transport will go on essentially unchanged. Returning to our calculations concerning the Mississippi drainage basin, a rate of sea-level rise on the order of 1 or 2 centimeters per 1000 years would presumably put us into the right range. Such a rate of rise of the lower end of the equilibrium profile would create average alluvial sedimentation rates on the order of 0.5 to 1.0 centimeter per 1000 years, representing approximately 10% of the total sedimentary load of the river system. If 90% of the present sedimentary load continued to come down the river into a rising sea level, surely the coastal processes would respond in a continuous fashion. In the delta area, the rate of vertical accumulation of swamp, marsh, and natural levee deposits would presumably be sufficient to confine the river to its channel and thereby allow deltaic sedimentation to continue to occur along the outer periphery of the preexisting delta (see Figure 12.18). Given this situation, longshore transportation of sediment away from the delta should also go on as usual, even though the sea level is rising slowly.

Turning now to the effects of slowly rising sea level on a prograding barrier bar (such as our Galveston Island model), we notice some things that will bear striking similarity to observations in the ancient record. The beach environment of the barrier bar more or less ignores the slowly rising sea level. From day to day, this beach environment simply goes on accumulating sediment and prograding seaward as long as there is adequate longshore sediment supply. On the other hand, topset deposits and lagoonal deposits take advantage of the rising water and expand with it. In our static sea-level model, marsh and swamp deposits behind barrier bars were thin because their accumulation was governed by a water table closely tied to sea level. Because of similar limitations, lagoonal deposits could only accumulate to some trivial thickness in our constant sea-level model. Now that we are developing a model that involves a slowly rising sea level, we observe that these deposits are given room to expand, both in a vertical sense and in a horizontal sense.

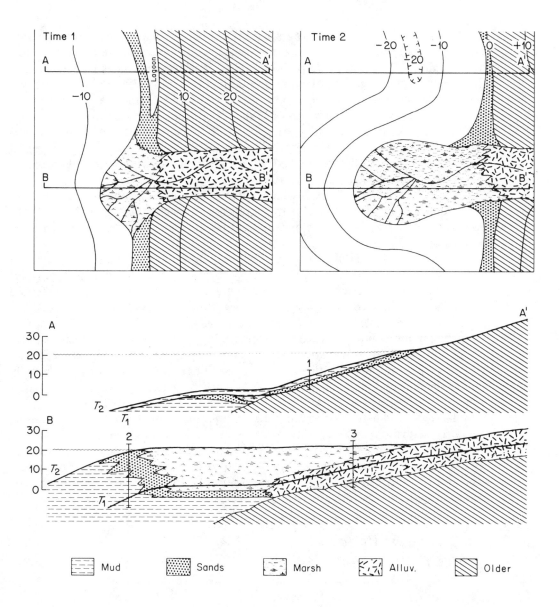

Figure 12.18 Theoretical model of stratigraphic relations generated by the interaction between coastal clastic sedimentation and rising sea level. In map view, time (1) shows the geographic distribution of major sediment types before submergence. Time (2) indicates one potential geographic distribution of major sediment types following an event of "moderately rapid" submergence. The delta has more or less maintained its position, whereas the beach environments to the north and south have migrated landward with rising water. Cross sections *AA'* and *BB'* demonstrate the vertical stratigraphies that may be generated by such an event. Note the "transgressive" sequences (1) and (3) are properly correlated with the "regressive" sequence (2). Observe also the thick accumulation of topset deposits in the delta area.

Thus, continued seaward progradation of a barrier beach during slowly rising sea level not only generates a "regressive sequence" during the time of rising water, but also allows thick and widespread marginal marine topset beds to accumulate. These beds will undergo lateral expansion both in the seaward direction (regressive) and in the landward direction (transgressive). Note the confusion that can arise from loose use of the words "transgression" and "regression" to describe the relationship of sediment to slowly rising sea level. Note further that the sea-level rise prescribed here (1 or 2 centimeters per 1000 years) is right in the range to be expected for subsiding basins (Chapters 8 and 9). We shall, therefore, expect this problem to be rampant in the geologic record.

Given a different balance between sediment supply and rate of sea-level rise, we could argue for the progressive transgression of beach facies onto the land surface as sea level rises. This relationship is the classical one that we used extensively in Chapter 2.

So long as there is a gently sloping topography along which the beach can extend, this model is perfectly reasonable. With rising water, beach facies trangress onto the land surface and former beach facies are overlain by fine-grained offshore sediments. As indicated in Figure 12.18, the model developed here suggests that it is perfectly reasonable to correlate a "transgressive" sequence in an area of low-sediment supply with a "regressive" sequence in an area of high-sediment supply.

In all models for continuous sedimentation during conditions of submergence, the preexisting sediments will be covered up with new sediment. The more rapidly a sediment becomes covered with new sediment, the better the chance is that that portion of sedimentation history will become incorporated into the permanent stratigraphic record. There is, therefore, a strong preservational bias for stratigraphic sequences deposited under conditions of submergence, as opposed to stratigraphic sequences deposited under conditions of emergence. Note that this statement does not use the terms "transgressive" and "regressive" sequences, which appear in classical stratigraphic terminology. Both transgressive and regressive sequences may be deposited under conditions of submergence.

An Ancient Example: Bottomset to Meandering-Stream Transition; Reedsville Shale–Oswego Sandstone, Upper Ordovician of Central Pennsylvania

The work of Horowitz (1966) provides a simple, yet instructive, test of our general model. First, let us discuss the section in classical stratigraphic terms.

Regional Setting and Earlier Stratigraphic Work. As indicated in Figures 10.14, 10.15, and 10.16, it has long been recognized that throughout the sedimentary Appalachians earlier Paleozoic limestone sedimentation gives way with time to marine clastic sedimentation, which in turn gives way gradually to nonmarine clastic sed-mentation. The timing of these events varies along the Appalachian front. In south central Pennsylvania, the transition from Reedsville shale to Oswego sandstone was recognized rather early as one of these transitions from marine clastics to continental clastic sedimentation.

The Reedsville shale is gray and silty, with occasional fine-grained sand beds, and generally contains marine fossils. Typically, the Oswego sandstone is green and red, high-angle, cross-stratified, and interbedded with shaly siltstone. The formation is generally unfossiliferous except for occasional bituminous fragments. Clearly, the distinction between Reedsville and Oswego lithologies provides the basis for the definition of separate and distinct formations. But where should we place the contact? Inasmuch as the Reedsville is predominantly shale and silt and the Oswego is predominantly sandstone, there is a certain logic to placing the formation contact so that the predominantly silt and shale lithology is below the predominantly sand lithology. Note that the formation boundary thus defined may have little paleogeographic significance.

Sedimentary Observations within the Reedsville-Oswego Contact Interval. Horowitz (1966) studied the contact between the Reedsville and Oswego formations in considerable detail (see Figure 12.19). Four of his measured sections are shown in Figure 12.20. Horowitz subdivides the transition from Readsville to Oswego into three units that he considers have environmental significance. The units of the transition zone are designated as *basal, medial,* and *upper;* their lithologic and paleontologic attributes are summarized in Table 12.1. In traditional stratigraphic terminology, the basal and medial units belong to the Reedsville shale. The upper unit, being predominantly sand, is designated as Oswego sandstone. A brief lithologic description of typical Oswego sandstone is also included in Table 12.1 for comparison.

Observe particularly the transition from medial unit to upper unit. Massively bedded, burrow-mottled, fossiliferous, clayey siltstones and very fine-grained sandstones give way upward to interbeds of evenly laminated very fine-grained sandstone and thick, cross-bedded, very fine-grained sandstone. These sandstones are poorly fossiliferous; those fossils that are present are poorly preserved. Note further that the upper unit is overlain by a sequence of unfossiliferous fining-upward units in which high-angle cross-stratified sandstone grades upward into shaly siltstone. The sequence from medial transition beds of the Reedsville into the fining-upward units of the Oswego presents us with three suites of lithologies that are quite familiar to us from our foregoing discussions of Recent sedimentation.

Sedimentological Interpretation. The basal unit of the transition zone presumably represents slump or storm-deposited material in essentially the marine bottomset position. Accumulation of fossil fragments at the base of sandstone layers and the generally graded texture of the sandstones suggest mass tranport and resedimentation of earlier marine deposits. This point will be discussed further in a later chapter. The medial unit has many of the attributes of the marine foreset deposits

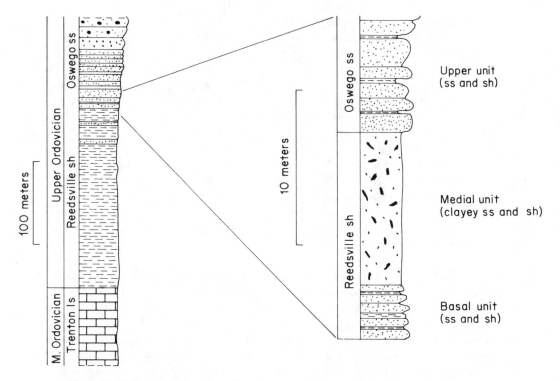

Figure 12.19 Generalized stratigraphic section of the contact zone between Reedsville shale and Oswego sandstone. (After Horowitz, 1966, with modification.)

that we discussed in some detail with respect to the Mississippi delta. The upper unit is the river-mouth bar-beach complex of the upper portion of the foreset beds. The fining-upward units, which constitute the bulk of the Oswego, are meandering-stream deposits that have prograded over the coastal clastic environments.

The environmental interpretation of the Reedsville-Oswego contact gives us a completely different unifying concept for the designation of formations. In particular, the upper unit of the transition zone has previously been considered part of the Oswego sandstone on the basis of the idea that "shale below is Reedsville and sandstone above is Oswego." If we could redesignate the formation boundary, we would probably prefer to place it at the point where coastal clastic sediments lie below and fluvial meandering-stream sediments lie above.

An Ancient Example: Coastal Clastic Sediments in the Subsurface Lower Oligocene of Southeast Texas

The work of Horowitz described earlier relies heavily on detailed examination of outcrops and hand specimens of the rock under consideration. Upon moving into subsurface problems, the geologist does not have these kinds of sample materials

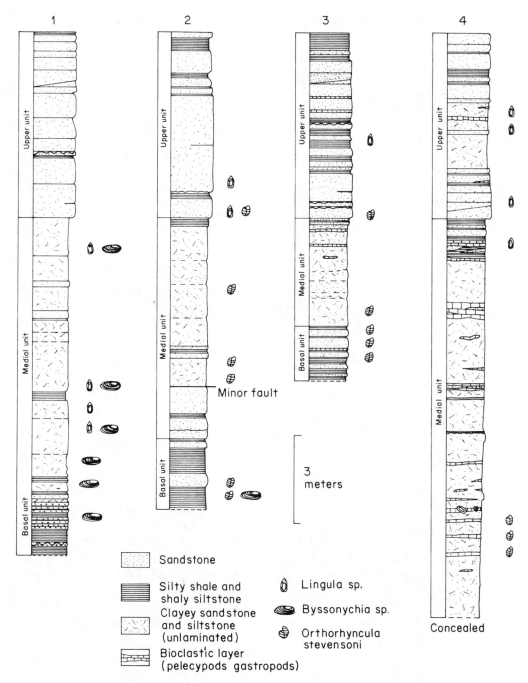

Figure 12.20 Four measured sections through the contact between Reedsville shale and Oswego sandstone. (After Horowitz, 1966, with modification.)

255

Table 12.1 Summary of Lithologic and Faunal Characteristics of Units of Deltaic Sequence

Unit	Fossils	Lithologies
Oswego (excludes basal 20 feet)	Unfossiliferous. Small bituminous coallike fragments scattered in a few shaly beds.	Cross-bedded, very fine-grained to fine-grained sandstone grades upward into shaly siltstone Beds were *red* before diagenetic reduction.
Upper	Fossils rare and fragmental; mostly *Lingula* and poorly preserved calcareous forms.	Evenly laminated, very fine-grained sandstone; thick, cross-bedded, fine-grained sandstone; unlaminated, clayey siltstone and very fine-grained sandstone; silty shale; fine-grained quartzite and other less common rock types.
Medial	Small pelecypods and gastropods concentrated in thin beds. *Lingula* occurs near top; brachiopod *Orthorhyncula stevensoni* and pelecypod *Byssonychia* sp. near base.	Mostly massive or thick-bedded, poorly sorted, clayey siltstone and very fine-grained sandstone. Internal structures suggest that sediment was churned or disturbed. Thin wavy fossiliferous beds are locally conspicuous features. Chert "balls" of probable slump origin are present.
Basal	Brachiopod *Orthorhyncula stevensoni* and pelecypod *Byssonychia sp.* locally abundant in shaly layers. Cephalopods, bryozoans, and small brachiopods rare.	Graded, very fine-grained to fine-grained sandstone interbedded with silty shale. Fossil fragments occur at base of sandstones.

After Horowitz, 1966.

to study. Where core is available, detailed examination of the lithology is possible. However, coring is an expensive way to drill a hole. The geologist is quite often left to carry out his or her investigations on the basis of electrical logs and well cuttings.

In the Gulf Coast, cuttings are of no value for sandstone petrology because the sands are so poorly cemented that only loose sand is recovered. Thus, all information concerning primary structures and systematic grain-size variation is lost in this drilling process. Cuttings from shale intervals, however, are valuable for the stratigraphic control provided by the foraminifera contained in the shales. Geologists studying the subsurface in the Gulf Coast, therefore, generally use (1) electrical logs, which provide their only lithology information, and (2) cuttings from the shale intervals, in which foraminifera provide paleontological control on the stratigraphy. The work of Gregory (1966) provides a good example of how a geologist proceeds to draw important paleogeographic conclusions from the study of these limited types of information.

Regional Setting and Earlier Stratigraphic Work. As noted in Chapter 10, the major stratigraphic subdivisions of Gulf Coast Tertiary stratigraphy are based on transgressive tongues of limestone or shale that give way vertically to regressive (progradational) deposits of marginal-marine and finally continental clastic sediments.

Figure 12.21 indicates how these major sedimentation units are further subdivided on the basis of benthonic foraminiferal zones. Recall that a formation is simply a mappable unit. If we choose to define our mappable unit on the basis of paleontology rather than lithology, that definition is perfectly acceptable.

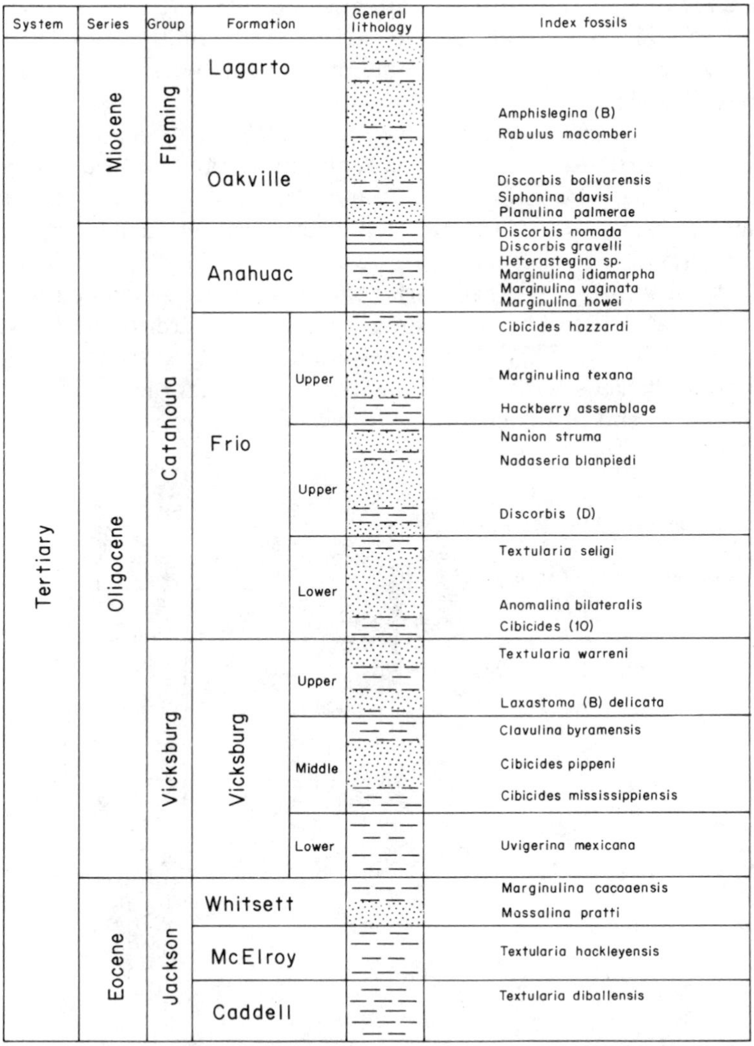

System	Series	Group	Formation		General lithology	Index fossils
Tertiary	Miocene	Fleming	Lagarto			
						Amphislegina (B)
						Rabulus macomberi
			Oakville			Discorbis bolivarensis
						Siphonina davisi
						Planulina palmerae
	Oligocene	Catahoula	Anahuac			Discorbis nomada
						Discorbis gravelli
						Heterastegina sp.
						Marginulina idiamorpha
						Marginulina vaginata
						Marginulina howei
			Frio	Upper		Cibicides hazzardi
						Marginulina texana
						Hackberry assemblage
				Upper		Nonion struma
						Nodaseria blanpiedi
						Discorbis (D)
				Lower		Textularia seligi
						Anomalina bilateralis
						Cibicides (10)
		Vicksburg	Vicksburg	Upper		Textularia warreni
						Laxostoma (B) delicata
				Middle		Clavulina byramensis
						Cibicides pippeni
						Cibicides mississippiensis
				Lower		Uvigerina mexicana
	Eocene	Jackson	Whitsett			Marginulina cocoaensis
						Massalina pratti
			McElroy			Textularia hackleyensis
			Caddell			Textularia diballensis

Figure 12.21 Generalized stratigraphic column, Vicksburg and adjacent formations, subsurface of southeastern Texas. (After Gregory, 1966.) Courtesy Houston Geological Society.

Sedimentological Observations in the Vicksburg Formation. Figure 12.22 presents a stratigraphic cross section of the Vicksburg formation oriented perpendicular to regional strike. There are several things in this cross section that should start a petroleum geologist thinking. To begin with, the Vicksburg formation abruptly thickens between wells (6) and (7). Abrupt thickenings of the section downdip are a general rule in Gulf Coast stratigraphy. Oil and gas accumulations are commonly associated with these *flexure zones.* Usually a large number of sands interlayered with the shale occur just downdip from the flexure zone. In the cross section we are looking at, the entire section downdip from the flexure zone is shale. This fact is disappointing, but at least we have one place on the map through which the flexure zone passes. Perhaps at some other place, we shall find sand downdip to the flexure zone.

Turning our attention to the Vicksburg formation updip from the flexure zone, we note particularly the seemingly continuous sandy zone developed within the Middle Vicksburg between wells (2) and (5). Sands such as these, in among a lot of shale, always interest the petroleum geologist. The sand can provide the reservoir for oil and gas; the shale can provide the seal on the trap. Thus, a little structure or some stratigraphic peculiarity can make these sands petroleum reservoirs. Observe that the sand body does not extend southward to the Vicksburg flexure zone. Well (6) indicates that a relatively thin Vicksburg section is all shale before the section thickens across the flexure zone. Note that the sand body also shales out on the updip end of the cross section [well (1)].

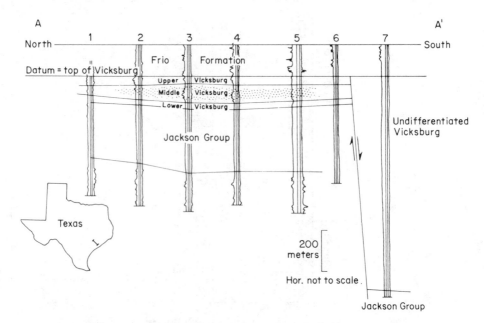

Figure 12.22 Stratigraphic cross section indicating the distribution of Middle Vicksburg sand development. (After Gregory, 1966.) Courtesy Houston Geological Society.

Making use of our first piece of paleoecological information provided by examination of the foraminifera content of the shales, we observe that these updip shales do not contain the marine benthonic foraminifera that are traced by biostratigraphers for stratigraphic control. Presumably, the absence of these marine benthonic foraminifera implies that this updip shale zone, stratigraphically equivalent to Middle Vicksburg sand, is not a marine shale but rather a freshwater or brackish-water shale.

Figure 12.23 is a map of the net thickness of sand contained in the Middle Vicksburg stratigraphic interval. Note that there are essentially two axes of thickening within the areal distribution of Middle Vicksburg sand. The thickening axis running from southwest to northeast more or less parallels regional strike; the thickening axis extending off to the northwest is almost perpendicular to regional strike. Note further the pronounced downdip bulge of net sand where these two axes of thickening come together. Surely this contour is beginning to look familiar to us.

Now we would like to relate the net sand map to the paleogeography in which it formed. Because of postdepositional tectonic deformation, we cannot construct a simple contour map of the Middle Vicksburg top that will reflect its initial sedimentary topography. Postdepositional tectonic activity has grossly distorted the depositional surface. A map showing the top of the Middle Vicksburg would be a map of structural deformation rather than of depositional topography.

A common way to get around this problem is to construct an isopach map of the sedimentary unit directly above the topographic surface that interests us. Clastic sediments do not generally pile up in big mounds; they tend to fill in the low spots. Thus, a map depicting the thickness of Upper Vicksburg sediments may be a crude reflection of the topography upon the Middle Vicksburg depositional surface. Where the Middle Vicksburg depositional surface was low, thicker accumulation of Upper Vicksburg material would be expected. Granted, this scheme is imperfect, but it is a step in the right direction and is probably the best that can be done on the basis of data generally available to the practicing petroleum geologist.

Figure 12.24 is an isopach map of Upper Vicksburg sediment within the study area. The tendency to abrupt thickening at or near the Vicksburg flexure is quite apparent. Observe that major sand accumulation in the Middle Vicksburg is considerably updip from the zone of abrupt thickening in Upper Vicksburg sediments. In short, Middle Vicksburg sands were deposited on a shallow-marine shelf marginal to deeper water.

Sedimentological Interpretation. On the basis of sand geometry, reconstructed paleobathymetry based on the thickness of overlying deposits, and paleontological data from associated shale, Gregory (1966) put together the environmental interpretation for the deposition of Middle Vicksburg sands, as indicated in Figure 12.25. If we briefly review his lines of evidence, we can recognize marine shales from electric log and micropaleontological data. Sands are shown to be sands from the electric log and no other evidence. Marshy, back-bay, mud-flat deposits and tidal-flat deposits are known from the electric log, which indicates shale, and from micro-

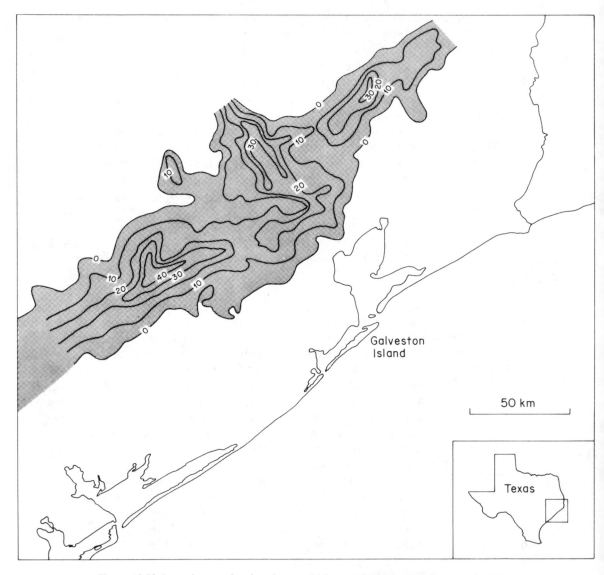

Figure 12.23 Isopach map showing the net thickness of Middle Vicksburg sand. (After Gregory, 1966, with modification.) Courtesy Houston Geological Society.

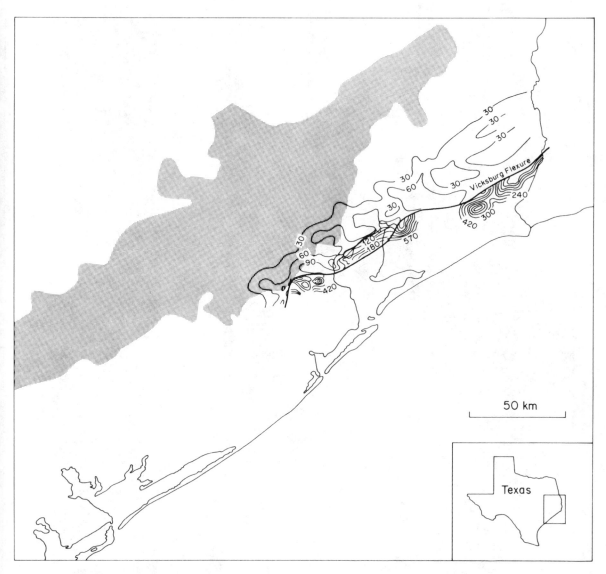

Figure 12.24 Isopach map depicting the total thickness of the Upper Vicksburg stratigraphic interval. Stippled area indicates distribution of Middle Vicksburg sand (Figure
12.23). Note that Middle Vicksburg sand accumulation lies tens of kilometers landward of
the Upper Vicksburg flexure. (After Gregory, 1966, with modification.) Courtesy Houston
Geological Society.

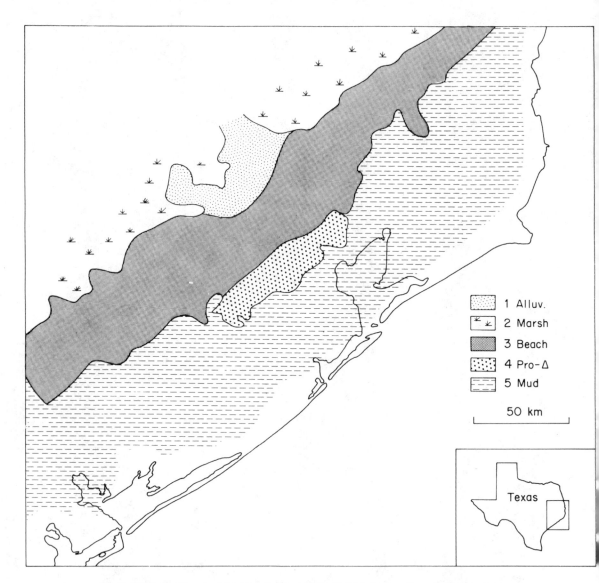

Figure 12.25 Environmental interpretation of Middle Vicksburg sediments. Sediment types are as follows: (1) meandering-stream deposits; (2) brackish-water to freshwater bay deposits, mud flats, and marshes; (3) beach and nearshore coastal clastics; (4) subtidal prodelta sands; (5) marine shales. (After Gregory, 1966, with modification.) Courtesy Houston Geological Society.

paleontological data, which indicate that they do not contain marine benthonic for-aminifera. The rest of the interpretation is based on the way in which these three general rock types fit together and fit the paleogeography provided by the Upper Vicksburg isopach map.

An Ancient Example: Stratigraphic Oil Trap in Coastal Clastic Sands of the Subsurface Upper Cretaceous of New Mexico

In both of the earlier Ancient examples, we have found a certain degree of reliability in the Ancient record as we have recognized sediment types and geometries that appeared in our study of Recent sedimentation. As we move on to look at Bisti field, New Mexico (Fig. 12.26), we shall begin to test our dynamic model of coastal clastic sedimentation. We shall be looking at a section that undoubtedly contains a lot of emergence, submergence, and progradation, if we can only sort them out.

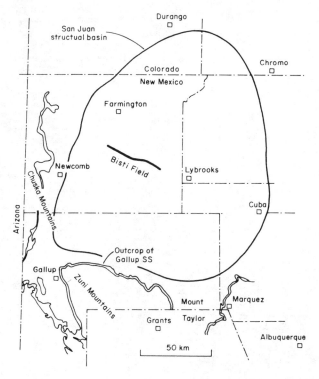

Figure 12.26 Index map of the San Juan basin, New Mexico, showing Bisti field and outcrops of the Gallup sandstones. (After Sabins, 1963, with modification.)

Regional Setting and Earlier Stratigraphic Work. The Bisti oil field is located near the center of the San Juan structural basin of northwestern New Mexico (see Figure 12.26). Upper Cretaceous stratigraphy of the area begins with the transgression of the Dakota sandstone over a major unconformity surface (Figure 12.27). From that time onward, the depositional history of the San Juan basin records a delicate balance between marine transgression from the northeast and continental regresssion from the southwest. Formation and member names are typically given to laterally extensive continental deposits, marginal-marine sandstones, and marine shales. The obvious stratigraphy of the area is decidedly lithostratigraphic.

Figure 12.28 is a structural contour map of a horizon just above the producing zone in Bisti field. The map shows only regional dip to the northeast. A structure map of a horizon below the producing zone would similarly show only northeast dip. Bisti field, therefore, is decidedly not a structurally controlled accumulation of petroleum. The oil occurs in that particular sand unit for some sedimentological-stratigraphic reason. Sabins (1963) attempts to explain the controlling factors on this stratigraphic trap.

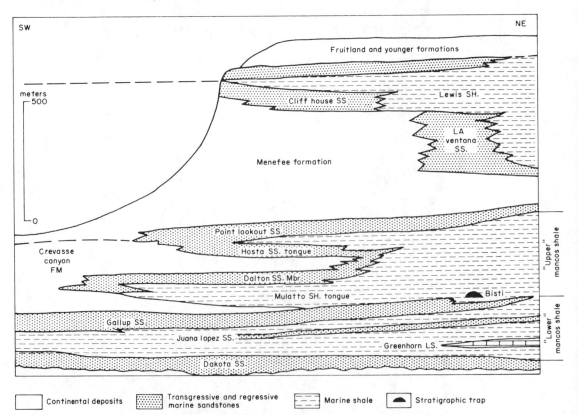

Figure 12.27 Upper Cretaceous stratigraphic diagram for the San Juan basin. No horizontal scale. (After Sabins, 1963, with modification.)

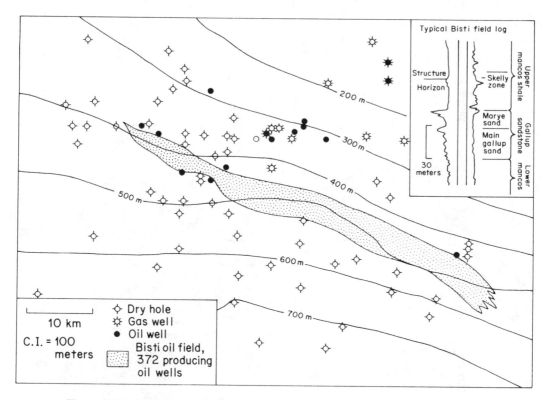

Figure 12.28 Structural contour map of the Bisti field area. Note that this occurrence of petroleum is not associated with structural closure. (After Sabins, 1963, with modification.)

Sedimentological Observations. Figure 12.29 presents Sabin's correlation of four wells crossing Bisti field from updip (south) to downdip (north). In brief, detailed electric log correlation has led Sabins to suspect that the sand bodies of Bisti field are petroleum reservoirs because they are physically separated from permeable Gallup sandstone updip by an impermeable zone such as occurs in well (2).

Cores are generally available within this field, so Sabins was able to make a detailed petrographic examination to test this theory. Consider first the main Gallup sandstone southwest of Bisti field (landward with respect to the paleogeography).

The basal contact of the main Gallup sandstone is gradational with Lower Mancos shale below. The lower portion of the Gallup sandstone is very fine-grained and contains abundant clay matrix. It tends to become coarsest toward the top, and/or clay matrix becomes less important. The upper 8 to 30 meters of the main Gallup sandstone are typically medium- to coarse-grained clean sands. Thus, it is reasonable to summarize this section southwest of Bisti field as a coarsening-upward cycle from marine Lower Mancos shale gradationally up into the Gallup sandstone and culminating in unfossiliferous medium to coarse-grained clean sand. The upper contact of this cycle with the Upper Mancos shale is sharp.

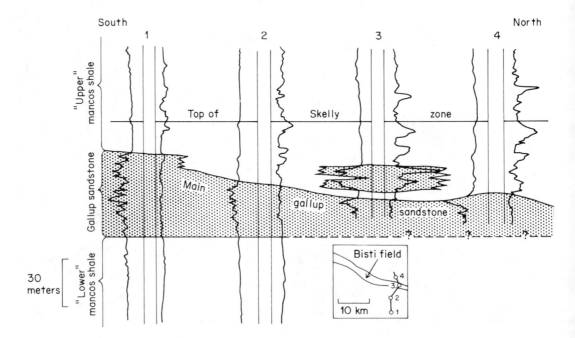

Figure 12.29 Stratigraphic cross section through Bisti field. Bisti field produces oil from sand bodies that are physically separated from the updip main Gallup sandstone by the occurrence of low-permeability material such as in well (2). (After Sabins, 1963, with modification.)

Next, let us consider a horizontal traverse of those facies that appear to correlate lithostratigraphically with the upper 20 meters of the main Gallup sandstone (see Figure 12.29). In the paleogeographic position represented by well (2), sediments of this lithostratigraphic interval are sandy and silty shales, highly burrowed, with benthonic foraminifera common. Moving on to well (3) (the producing horizon of Bisti field), we see that this field is occupied by relatively clean, medium to fine sand, with abundant glauconite grains and a few pyrite-filled foraminifera but no other elements of a marine fauna. In well (4) (seaward from the producing zone in Bisti field), the sediments of this lithostratigraphic interval are sandy and silty shales, which contain an abundant marine fauna dominated by the pelecypod *Inoceramus,* by fish bones and teeth, and by planktonic foraminifera. Similar sediment types predominate in the Upper Mancos shale, which overlies the Gallup sandstone and related sandbars.

Sedimentological Interpretation. Throughout the preceding lithologic descriptions, we avoided the use of genetic terms such as beach, offshore shales, and coastal sandbars. Now let us go back and see whether some of those words might fit. To begin with, we almost said "beach deposits" when we looked at the coarsening-upward cycle in well (1) (Figure 12.29). Sabin's lithologic description of a transition from Lower Mancos shale to the culmination of the Gallup sandstone certainly appears like the kind of sequence that would be generated by prograding, coastal clastic, longshore sedimentation. Indeed, when we look back at the regional stratigraphy (Figure 12.27), we see the Green Horn limestone as the maximum transgressive facies in this section. The constant thickness of the main Gallup sandstone and the gradual thickening of the upper portion of the Lower Mancos shale are certainly compatible with progradation of a Gallup sandstone beach building outward into the originally deeper portions of this shallow sea.

Clearly, the Gallup sandstone progradation was ended rather abruptly by submergence. That the submergence was rapid rather than gradual is suggested by the Mulatto shale tongue overlying the lower portion of the continental Crevasse Canyon formation with no intervening marine sandstone member. According to our generalized model for coastal clastic sedimentation under conditions of submergence, rapid submergence may be expected to temporarily trap the sediment supply in alluvial deposits and in newly constructed deltas, thus allowing water to rise over the former land area without the deposition of associated coastal coarse-grained clastics.

Bisti field, therefore, is the sandbar complex along the front edge of a major progradational sand beach. The reason we see it preserved today is that progradation was abruptly terminated by renewed submergence. As indicated in Figure 12.30,

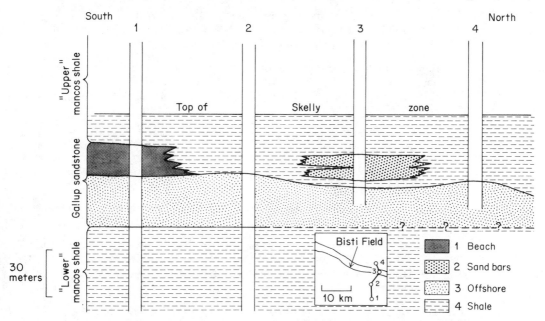

Figure 12.30 Environmental interpretation of the stratigraphic cross section presented in Figure 12.29. Sediment types are as follows: (1) beach sand; (2) forebeach sandbars; (3) offshore marine sands and silts; (4) low-permeability marine deposits, shales, and shaly siltstones. (After Sabins, 1963, with modification.)

Sabins concludes that the lead edge of the Gallup sandstone progradational surface held its position for a time during the initiation of submergent conditions. Beach sediments accumulated to thicknesses of 8 to 30 meters in the area southwest of Bisti field. In the area of Bisti field itself, at least two generations of sandbar development took place before submergence finally dominated the area, driving coarse clastic sedimentation far to the southwest and replacing it with Upper Mancos shale sedimentation over the Bisti field.

Note that a stratigraphy of coastal clastic sediments such as the one indicated in Figure 12.27 affords magnificent opportunity for the development of (1) numerous other examples showing the delicate balance between progradation and submergence and (2) other stratigraphic traps like Bisti field.

Citations and Selected References

ALLEN, J. R. L. 1965. Late Quaternary Niger delta, and adjacent areas: sedimentary environments and lithofacies. Bull. Amer. Assoc. Petrol. Geol. 49: 547–600.

———. 1970. Sediments of the modern Niger Delta: A summary and review, p. 138–151. *In* J. P. Morgan and R. H. Shaver (eds.), Deltaic sedimentation: modern and ancient. Soc. Econ. Paleont. Mineral. Tulsa, Okla., Spec. Pub. 15.

ASQUITH, D. O. 1970. Depositional topography and major marine environments, late Cretaceous, Wyoming. Bull. Amer. Assoc. Petrol. Geol. 54: 1184–1224.
Detailed discussion of a complicated stratigraphy of shelf, slope, and basin clastic sediments.

BEALL, A. O., JR. 1968. Sedimentary processes operative along the western Louisiana shoreline. J. Sed. Petrology 38: 869–877.

BERG, ROBERT R. 1979. Oil and gas in delta-margin facies of Dakota sandstone, Lone Pine Field, New Mexico. Bull. Amer. Assoc. Petrol. Geol. 63 (6): 886–904.

COLEMAN, J. M. 1976. Deltas: processes of deposition and models for exploration. Continuing Education Pub. Co., Inc., Champaign, Ill. 102 p.

CURRAY, J. R. 1960. Sediments and history of Holocene transgression, continental shelf, northwest Gulf of Mexico, p. 221–266. *In* F. P. Shepard, F. B. Phleger, and T. H. van Andel (eds.), Recent sediments, northwest Gulf of Mexico. Amer. Assoc. Petrol. Geol., Tulsa, Okla.

DAILLY, G. C. 1976. Pendulum effect and Niger Delta prolific belt. Bull. Amer. Assoc. Petrol. Geol. 60 (9): 1543–1550.

DOLAN, ROBERT, BRUCE HAYDEN, and CONRAD JONES. 1979. Barrier island configuration. 204: 401–402.

EDWARDS, MARC B. 1981. Upper Wilcox Rosita delta system of south Texas: growth-faulted shelf-edge deltas. Bull. Amer. Assoc. Petrol. Geol. 65 (1): 54–73.

EJEDAWE, J. E. 1981. Patterns of incidence of oil reserves in Niger Delta Basin. Bull. Amer. Assoc. Petrol. Geol. 65 (9): 1574–1585.

ELLIOTT, T. 1977. The variability of modern river deltas. Scientific Progress 64: 215–227. Organizes delta variability as function of interaction among fluvial, wave, and tidal processes.

EMERY, K. O., and A. S. MERRILL. 1979. Relict oysters on the United States Atlantic continental shelf: a reconsideration of their usefulness in understanding late Quaternary sea-level history: discussion and reply. Bull. Geol. Soc. Amer. 90 (pt.1): 689–694.

EVAMY, D. D., J. HAREMBOURE, P. KAMERLING, W. A. KNAAP, F. A. MOLLOY, and P. H. ROWLANDS. 1978. Hydrocarbon habitat of Tertiary Niger Delta. Bull. Amer. Assoc. Petrol. Geol. 62: 1–39.

EVANS, G. 1965. Intertidal flat sediments and their environments of deposition in the wash. Geol. Soc. London Quart. J. 121: 209–245.

———. 1970. Coastal and nearshore sedimentation: a comparison of clastic and carbonate deposition. Geol. Assoc. Canada Proc. 81: 493–508.

FISHER, W. L., and K. H. MCGOWEN. 1969. Depositional systems in Wilcox Group (Eocene) of Texas and their relation to occurrence of oil and gas. Bull. Amer. Assoc. Petrol. Geol. 53: 30–54.
Recognition of deltaic systems in the subsurface and their importance to petroleum exploration.

FISK, H. N., and E. MCFARLAN, JR. 1955. Late Quaternary deltaic deposits of the Mississippi River, p. 279–302. *In* A. Poldervart (ed.), Crust of the earth. Geol. Soc. Amer. Spec. Paper 62.

FISK, H. N., E. MCFARLEN, C. R. KOLB, and L. J. WILBERT, JR. 1954. Sedimentary framework of the modern Mississippi delta. J. Sed. Petrology 24: 76–99.
Basic sediment geometries of the Mississippi delta.

FRAZIER, D. E., and A. OSANIK. 1969. Recent peat deposits—Louisiana coastal plain, p. 63–86. *In* E. C. Dapples and M. E. Hopkins (eds.), Environments of coal deposition. Geol. Soc. Amer. Spec. Pub. 14.
Note especially the detailed stratigraphy of the St. Bernard delta complex.

GOULD, H. R. 1970. The Mississippi delta complex, p. 3–30. *In* J. P. Morgan (ed.), Deltaic sedimentation, modern and ancient. Soc. Econ. Paleont. Mineral. Spec. Pub. 15.
Good general summary.

GOULD, H. R., and E. MCFARLAN, JR. 1959. Geologic history of the Chenier Plain, southwest Louisiana. Gulf Coast Assoc. Geol. Soc., Trans. 9: 261–270.

GREGORY, J. L. 1966. A Lower Oligocene delta in the submarine of southeastern Texas, p. 213–228. *In* M. L. Shirley and J. A. Ragsdale (eds.), Deltas in their geologic framework. Houston Geol. Soc. Houston, Tex.

HALBOUTY, M. T. 1969. Hidden treasures and subtle traps in the Gulf Coast. Bull. Amer. Assoc. Petrol. Geol. 53: 3–29.
A successful petroleum geologist discusses sedimentation.

HARRISON, ELLEN Z., and A. L. BLOOM. 1977. Sedimentation rates on tidal salt marshes in Connecticut. J. Sed. Petrology 47 (4): 1484–1490.

HINE, ALBERT C. 1979. Mechanisms of berm development and resulting beach growth along a barrier spit complex. Sedimentology, 26 (4): 333–352.

HOLEMAN, J. N. 1968. The sediment yield of major rivers of the world. Water Resources Res. 4: 737–747.

HOROWITZ, B. H. 1966. Evidence for deltaic origin of an Upper Ordovician sequence in the Central Appalachians, p. 159–170. *In* M. L. Shirley and J. A. Ragsdale (eds.), Deltas in their geologic framework. Houston Geol. Soc. Houston, Tex.

JOHNSON, K. G., and G. M. FRIEDMAN. 1969. The Tully clastic correlatives (Upper Devonian) of New York State: a model for recognition of alluvial, dune (?), tidal, nearshore (bar and lagoon), and offshore sedimentary environments in a tectonic delta complex. J. Sed. Petrology 39: 451–485.

KLEIN, G. DEV. 1970. Depositional and dispersal dynamics of intertidal sand bars. J. Sed. Petrology 40: 1095–1127.
Well-studied spectacular examples of intertidal primary structures in clastic sediments of Nova Scotia.

———. 1971. A sedimentary model for determining paleotidal range. Bull. Geol. Soc. Amer. 82: 2585–2592.

KOLB, C. R., and J. R. VAN LOPIK. 1966. Depositional environments of the Mississippi River deltaic plain—southeastern Louisiana, p. 17–62. *In* M. L. Shirley and J. A. Ragsdale (eds.), Deltas in their geologic framework. Houston Geol. Soc. Houston, Tex.

KRAFT, JOHN C., and CHACKO J. JOHN, 1979. Lateral and vertical facies relations of transgressive barrier. Bull. Amer. Assoc. Petrol. Geol. 63 (12): 2145–2163.

LANE, D. W. 1963. Sedimentary environments in Cretaceous Dakota sandstone in northwestern Colorado. Bull. Amer. Assoc. Petrol. Geol. 47: 229–256.
Primary structures suggest a transgressive sequence within the formation.

LEBLANC, R. J., and W. D. HODGSON. 1959. Origin and development of the Texas shoreline. Gulf Coast Assoc. Geol. Soc. Trans. 9: 197–220.

McCUBBIN, D. G. 1969. Cretaceous strike-valley sandstone reservoirs, northwestern New Mexico. Bull. Amer. Assoc. Petrol. Geol. 53: 2114–2140.
Stratigraphic analysis in an area approximately 30 miles northwest of Bisti field produces an interpretation of linear sand bodies that is grossly different from the interpretation of Sabins (1963). It is interesting to consider whether or not the two papers are talking about the same type of sand body.

McGregor, A. A., and C. A. Biggs. 1968. Belle Creek field, Montana: a rich stratigraphic trap. Bull. Amer. Assoc. Petrol. Geol. 52: 1869–1887.
Paleogeography and facies relationships were important in the discovery of this 200-million-barrel oil field.

Morgan, J. P. and R. H. Shaver (eds.). 1970. Deltaic sedimentation, Modern and Ancient. Soc. Econ. Paleont. Mineral. Spec. Pub. 15. 312 p.
Symposium volume.

Reineck, H. E. 1967. Layered sediments of tidal flats, beaches and shelf bottoms, p. 191–206. *In* G. H. Lauff (ed.), Estuaries. Amer. Assoc. Adv. Sci. Pub. 83.

Rice, Dudley D. 1980. Coastal and deltaic sedimentation of upper Cretaceous Eagle sandstone: relation to shallow gas accumulations, north-central Montana. Bull. Amer. Assoc. Petrol. Geol. 64 (3): 316–338.

Rich, J. L. 1951. Three critical environments of deposition and criteria for recognition of rocks deposited in each of them. Bull. Geol. Soc. Amer. 62: 1–20.
A landmark paper concerning the conceptual value of recognizing bottomset, foreset, and topset sediments in the stratigraphic record.

Roberts, H. H., R. D. Adams, and R. H. W. Cunningham. 1980. Evolution of sand-dominant subaerial phase, Atchafalaya delta, south-central Louisiana. Bull. Amer. Assoc. Petrol. Geol. 64 (2): 264–279.

Sabins, F. F., Jr. 1963. Anatomy of stratigraphic trap, Bisti field, New Mexico. Bull. Amer. Assoc. Petrol. Geol. 47: 193–228.

Scruton, P. C. 1960. Delta building and the deltaic sequence, p. 82–102. *In* F. P. Shepard, F. B. Phleger, and T. H. van Andel (eds.), Recent sediments, northwest Gulf of Mexico.

Shepard, F. P., F. B. Phleger, and T. H. van Andel (eds.). 1960a. Recent sediments, northwest Gulf of Mexico. Amer. Assoc. Petrol. Geol., Tulsa, Okla. 394 p.
Summary of extensive Recent sediment studies carried out between 1951 and 1958.

————. 1960b, Mississsippi delta: marginal environments, sediments and growth, p. 56–81. *In* F. P. Shepard *et al.* (eds.), Recent sediments, northwest Gulf of Mexico.

————. 1960c, Rise of sea level along northwest Gulf of Mexico, p. 338–344. *In* F. P. Shepard *et al.* (eds.), Recent sediments, northwest Gulf of Mexico.

Shirley, M. L., and J. A. Ragsdale (eds.). 1966. Deltas in their geologic framework. Houston Geol. Soc., Houston, Tex. 251 p.
Eleven papers dealing with Recent and Ancient deltaic sedimentation. Also contains vital statistics on 24 modern deltas.

Van Straaten, L. M. J. U. 1961. Sedimentation in tidal flat areas. J. Alberta Soc. Petrol. Geol. 9: 203–226.
Summary article with extensive bibliography.

Wilkinson, B. H., and R. A. Basse. 1978. Late Holocene history of the central Texas Coast from Galveston Island to Pass Cavallo. Bull. Geol. Soc. Amer. 89: 1592–1600.

WITHROW, P. C. 1968. Depositional environments of Pennsylvania red fork sandstone in northeastern Anadarko Basin, Oklahoma. Bull. Amer. Assoc. Petrol. Geol. 52: 1638–1654.
Oil exploration based on paleogeography and the depositional history of alluvial and coastal clastic sand bodies.

Carbonates
of the Shelf Margin
and Subtidal Shelf Interior

13

In the previous chapter, we dealt with clastic sedimentation from the coastline out onto the continental shelf. In the next two chapters, our discussion of chemical and biochemical sedimentation will deal with this same geographic area: the relatively shallow waters adjacent to land masses. The fundamental difference will be that *in situ* origin allows a greater variety of chemical and biochemical sediments than does clastic sedimentation, which depends on a source of sediment supply. Whereas clastic sediments are by and large inert objects *brought to* the depositional environment, chemical and biochemical and biochemical sediments are generally the *product of* the depositional environment.

When comparing carbonate sedimentation to clastic sedimentation, we may find it convenient to think of the carbonate sediments as being, in a very real sense, "alive." Clastic sediments are inanimate objects that must be acted upon by external forces; but chemical and biochemical sediments are self-generating, thereby needing no external energy source to deliver them to the sedimentary environment. Although external forces are still an important agent in carbonate sedimentation, local production of the sediment upon which these forces act is more important.

Perhaps the most striking example of the contrast between chemical and clastic sediments occurs at the outer margin of the continental shelf. Clastic sediments require a source of sediment and external forces to deliver that sediment to the depositional site. For clastics in general, therefore, the continental-shelf margin is a long way from the modern coastline, and the water depth over the major physiographic break is from continental shelf to continental slope is at least 100 meters. Thus, the shelf break is far from the source of clastic material. Clearly, the shelf break is occupied by relict sediments, that is, sediments deposited during the Wisconsin lowstand of the sea (from 10,000 to 20,000 years ago).

In contrast, carbonate sediments in this same physiographic position are self-generating. Coral reefs and oolite shoals commonly occupy the physiographic break from continental shelf to ocean basin. Because of this self-generating character of carbonate deposits, our study of carbonate sedimentation is in large part set free from considerations of sediment source and external forces. The water itself is a sufficient "source." Nor is there any need for physical forces to transport the material, because very large particles can accumulate essentially *in situ.*

Recent Carbonate Sedimentation

As in previous chapters, this discussion will be organized around the development of general principles rather than a description of regional sedimentation.

Because carbonate sediments form the major petroleum reservoirs of the world and because many complications have been encountered in attempting to locate new carbonate petroleum reservoirs, Recent carbonate sediments have been studied in great detail. By way of introduction, therefore, several of the principal areas of study of Recent carbonate sedimentation do deserve brief regional description.

The Bahama platform is a vast area of shallow subtidal carbonate sedimentation (Figure 13.1). Platforms rise abruptly out of several hundred meters of water and are constructed by the accumulation of carbonate and evaporite sediments at or near sea level during the continuing subsidence of this area since Cretaceous time. Oolite sedimentation predominates at the shelf margin, where daily tides cause rapid flow of water onto and off of the bank. The subtidal interior of the platform is the site of extensive carbonate-mud sedimentation. Portions of many of the larger islands, western Andros Island for example, are supratidal accumulations of Recent carbonate sediments.

South Florida is another area where carbonate sedimentation has beeen studied extensively (see Figures 10.7 and 13.2). The shelf margin facing the Florida Strait is the site of discontinous coral-reef development. Fifteen to 30 kilometers interior from, and parallel to, the platform margin, outcrops of Pleistocene limestone form the Florida Keys. The Florida Keys separate the south Florida reef tract from Florida Bay and appear to exert a pronounced effect on carbonate sedimentation in Florida Bay.

Along the Caribbean coast between Mexico and Honduras in Central America, the coral reefs of Belize (Figure 13.3) are the most extensive development of barrier reefs and normal salinity backreef lagoons to be found anywhere in the Western Hemisphere. Well-developed barrier reefs exist on the eastern edge of a shallow shelf marginal to the deep water of the Caribbean. Behind the barrier reefs is an extensive development of *in situ,* shallow-water, carbonate-skeletal sands referred to as the Barrier Platform. Interior to the shallow waters of the Barrier Platform, the Main Channel Lagoon is the site of moderately deep-water lime-mud and clay-mud sedimentation. The patch reefs rising out of 30 to 60 meters of water in the southern

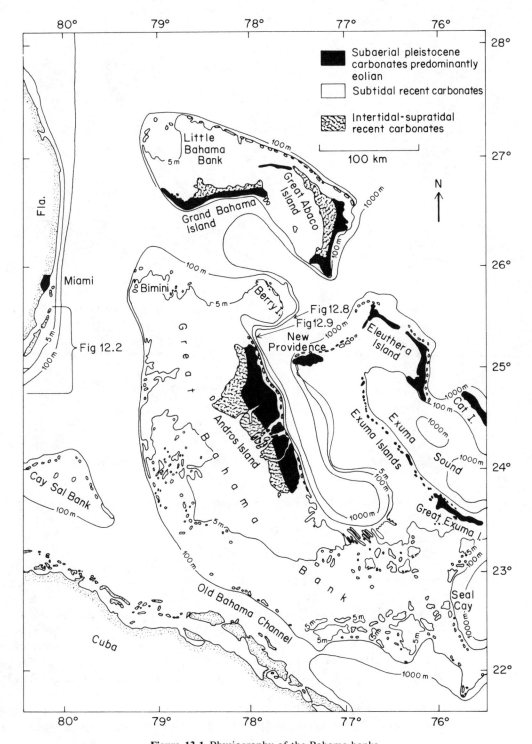

Figure 13.1 Physiography of the Bahama banks.

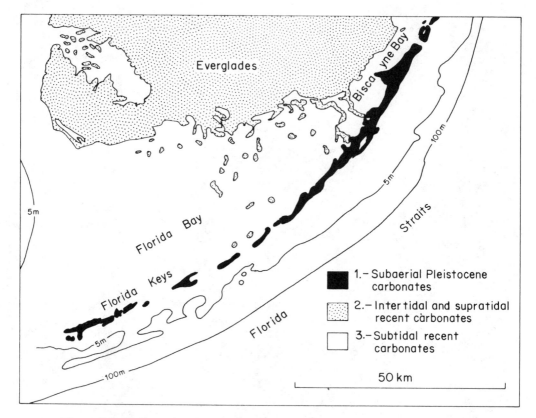

Figure 13.2 Physiography of environments with Recent carbonate sedimentation in south Florida. (After Ginsburg, 1964.)

portion of the Main Channel Lagoon are unique in the Western Hemisphere and have been carefully studied.

The Great Barrier Reef of eastern Australia is the largest such feature in the world, extending for 2000 kilometers along the eastern margin of the Australian continent and having a backreef province as much as 300 kilometers wide. Because of its immense size, sedimentation studies in this area have been predominantly reconnaissance in nature. Logistical problems concerning the study of such a large offshore area are substantial. For these reasons, we shall not incorporate data from the Great Barrier Reef into our general considerations of recent carbonate sedimentation. Rather, the area can be reserved as a test for models constructed from areas that have been studied in greater detail. For further discussion of the Great Barrier Reef, see especially Maxwell and Swinchatt (1970), and Stoddart and Yonge (1978a, 1978b).

In contrast, coral reefs off the north coast of Jamaica, though volumetrically insignificant, are of special interest because they have been examined in detail. This

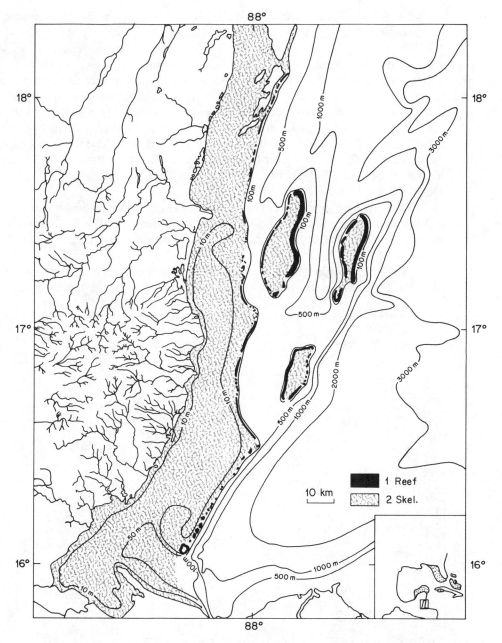

Figure 13.3 Physiography of regions with Recent carbonate sedimentation in Belize. Sediment types are as follows: (1) barrier reefs and associated skeletal sands; (2) subtidal shelf-interior sediments, predominantly skeletal carbonates, with significant clastic input along the mainland coast throughout the area shown in the southern two-thirds of the map.

region is where Tom Goreau pursued his pioneering studies on coral-reef communities and physiology (Goreau, 1959; Goreau and Land, 1972).

The Persian Gulf (Figure 13.4) interests us primarily because of its shallow water and increased salinities. In all of the areas outlined above, shallow-water carbonate sedimentation begins abruptly at a shelf margin that is adjacent to deep water. In the Persian Gulf, the transition from moderately deep water to shallow water is exceedingly gradational. Here we see a geometric condition that has undoubtedly occurred numerous times in the Ancient record but has not been duplicated in any of the well-studied areas just outlined. In addition, high evaporation rates and the virtual absence of rainfall make the entire Persian Gulf an area of increased salinities. This fact is a sharp contrast to high-rainfall areas such as south Florida.

With this brief and general introduction to several areas where carbonate sedimentation has been studied, let us take a more detailed look at the various components that we shall try to fit together into a general model.

The Shelf Margin Shoal. Our discussion of coastal clastic sediments logically began at the delta, where clastic sediments are brought to the marine environment.

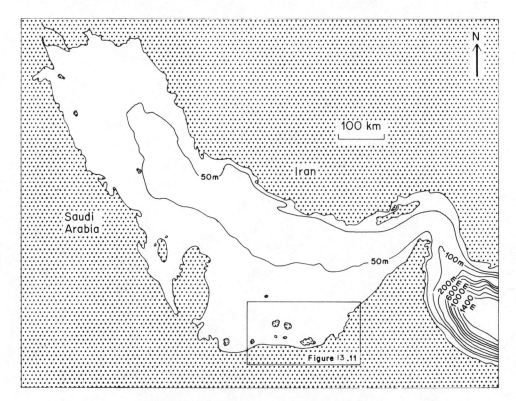

Figure 13.4 Physiography of the Persian Gulf.

Our discussion of carbonate sedimentation logically begins at the shelf margin, where the water from which the carbonates are to be precipitated first comes onto the shelf.

Two sediment types dominate the carbonate bank-margin environment: coral reefs and oolite mobile sand belts. We can gain important input for our general model from southern Belize, northern Jamaica, south Florida, the Bahamas, the Persian Gulf. The shelf margin of southern Belize provides a striking example of how coral-reef development can dominate a shelf margin (Figure 13.5). Let us begin at reef crest, first looking at the forereef environment and then looking at the back-reef environment.

Goreau (1959) provides an authoritative discussion of forereef coral zonation in Jamaica. Extensive reconnaissance of the Belize forereef region (James *et al.,* 1976; James and Ginsburg, 1979; and Rützler and Macintyre, 1982) confirm his observations and provide further detail concerning the deep forereef environment. From surf zone to approximately 5 meters, modern Caribbean coral reefs are dominated by *Acropora palmata,* the moose-horn coral. From 5 to 15 meters, forereef coral developments are predominantly *Acropora cervicornis,* the stag-horn coral, or *Montastrea annularis.* Within this zone, it is common for thick coral growth to alternate with channels through which reef-derived sands are transported to the deep-water forereef environment. Such an arrangement is shown in Figure 13.6. The elongate patches of reef development (dark in Figure 13.6) are topographically high

Figure 13.5 Oblique aerial photograph of the Belize barrier reefs. Deep water of the Caribbean is to the right and the shallow carbonate environments of the Barrier Platform are to the left.

Figure 13.6 Oblique aerial photograph showing spur and groove structures in the forereef of the Belize barrier reef.

and are referred to as *spurs*. The channels down which reef-derived carbonate-skeletal debris is transported (white streaks in the middle of Figure 13.6) are called *grooves*.

At a water depth of approximately 15 meters, there is a transition from the *A. cervicornis* and/or *M. annularis* buttress zone to the deep-water, coral-head zone. Whereas shallower zones of the reef are dominated by single species or few species, the deep-water coral-head zone accommodates a great variety of corals. Approximately 60 meters water depth is the lower limit of active coral growth in the forereef of Belize.

From approximately 60 meters to 110 meters, there is a vertical escarpment, largely cut into Pleistocene coral-reef materials by the 18,000 B.P. lowstand of the sea. Below approximately 110 meters, there begins a reef-debris talus slope that extends down to hundreds of meters. Coarse rubble dominates the upper portions of the talus slope; coral-derived sands, the mid-portion; and lime mud, below 200 meters. This region is presumably the final resting place for the coral sands that we saw moving down the grooves of the upper forereef region.

While it is technologically difficult to estimate the volume of reef-derived skeletal debris that has accumulated on the talus slope since the beginning of the Holocene, the number is clearly rather large. The mere presence of the sand-floored grooves indicates a rather active process. If transportation of sediment down these grooves were not significant, coral growth would colonize the channel floors. A bit later in this chapter, we shall derive some estimates of potential shallow-water vertical carbonate accumulation rates. It is tempting to suggest that these rates of sediment production, in a volumetric sense, continue whether sea level is rising or not. If sea level is rising, the material accumulates *in situ;* if sea level remains constant, the calcium carbonate deposited by coral skeletal growth simply becomes ground up into sand and mud and is transported down the forereef.

Turning our attention now to the area behind the reef crust, we note abundant evidence of reef-derived material that has been transported into the immediate back-reef area. In Figure 13.5, the predominantly white area nearest the bank margin is the reef flat. This zone is composed almost entirely of cobbles and boulders of *A. palmata,* ripped loose from the reef-crest zone and thrown back by wave action. Immediately behind the reef flat is the sand apron, which consists almost entirely of reef-derived skeletal sand that is transported only a slight distance back onto the Barrier Platform. The sand bottom in this environment is constantly shifting. Thus, no marine grasses or other attached forms can become established.

The shelf margin in southern Florida interests us because it is somewhat different from the Belize example and because available data allow us to say more about the forereef environment.

Coral reefs occupy the shelf-margin position in southern Florida, but they are by no means as continuous as the barrier reefs of southern Belize. Let us look first at forereef sedimentation.

In south Florida, limited data are available concerning sediment in the forereef. There is a pronounced tendency for south Florida forereef samples to become muddy with increasing depth (see Figure 13.7). Indeed, the absence of carbonate mud appears to be a striking characteristic of the shelf-margin environment. In front of the shelf margin, forereef sediments contain mud. Behind the shelf-margin environment, shelf-interior sediments tend to contain considerable mud. The south Florida reefs are not such an obvious source of sediment supply for the backreef area as the reefs are in Belize. At Key Largo Dry Rocks, for example, a *A. cervicornis* thicket grows profusely immediately behind the *A. palmata* reef-crest community. In Belize, we saw this geographic position occupied by the reef flat, a debris pile derived from the *A. palmata* community. Thus, in the absence of a large accumulation of reef-derived detritus, quiet-water coral communites may predominate in the area immediately behind the reef-crest community.

Similarly, reef-associated sands show no obvious geometric relationship to shelf-margin reefs in south Florida. Sand composition indicates a significant portion of the skeletal material probably originated on the reef (Figure 13.7). Yet sediment supply is not great, and there is ample time for reworking of sand bodies into their own discrete geometries.

Most of the Great Bahama Bank (Figure 13.1) exemplifies a completely different type of shelf-margin sedimentation. In many areas, coral reefs are substantially absent from the shelf margin. Usually, there are only occasional coral heads and coralline algae on the foreslopes of the shelf.

The predominant sediment type of this shelf-margin environment is the oolite shoal (Figure 13.8). The daily tides have currents that flow rapidly onto and off of

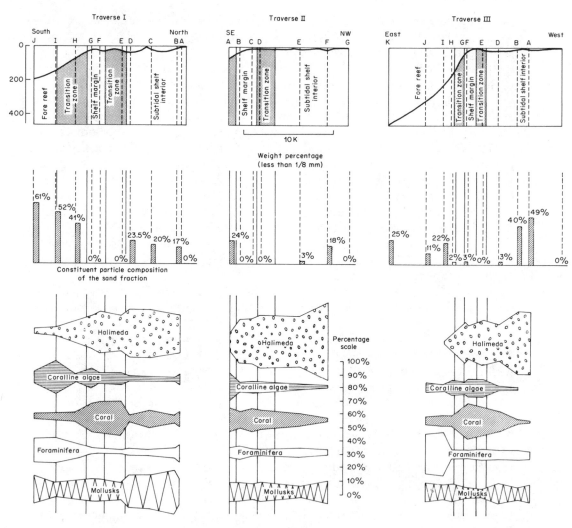

Figure 13.7 Composition by grain size and constituent particles of carbonate sediments in the Florida reef tract. (After Ginsburg, 1956.)

the shelf margin. The oolite sand is locally transported by these currents. As a result, the oolite-shoal environment has two characteristics. First, continual resuspension of sand grains provides opportunity for additional oolitic laminae to grow around each grain by chemical or biochemical precipitation of aragonite. Second, the constant movement of sand results in such bottom instability that benthonic organisms cannot flourish. Sessile forms find no place to attach themselves; burrowing organisms are continually uprooted by the erosion of sand from around their burrows. Thus, it is very difficult for any sedimentological succession to occur in the oolite-shoal environment. So long as new fauna and flora cannot establish themselves, the

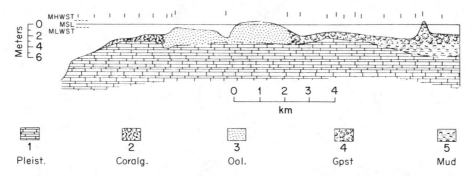

Figure 13.8 Cross section from shelf margin to shelf interior. Frazer's Hog Cay area, Great Bahama Bank. Sediment types are as follows: (1) Pleistocene and older carbonate rocks, (2) coralgal facies, (3) oolite sandbars, (4) grapestone facies,(5) pellet mud facies. (After Purdy and Imbrie, 1964, and Buchanan, 1970, with modification.)

environment remains a mobile sand belt. So long as the environment remains a mobile sand belt, new fauna and flora cannot establish themselves. This system will not change unless something changes in front of the mobile sand belt.

Observations in the Jolters Cay area of the Bahamas (Figure 13.9) indicate that changes can ocur in front of oolite sand belts that may have significant potential for breaking the mobile sand belt syndrome described above. Specifically, if there is enough shallow platform in front of the oolite mobile sand belt, coral reefs may establish themselves seaward of the oolites and dissipate some of the wave and tidal energy that formerly went into maintaining the mobile sand belt. Reef-building corals have a greater depth tolerance than do oolite shoals. Coral communities can establish themselves in 15 meters of water and grow upward to sea level. This growth appears to be happening in the Jolters Cay area today, for example. Oolite shoals are extensive and located in very shallow water where the shelf-margin topography closely approaches modern sea level. Coral reefs out in front seem to be just getting started; there is no reef-flat debris accumulation nor obvious abundance of reef-associated sand. With continued reef development, the oolite-shoal environment will become a relatively quiet-water environment, thus allowing sedentary benthic organisms to colonize the region and thus generate a sedimentological succession capping the mobile sand belt.

Another interesting example of the interplay between a coral reef and mobile sand belt has been documented on the western margin of the Little Bahama Bank (Hine and Neumann, 1977). Here, coral reefs established themselves 40 meters below present sea level at approximately 10,000 B.P., during the Holocene sea-level rise. So long as the Little Bahama Bank was an exposed island of Pleistocene rock, conditions on the west side of the bank were suitable for coral-reef growth. However, when the bank became flooded, the reefs were killed and off-bank sand transport came to dominate the region. Precisely what killed the reefs is a subject for speculation. Intermittently high salinity conditions in the shallow-water bank interior might produce a water inhospitable to corals; likewise, winter temperatures

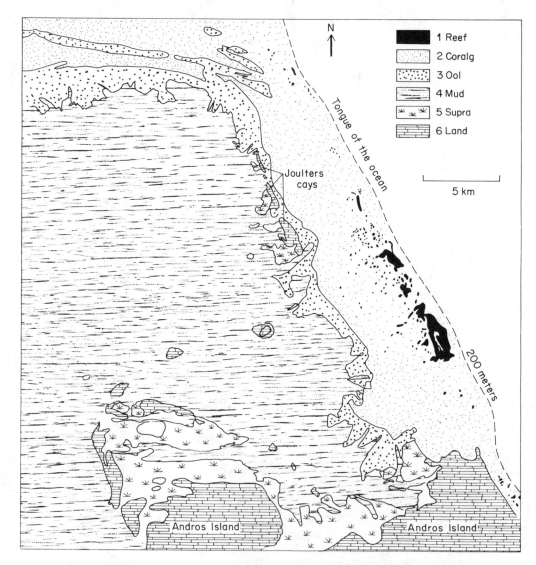

Figure 13.9 Map showing Recent sedimentary facies in the Joulters Cay area, Great Bahama Bank. Sediment types are as follows: (1) coral reefs, nearly awash at low tide; (2) thin veneer of coralgal sand over Pleistocene rock; (3) oolite shoals, awash at low tide; (4) subtidal platform-interior deposits, mostly carbonate mud; (5) supratidal mud flats; (6) land, Recent or Pleistocene in age. (After Purdy and Imbrie, 1964, with modification.)

of bank water fall considerably below that of open-marine water nearby. Alternatively, sediment produced on the interior shelf may have rendered the bank-margin environment too turbid for continued reef growth.

In summary, shelf-margin mobile sand belts and coral reefs have different sets of requirements for their origin. However, there appear to be feedback mechanisms that allow for stratigraphic interplay of these two types of shelf-margin shoals. If vigorous reef growth is established on rising sea level, it will dominate the shelf margin unless something happens to kill it. In the absence of shelf-margin reefs, mobile sand belts will likely dominate the shelf margin; the mobility of the sand bottom will preclude direct colonization by attached organisms. However, with the passage of time, the mobile sand belts may become a sufficient barrier to outflow of shelf-interior water that coral reefs may reestablish themselves in front of the mobile belts. Such a scenario then decreases the energy available to the former mobile sand belt and with time the shelf-margin coral reefs are once again at risk from exposure to shelf-interior salinity-temperature hazards.

The Shelf-Margin Ramp. In all of the above examples, the post-Wisconsin sea-level rise has flooded on to a shallow carbonate shelf; the shallow shelf margin is juxtaposed to relatively deep water. Presumably, the existence of well-developed topographic prominence at the shelf margin is itself related to shelf-margin sedimentation associated with previous highstands of the sea. The Campeche Bank of southeastern Mexico (Figure 13.10) provides an interesting example of a shelf margin on which carbonate sedimentation has not kept pace with highstands of the sea.

Here the major topographic break from shelf to slope lies at about 70 to 100 meters below present sea level. With the post-Wisconsin eustatic sea-level rise, only a few isolated coral banks managed to keep pace with the rising water. Over the majority of the shelf, shallow-water carbonate sedimentation could not develop as sea level rose. Consequently, the bottom layer over most of the bank today is a thin veneer of relict shallow-water carbonate sediments combined with deep-water carbonate sediments of the modern environment. The result is some rather unusual sediment types. Unless we have studied an area like Campeche Bank, we may find it rather difficult to conjure up an explanation for an oomicrite with abundant pelagic foraminifera, for example.

To round out our discussion of the shelf-margin environment, let us now look at the southern shore of the Persian Gulf (Figures 13.4 and 13.11). First, note that the shelf-margin geometric relationships are completely different from anything we have looked at earlier. In our previous examples, shelf-margin facies are very close to deep water. In the Persian Gulf, the shelf margin is just a slight wrinkle in a bathymetric surface where shallow water extends basinward for tons of kilometers. In short, the steep geometries of our previous forereef environments have been replaced by the very low-relief, shallow, subtidal, open-marine environment. Such a forereef geometry must have existed numerous times in the stratigraphic record however. There are a lot of places in the continental United States where there simply could not have been enough space available for the development of the precipitous forereefs that characterize most carbonate shelves today.

In the Persian Gulf, reefs and subtidal oolite shoals blend together to form a shelf-margin complex. Reefs predominate in front of island, and oolite shoals pre-

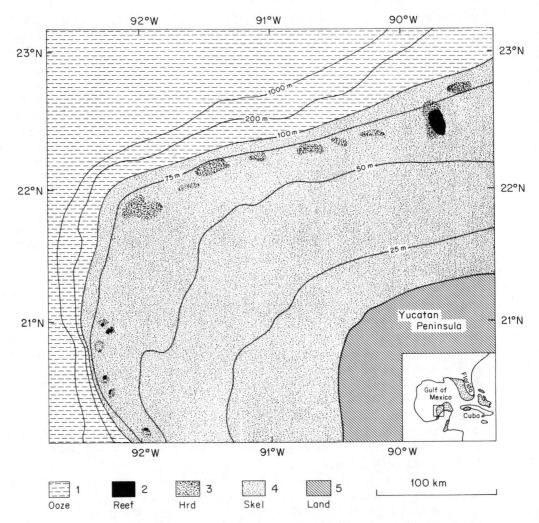

Figure 13.10 General facies and physiography of the western portion of Campeche Bank. Sediment types are as follows: (1) pelagic ooze; (2) shallow-water coral reefs; (3) deeper-water hardbank communities; (4) thin veneer of Recent carbonate sediment, predominantly skeletal sand, (5) land. (After Logan *et al.,* 1969.)

dominate in the tidal channels between islands. Eolian dunes on many of the islands are composed of Recent oolite with a marine origin. The dunes are an interesting example of a topset facies in close association with shelf-margin sediments.

The Subtidal Shelf Interior. The shelf-margin environment marks the transition from relatively deep water to the relatively shallow water of the carbonate shelf. Sedimentation on the shelf interior varies greatly as a function of water circulation over the shelf. The very fact that the shelf is generally shallow impedes circulation

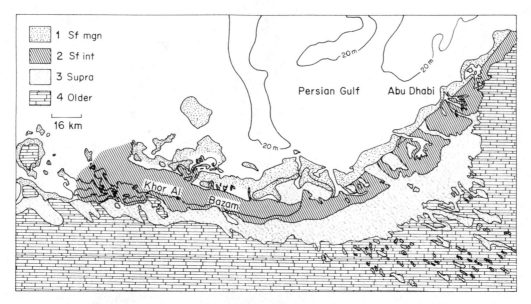

Figure 13.11 Physiography and sedimentary facies of a portion of the southern Persian Gulf. Sediment types are as follows: (1) high-energy shelf-margin facies, includes coral reefs, oolites, and eolian sands; (2) subtidal shelf-interior, predominantly pelletal, lime mud with some nonskeletal carbonate sands; (3) intertidal-supratidal deposits, from algal stromatolites to sabkha evaporites; (4) older rock. (After Kendall and Skipwith, 1969, with modification.) Courtesy Geological Society of America.

to some extent. Furthermore, the shallow shelf-interior water may be altered considerably by either freshwater input or by net evaporation.

Four examples from the Recent epoch will help us understand the variability of subtidal sedimentation of the shelf interior. In southern Belize, normal Caribbean seawater passes over the Barrier Platform and produces a net flow to the south in the Main Channel Lagoon. Reduced salinities are restricted to coastal areas; the majority of the Main Channel Lagoon and all of the Barrier Platform are left with normal salinity environments. In south Florida, the existence of the Forida Keys restricts circulation both in the inner reef tract and in Florida Bay. Furthermore, net evaporation alternates with the input of fresh water from the Everglades and so causes large fluctuations in Florida Bay salinity. In the Bahamas, shelf-interior circulation is somewhat impeded by both the sheer size of the platform and the "shadow effect" of Andros Island. Salinities that are 20% or 30% above normal marine are common here. The shallow lagoons along the southwest coast of the Persian Gulf are the site of gross net evaporation. Salinities two or three times normal marine are common in the subtidal environments. Let us examine each of these examples in more detail.

Figure 13.12 presents a map and cross section of a portion of the southern Belize shelf interior. The shelf interior of Belize exhibits a large amount of topo-

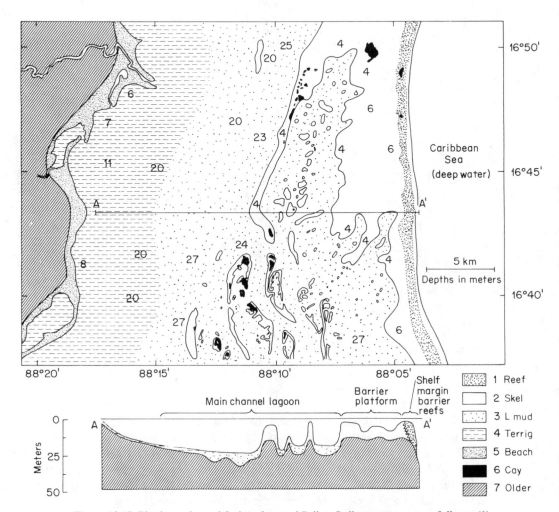

Figure 13.12 Physiography and facies of central Belize. Sediment types are as follows: (1) shelf-margin coral reefs and clean reef-derived sands; (2) shelf-interior, *in situ,* skeletal sands; (3) shelf-interior, deep-water, lime muds; (4) terrigenous mud; (5) near shore to subaerial coastal clastic sediments; (6) cays, predominantly mangrove swamp; (7) older rock.

graphic relief. In particular, the deep water of the Main Channel Lagoon is a feature that will be lacking in examples of more restricted shelf interiors.

Undoubtedly, the existence of the Main Channel Lagoon plays an important role in maintaining normal marine salinities throughout the southern Belize shelf interior. This lagoon serves as a conduit for net southward flow of water brought onto the shelf during flood tide. Furthermore, net southward flow in the Main Channel Lagoon serves to hold freshwater discharge from the mainland close to the shore. Only in the extreme southern portions of the Belize reef tracts do freshwater wedges begin to impinge on the reef environments. Finally, the Main Channel Lagoon serves as a settling basin to separate carbonate sedimentation of the Barrier Platform from clatic sedimentation of the coastal areas. Without this settling basin, clastic sediments would undoubtedly exert a harmful influence on Barrier Platform carbonate sedimentation.

Distribution of sediment types on the southern Belize shelf interior is related to bathymetry, proximity to carbonate shoals, and proximity to the mainland coast. The Barrier Platform sediments are predominantly clean skeletal sands composed in large part of the green algae *Halimeda,* molluscs, porcellaneous foraminifera, hyaline foraminifera, and lesser amounts of coraland coralline algae. This sediment is apparently forming *in situ* on the Barrier Platform and contains only minimal contributions of transported skeletal detritus from the barrier-reef shelf-margin environment. Significant accumulation of skeletal detritus from the barrier reef seems to be limited to the reef flat and sand apron (discussed earlier).

Sediments of the Main Channel Lagoon are predominantly muds: carbonate mud near the Barrier Platform, sandy clay mud near the mainland coast, and a spectrum of carbonate-clay mud in between. The Main Channel Lagoon has its own distinctive fauna, consisting primarily of thin-shelled molluscs and hyaline benthonic foraminifera. By and large, the water depth precludes a benthonic flora. The bulk of the lime mud is apparently generated by skeletal comminution in the barrier-reef and Barrier Platform environments and is transported westward to its final resting place in the deeper-water settling basin that is the Main Channel Lagoon (Matthews, 1966).

Platform-interior sediments of south Florida offer an interesting contrast to those of southern Belize (see Figure 13.13). Physiographic differences between the two areas undoubtedly sets the stage for the sedimentological differences. To begin with, the south Florida shelf possesses no main channel lagoon, nor does it have any built-in mechanism to control the net flow of water across the shelf. Replenishment of the very shallow water of the shelf interior under normal circumstances must be accomplished by oscillatory tidal flow from the shelf margin. The sheer size of the shelf may place serious limitations on this process.

Communication between the shelf interior and the Florida Strait is further inhibited by the existence of the Florida Keys. For instance, faunas of the inner reef tract are found in Florida Bay only at the various passes through the Florida Keys (Figure 13.13 and 13.14). If the keys were not there, inner reef tract faunas and Florida Bay faunas would seek their own equilibrium positions, which would undoubtedly be different from the positions now governed by the existence of the keys.

It is convenient to discuss southern Florida sedimentation in terms of three major environments. First, the outer reef tract is the shelf-margin environment just discussed. Second, the inner reef tract is that belt of sediment between the shelf-margin environment and the Florida Keys. Third, Florida Bay is that triangular area of shallow-water sedimentation between the Florida Keys on the southeast, the Gulf of Mexico on the west, and the Florida Everglades on the north.

The south Florida inner reef tract is more or less analogous to the barrier-

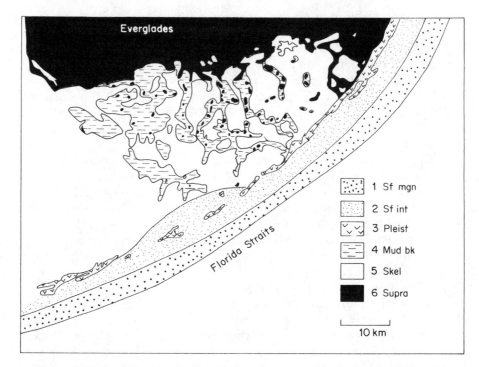

Figure 13.13 Generalized facies of south Florida. Sediment types are as follows: (1) shelf-margin coral reefs and clean reef-derived sands; (2) shelf-interior muddy skeletal sands; (3) Pleistocene key rock; (4) grass-covered, carbonate mudbanks, awash at low tide; (5) thin veneer of skeletal sand; (6) intertidal-supratidal deposits, commonly mangrove swamp.

platform environment of southern Belize. The major distinction is the tendency of lime mud to build up in the shallow waters of the inner reef tract; similar environments in southern Belize are predominantly clean skeletal sands. The accumulation of mud within the inner reef tract results simply from the fact that there is no place for the mud to go. In the absence of a well-developed current system on the shelf, mud transported to this area by flood tides simply accumulates here.

Some consideration should also be given to the role of marine grasses *(Thallasia)* in the stabilization of mud in shallow water. Scientists have observed that sediment samples from patches of *Thallasia* generally have more mud than samples from sediment-floored areas nearby. It can be argued that the blades of *Thallasia* tend to create a less agitated environment near the sediment-water interface and thereby encourage the accumulation of fine-grained sediments on the *Thallasia*. Note, however, that the inevitable question arises: Is the mud there because the *Thallasia* came first, or is the *Thallasia* there beacause it prefers to grow on a muddy bottom?

An alternate proposal for the sedimentology of carbonate-mud accumulation simply considers pelleted mud to be a soft sand particle (Ginsburg, 1969). The stuff is as gooey as we would want a mud to be, yet the hydrologic behavior of the

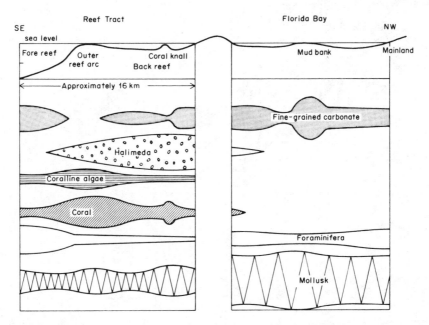

Figure 13.14 Generalized variations in sediment grain size and constituent composition from the south Florida reef tracts into Florida Bay. (After Ginsburg, 1956.)

individual soft pellets is that of a sand-sized particle. In this fashion, we can argue that a shallow-water mudbank accumulating under agitated conditions is simply an accumulation of coarse-grained "soft sand."

Crossing over the Florida Keys and entering Florida Bay, we enter a totally different situation (Figures 13.13 and 13.14). This region is a restricted shallow-water shelf interior with a large potential source of fresh water along its northern boundary. The combination of a restricted shallow-water shelf interior and a potential freshwater input makes Florida Bay a highly variable environment (Enos and Perkins, 1979). As indicated in Figure 13.15, salinity within the bay may vary from hypersaline to brackish. Salinity variations within the bay render this environment hospitable only to certain molluscs, foraminifera, incrusting red algae and soft-bodied bryozoa. Water depths are generally 2 meters or less and an intricate pattern of *Thallasia* mudbanks rise essentially to sea level. Intervening areas between mudbanks are flooded by approximately 15 centimeters of mollusc-foraminifera skeletal sand.

The origin of carbonate mud in shelf-interior environments has been a subject of considerable controversy. On the Florida shelf, it has been rather conclusively demonstrated that a single genus of calcified green algae, *Penicillus,* can produce the vast majority of the carbonate mud observed in these environments (Stockman *et al.,* 1967). By gathering data on the weight of aragonite needles in single specimens of *Penicillus,* the population density of *Penicullus,* and the turnover rate (life cycle) of *Penicullus*, we can estimate the rate of production of aragonite-needle

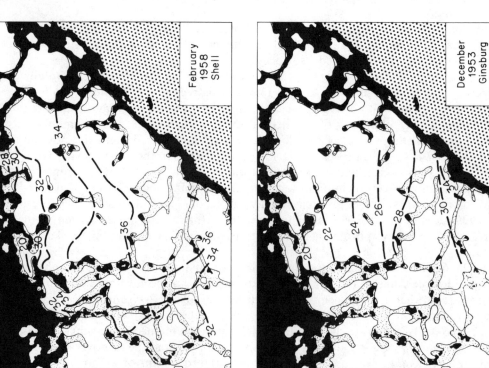

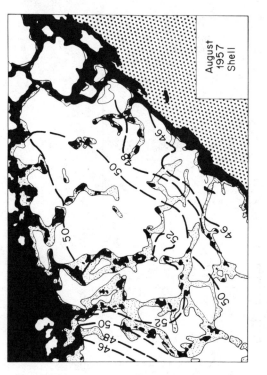

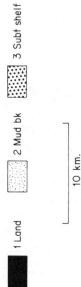

Salinity distribution in Florida Bay

30 — Isohaline (in parts per thousands)
Normal salinity of sea water in this area is 36–37 0/00

1 Land   2 Mud bk   3 Subt shelf

10 km.

August 1957 Shell

February 1958 Shell

December 1953 Ginsburg

Figure 13.15 Salinity variations in Florida Bay. Sediment types are as follows: (1) Pleistocene rock and supratidal swamp areas; (2) shallow, subtidal, grass-covered mudbanks; (3) subtidal platform-interior, Atlantic province. Open areas are Pleistocene rock surface overlain by less than 6 inches of mollusc foraminiferal sand. (After Ginsburg, 1964, with modification.)

carbonate mud. This rate, applied uniformly since the time when this area became flooded by the post-Wisconsin sea-level rise, is sufficient to account for the vast majority of mud accumulated in the shelf-interior environments of southern Florida.

In the Great Bahama Bank, (Fig. 13.16) we can examine shelf-interior sedimentation in a physiographic setting quite similar to south Florida but with two important differences: (1) Shelf-margin oolite shoals and Pleistocene outcrops (like Andros Island) provide greater restriction of the platform interior; and (2) there is

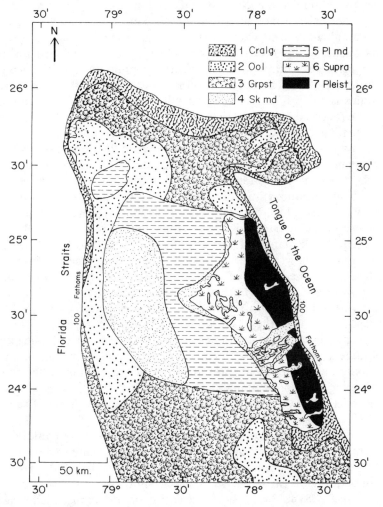

Figure 13.16 Generalized facies map of the Great Bahama Bank. Sediment types are as follows: (1) coralgal sand facies; (2) oolitic facies, mostly oolite tidal bars; (3) grapestone sand facies; (4) skeletal-mud facies; (5) pellet-mud facies; (6) supratidal carbonates; (7) lithified Pleistocene eolian sands. (After Imbrie and Purdy, 1962, with modification.)

no large source of fresh water, such as the Florida Everglades. Consequently, there is a tendency for the shelf-interior environment to have increased salinity.

Recent sediment thickness over the shelf interior is thin, averaging around 1 meter. The sediment of the skeletal-mud facies is similar to the muddy skeletal sands of the inner reef tract of south Florida. However, sediment types that predominate in the shelf interior are rather different from anything discussed thus far. In the pellet-mud facies, a large percentage of the sand fraction consists of lithified pelleted mud. These pellets are presumably similar in origin to the "soft sand" discussed earlier. Lithification resulted from some combination of increased salinity and a relatively long residence time at the sediment-water interface.

The grapestone facies is a relatively clean sand in which grapestone grains are predominantly sand-sized particles. Grapestone grains are aggregates of preexisting carbonate grains held together by chemical or biochemical carbonate cement. The original sedimentary particles are most commonly recrystallized oolite grains or indurated pellets. The cementation process by which they are united into grapestone grains is a combination of biochemical cementation by foraminifera of blue-green algae followed by physical-chemical recrystallization and precipitation of additional cementing material. The abundance of these sediment types in the shelf interior of the Bahamas can be attributed to some combination of increased salinity and long residence time at the sediment-water interface. Indeed, it has been proposed that the grapestone facies represents a surface of nondeposition on which no new grains are being produced; just old grains are being cemented together (Winland and Matthews, 1974). Thus, the grapestone facies may represent a modern, shallow, subtidal environment of essentially zero sedimentation rate.

In the Persian Gulf, the extremely low-relief profile of coastal sedimentary environments has resulted in substantially reducing the areal importance of subtidal shelf-interior sediments. The large area that was once shallow, subtidal shelf interior is now occupied by supratidal sediments of the sabkha environment. Subtidal shelf-interior sedimentation in this area today restricted to relatively small, shallow lagoons. Salinity of these lagoons is typically two or three times normal marine values. In some lagoons, aragonite-needle lime mud and pelleted lime mud are predominant sediment types. In other lagoons, skeletal sands have become well lithified in the submarine environment, presumably by a process somewhat akin to the formation of grapestone in the Bahamas. Here, the process appears to have gone to ultimate completion: the formation of laterally continuous, lithified sand layers.

Topographic Control of Recent Carbonate Sedimentation. We have already noted a large-scale topographic control of carbonate sedimentation, the shelf-margin environment. Let us now look closer at some of the second-order effects of topography on carbonate-sediment accumulation.

In Biscayne Bay, Florida, scientists have demonstrated that preexisting topography on the underlying Pleistocene carbonate rock serves to modify patterns of tidal flow and thereby accounts for the geometric configuration of Recent carbonate-bank deposits (Wanless, 1970). Where Pleistocene topographic highs form islands, strong tidal jets exist between them. Elongate carbonate banks are thus generated perpendicular to the shelf margin. Elsewhere, the Pleistocene surface forms a continuous subtidal lip at the eastern margin of the bay. Here tidal velocities are increased over the shallow water of the lip and abruptly decreased in the deeper water immediately behind the lip. Sediment dumping associated with this decrease in velocity has resulted in the accumulation of a sediment bar running more or less parallel to the Pleistocene topographic prominence.

In southern Belize, the distribution of carbonate shoals within the Main Channel Lagoon is controlled by preexisting topography. As indicated in Figure 13.12, there is a good correlation between carbonate-shoal areas in the lagoon and preexisting topography. Figures 13.17 and 13.18 show that much of this preexisting to-

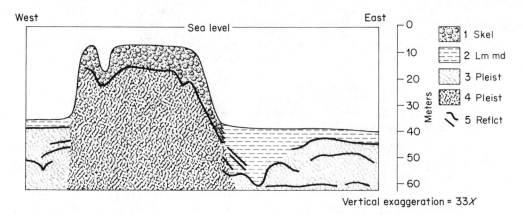

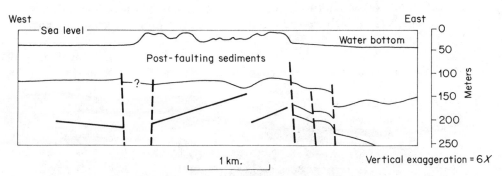

Figure 13.17 Reflection seismic profiles across a shoal area in Main Channel Lagoon, southern Belize shelf. The upper profile records relatively shallow information, and the lower profile records deeper information in approximately the same area. Sediment types are as follows: (1) shelf-interior skeletal sands and patch reefs; (2) shelf-interior, deep-water, lime muds; (3) Pleistocene clastic sediments; (4) Pleistocene limestone; (5) prominent reflecting horizons. In the upper profile, note that the preexisting topography determines Recent distribution of sand facies and mud facies. Data in the lower profile suggest that the preexisting topography was in turn the result of previous differential sedimentation ultimately initiated on fault blocks. (After Purdy and Matthews, 1964; and Purdy, 1974, with modification.)

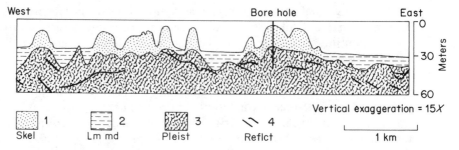

West   Bore hole   East

Figure 13.18 Additional example of topographic control of Recent carbonate sedimentation, Main Channel Lagoon, southern Belize. Sediment types are as follows: (1) shelf-
interior skeletal sands and patch reefs; (2) shelf-interior, deep-water, lime mud; (3) Pleistocene or older clastic sediments; (4) prominent reflecting horizons. In this case, carbonate
shoals have developed on the topography related to a breached anticlinal fold. The borehole recovered soft, orange, terrestrial clay from approximately 40 meters. Other seismic
data in the area confirm the existence of this broad anticlinal feature. (After Purdy and
Matthews, 1964; and Purdy, 1972, with modification.)

pography is directly related to underlying structural features that have been
subsequently modified by subaerial geomorphologic processes.

It is interesting to note that preexisting topographic highs and carbonate-shoal
sedimentation apparently become interrelated during times of sea-level fluctuation,
such as exemplified by the Pleistocene and Recent epochs. During a period of highstand sedimentation, the carbonate shoals generate more sediment than the surrounding deep-water environment. With subaerial exposure during a lowstand, the
carbonate-bank deposits of the topographic highs are permeable and thus allow
rainfall to pass through them freely. Cavernous porosity may develop, but the carbonate-bank facies will nevertheless withstand the subaerial exposure and become
a well-cemented cavernous limestone. With the next highstand of the sea, the hard
cavernous limestones of the bank facies remain as topographic highs for rejuvenation of bank facies sedimentation (Purdy, 1974; Bloom, 1974).

Recent Carbonate Sedimentation Rates. Estimation of Recent carbonate sedimentation rates can be made from three general sources: (1) from the study of living
organisms in Recent environments, (2) from carbon-14 dating of samples buried
beneath a known thickness of sediment, or (3) from the thickness of Recent sediments overlying the Pleistocene surface, wherever the contact between Recent and
Pleistocene can be well defined.

Studies of lime-mud production by the green algae *Penicillus* provide the most
complete example of directly determining sedimentation rates from living organisms
(Stockman *et al.,* 1967; Newman and Land, 1975). *Penicillus* colonies stand 5 to
7.5 centimeters tall and are easily visible to the skin diver surveying the bottom of
a carbonate environment. Measurement of sediment production rates for *Penicillus*
consists of three steps. First, establish the weight of aragonite sediment produced
by a single specimen. Simply collect a large number of specimens and determine the

average weight of aragonite per specimen. Next, estimate the population density of *Penicillus* within the environment. At any one time, how many plants are growing per square meter of sea floor? Finally, establish the life cycle of *Penicillus*. How long does it take an average specimen to grow to full size and die, thus releasing its aragonite to the sediment? Having these numbers, calculate the weight of aragonite sediment produced by *Penicillus* per square meter per year. Assumptions concerning porosity of the sediment allow translation of these weight figures into sediment volume figures. Resultant estimates of net lime-mud sedimentation rate range from 0.3 to 3.0 centimeters per 1000 years in the southern Florida area. In Bight of Abaco, Bahamas, the observed rate of accumulation is 12 centimeters per 1000 years, but the calculated production rate is 1.5 to 3 times the mass of aragonite sediment presently observed in Holocene deposits on the bank. Presumably, the bank is an important source of sediment to the nearby off-bank deep-water environment.

A similar calculation should be possible for sediment production by corals. Calcification rates are fairly well-known and indicate a potential for extremely rapid sediment accumulation (see Hoffmeister and Multer, 1964, for example). However, the population density of corals and the life cycle of individual colonies are not known in sufficient detail to complete calculation.

An alternative approach to estimation of carbonate sedimentation rates is to obtain a sample buried beneath a measured thickness of sediment and determine its age by carbon-14 analysis. A comprehensive study of this sort is contained in Purdy (1974). Analyzing fragments of coral taken from boreholes through islands in Belize, minimum estimates of barrier platform sedimentation rate range from 0.7 to 4.2 meters per 1000 years. Similarly, by analyzing peats recovered from the base of Belize shelf-lagoon sediments by piston cores rates of 10 to 30 centimeters per 1000 years were obtained. In the forereef of Jamaica, Goreau and Land (1972) report that numerous attempts to quarry into the face of forereef sediments yield ages of approximately 3000 B.P., about 1 meter below the present sediment-water interface. Thus, it appears that the forereef slope in building outward at a rate of approximately 30 centimeters per 1000 years.

As a third alternative to the estimation of Holocene carbonate sedimentation rates, the contact between Recent sediment and the underlying Pleistocene surface can be mapped by reflection seismic techniques or by probing through the soft sediment with a steel rod. These methods give the thickness of Recent sediment accumulation over the Pleistocene surface and the depth to the Pleistocene surface. This data can be combined with estimates of the post-Wisconsin sea-level rise to give an indication of the amount of sediment that has accumulated since the post-Wisconsin sea-level rise first flooded this area. Such estimates for Florida Bay and the inner reef tract indicate net sedimentation rates of 20 to 35 centimeters per 1000 years (Stockman *et al.,* 1967). Similar techniques in the Bahamas yield sedimentation rates from 45 centimeters per 1000 years for oolite shoals down to 10 centimeters per 1000 years for subtidal grapestone and mud.

In southern Belize (Figures 13.12, 13.17, and 13.18), reflection seismic data indicate that the Pleistocene surface is deeper in the southern part of the study area than it is in the northern part of the study area. In the northern portion of Ambergris Cay, a Pleistocene reef facies outcrops on the island. In southern Belize, this horizon is as much as 30 meters below present sea level. Presumably, the lowering of this surface reflects postdepositional tectonic downwarping into the Cayman Trench. However, the exact cause of the lowering of the Pleistocene surface is not important to our observation; we care only that the Pleistocene surface has subsided. If we assume that this Pleistocene topographic feature is 120,000 years old, as is most of the Pleistocene topography beneath Recent shallow-water carbonates around the world, an average rate of subsidence on the order of 30 centimeters per 1000 years is indicated. Thus the Belize reef tract records the results of an experiment that nature has run for us concerning sediment accumulation capabilities of these Recent carbonate environments.

As sea level passed on to the Pleistocene topographic highs, sedimentation was initiated. Has sedimentation kept pace with the rising water? In central Belize (just north of the area pictured in Figure 13.12), approximately 10 to 13 meters of Recent sediment overlie the Pleistocene surface that is situated 13 meters below present sea level, indicating sedimentation rates of approximately 150 centimeters per 1000 years. The shelf-margin facies and Barrier Platform clean sands are of approximately equal thickness, even though these two facies do not have a common sediment source. Apparently both facies have sediment accumulation rates sufficient to maintain shallow-water depths during rising sea level.

In contrast, to the south of the area depicted in Figure 13.12, where the Pleistocene surface lies 25 to 30 meters below sea level, the shelf-margin facies is discontinuous, and the Barrier Platform clean-sand facies does not completely occupy the preexisting topographic platform. Only occasional shoals rise out of deep water. The shelf-margin facies is apparently accumulated at a sedimentation rate of about 275 centimeters per 1000 years. At this sediment accumulation rate, no truly continuous barrier reef has been maintained. Barrier Platform clean-sand sedimentation apparently has been unable to keep pace with rising water in this area.

Note that sedimentation rates for southern Belize are considerably greater than sedimentation rates for related environments in southern Florida and the Bahamas. This point will be of considerable interest to us in the formulation of general models.

Mineralogy of Recent Carbonates. The vast majority of shallow-marine Recent carbonates are composed of aragonite and high-magnesium calcite. Corals, calcified green algae, most oolites, and most tropical molluscs are composed of aragonite. Coralline algae, echinoderms, and benthic foraminifers are commonly high-magnesium calcite. In sharp contrast, most Ancient carbonates consist of low-magnesium calcite and dolomite. To a first approximation, it is convenient to regard aragonite and high-magnesium calcite as *unstable minerals* whereas low-magnesium calcite and dolomite are stable minerals. As a general rule, we shall anticipate that unstable minerals recrystallize to stable minerals early in the history of the sediment—a point to which we shall return later.

A Basic Model for Shelf-Margin and Subtidal Shelf-Interior Sedimentation

With the preceding examples of Recent environments serving as our guide, let us make some generalizations about what we would expect to see if we encountered similar deposits in the stratigraphic record.

Areal Distribution of Contemporaneous Sedimentary Facies. Generalities are summarized in Figure 13.19. The foreslope of the shelf margin is usually the site of accumulated skeletal sand and carbonate mud. Because access is difficult, this environment in the Recent epoch is not well studied.

The shelf-margin environment is the site of coral-reef growth. In addition, current-transported skeletal sands or oolitic sands can accumulate here. Modern coral reefs of the shelf-margin environment have a pronounced zonation of coral communities; we would expect Ancient reefs to also show ecological zonation. The reef-crest community, therefore, may form an environmental datum upon which we can reconstruct the environment and the history of sea-level fluctuation. For example, the modern *A. palmata* community occurs from reef crest to water depths of 5 meters. Thus, if we encounter a vertical section of *A. palmata* facies 20 meters thick, we would suspect that the reef-crest community grew upward with the rising sea level.

Sands of the shelf-margin environment should show evidence of current trans-

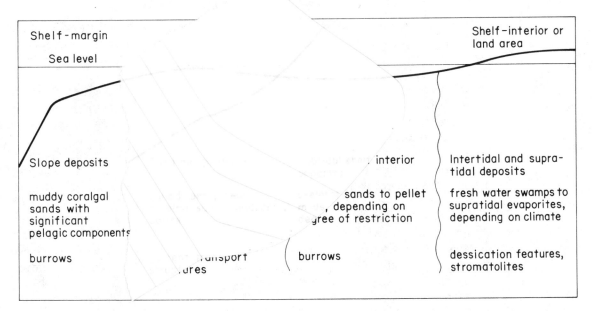

Figure 13.19 Generalized model for carbonate facies distribution from the shelf margin inward.

portation (Imbrie and Buchanan, 1965; Ball, 1967). Skeletal sands may predominate around shelf-margin coral reefs. In the absence of sediment supply from coral reefs, oolitic sands may typify the shelf-margin environment.

A generalization concerning the sediments of the shelf interior is more difficult than a generalization concerning shelfmargin sedimentation. Indeed, the farther we get from the open-marine, sea-water conditions of the shelf margin, the more chances there are for one variable or another to predominate in the environment and change sediment type. In the shelf interior, circulation patterns, evaporation rates, and the presence or absence of freshwater input from nearby landmasses become very important attributes of the paleogeography. Given open circulation, such as over the Barrier Platform of southern Belize, the shelf interior may be the site of widespread production of skeletal sands and carbonate mud. Sands may accumulate *in situ*, whereas muds may be transported by the throughgoing circulation pattern.

Given a somewhat more restricted circulation pattern, such as in the inner reef tract of south Florida, both skeletal sands and muds may accumulate *in situ* in the shelf-interior, shallow-water environment.

The Bahamas apparently represent the next step in the restriction of the shelf interior. Water flow is limited, thus allowing the accumulation of carbonate mud in shallow-water environments. In addition, increased salinities restrict the fauna and therefore the skeletal constituents of the shelf-interior sediments. Lithification of mud pellets and the formation of grapestone probably reflect some combination of increased salinity and reduced sedimentation rate.

The Concept of Bottomset, Foreset, and Topset Beds in Carbonate Sedimentation. In our discussion of coastal clastic environments, we found it convenient to use the top of the foreset (on a beach or a river-mouth bar) as our single, best environmental datum in the coastal clastic model.

In carbonate sedimentation, we shall continue to use the concept of foreset sedimentation, but with at least two important modifications. First, in any carbonate paleogeography, there may be numerous simultaneously active foresets: the shelf margin and the subtidal-supratidal transition, for example (see Figure 13.20). Secondly, the concept of systematic decrease in sedimentation rate from upper foreset to bottomset deposits seldom applies to carbonates because of *in situ* production of sediment within the local environment.

Vertical Accumulation in Carbonate Sediments. In our discussion of coastal clastic sedimentation, we observed that virtually all of the Recent sediment accumulates within the foreset slope enviroment. Only very fine-grained clays achieve sufficiently wide distribution from the source area to be regarded as truly bottomset. *In situ* production, and therefore vertical accumulation of carbonate sediments, require that we reconsider this generality and the facies relationships that may be predicted by it.

In a very real sense, the Barrier Platform, carbonate-skeletal sands of southern

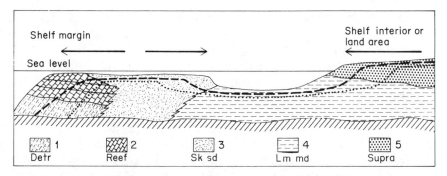

Figure 13.20 Schematic cross section indicating the various possibilities for progradation within carbonate environments. Sediment types are as follows: (1) forereef detritus; (2) shelf-margin reef or similar facies, with reef zonation indicated by solid lines; (3) shelf-interior skeletal sands; (4) shelf-interior subtidal muds; (5) intertidal-supratidal deposits, typically algal stromatolites, supratidal dolomites, and evaporites, or alternatively, coastal clastics. Dotted line and dashed line are time lines. Solid arrows indicate directions in which progradation may be expected.

Belize are bottomset deposits to the prograding coastal clastics of the mainland (see Figure 13.12). Note, however, that the Barrier Platform sands have such a high sedimentation rate that the application of the term "bottomset" is a purely academic exercise. Indeed, the Barrier Platform is itself a source of sediment supply westward to the Main Channel Lagoon. Thus, the concept of a foreset to bottomset transition has equal application to sedimentation from both margins of the Main Channel Lagoon.

Because of *in situ* production of sediment, it is quite possible for a carbonate basin to be filled by vertical accumulation. According to our former terminology, this process means that bottomset beds may be directly overlain by topset beds with no intervening foreset deposits.

Progradation in Carbonate Sedimentation. In coastal clastic sedimentation, the ability of a shoreline to prograde was tied directly to sediment supply. Once again, *in situ* production from carbonate sediment opens several new possibilities. In southern Belize, for example (see Figure 13.12), progradation is occurring in three different geographic positions, each having their own separate sediment supply. Coastal clastics are prograding eastward into the lagoon; the Barrier Platform is prograding westward into the lagoon; and the barrier reefs of the shelf margin are prograding eastward into the Caribbean. Simultaneous progradations from independent sediment sources must be expected to produce complicated stratigraphies.

Seaward progradation of the shelf-margin facies is fundamentally important in carbonate sedimentation. The establishment of a well-developed shelf-margin facies creates the other carbonate environments, just as the development of a clastic barrier island creates the bay environments. Continued maintenance and progradation of the bank-margin facies enlarges the shelf-interior environment. Indeed, magnetic data indicate that the carbonate platform of Florida was initiated on discrete volcanic cones separated from each other by as much as 100 kilometers. Maintenance and progradation of the shelf-margin facies eventually led to coalescence and development of a single Florida platform.

If we look at a more detailed level, we can expect that progradation of reef facies during times of constant sea level will result in sheetlike deposits with a vertical zonation of coral similar to the depth zonation of the forereef at any one instant in time (see Figure 13.20).

Furthermore, we can see that the rate of seaward progradation of the shelf-margin facies may be governed in large part by the preexisting topography of the foreshelf environment (Figure 13.21). Later, we shall study the relationships between progradation capabilities and sea-level fluctuations.

Progradation of supratidal deposits over platform-interior deposits can also be expected to produce laterally extensive sheet deposits in which the vertical stratigraphy mimics map-view relationships seen at any one instant in time. These deposits are the subject of the next chapter.

Continued Sedimentation at Constant Sea Level. In looking at various examples of Recent sedimentation, we have noted varying degrees of restriction on the shelf interior. In Recent sedimentation, these distinctions are primarily a function of local geometry, tectonic activity, and climate. For example, it is almost certainly no accident that the carbonate shelf exhibiting the best circulation (southern Belize) is also the carbonate shelf that is actively subsiding.

Clearly, continuing sedimentation will lead to progressively greater and greater restriction of water circulation. Because the Recent epoch provides only a brief glimpse into this process, we must now combine several of our Recent models into a temporal succession of sedimentary facies that we would expect to find if a carbonate shelf continued to accumulate sediment over an extended period of constant sea level.

Shelf-interior sedimentation of the Belize Barrier Platform and Main Channel Lagoon, the inner reef tract of south Florida, and the interior of the Great Bahama Bank offer us glimpse of three stages that might be superimposed one upon the other.

With open circulation, sediments are predominantly skeletal and there is good separation of *in situ* skeletal sands and transported carbonate muds (Belize). With increasing restriction, as sediments begin to fill the space between the shelf and the level of the sea, there will likely come a time when water chemistry is still satisfactory for the production of skeletal sands but currents are sufficiently weak that lime mud accumulates with the sand (south Florida inner reef tract). As restriction continues, water chemistry and nutrient supply become unfavorable for normal marine benthonic organisms. Thus, fauna is limited, and pelleted lime-mud grapestone predominate (Great Bahama Bank). Finally, we might expect supratidal or eolian deposits to prograde out over much of the shelf interior.

Note that the vertical profile generated in this discussion is not duplicated in any single example of Recent carbonate sedimentation. In all cases, the model has

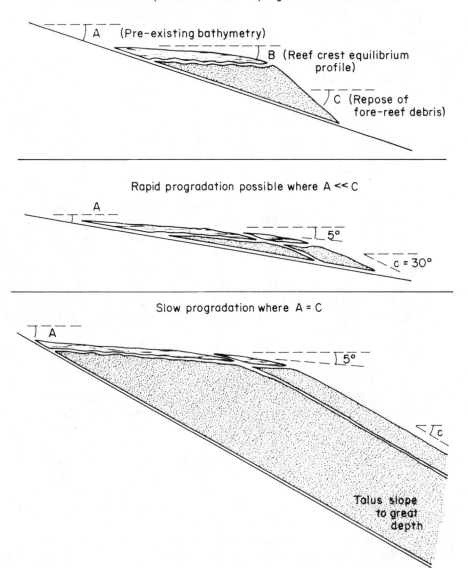

Geometries important in seaward progradation of coral reefs

A (Pre-existing bathymetry)

B (Reef crest equilibrium profile)

C (Repose of fore-reef debris)

Rapid progradation possible where A << C

A

5°

c = 30°

Slow progradation where A = C

A

5°

c

Talus slope to great depth

Figure 13.21 Geometries important in the seaward progradation of coral reefs. The upper diagram summarizes the three angles important to the maintenance of a coral reef during time of emergence. The reef has developed on some preexisting bathymetric surface that slopes seaward (angle A). If the living reef is to prograde seaward and remain unharmed by emergence, the reef crest equilibrium profile (angle B) must be maintained. Maintenance of angle B during seaward progradation will require volumes of forereef sediment determined by preexisting forereef bathymetry. (After Matthews, 1969.)

either not run long enough, or the interaction between preexisting topography and the Recent sea-level rise has set the model running in an already constricted configuration.

Response of Carbonate Sedimentation to Conditions of Submergence. In the discussion of coastal clastic sedimentation, we observed that rising water may tend to trap clastic sediment supply at the heads of estuaries, therefore forcing coastal areas into a condition of nondeposition. In carbonate sedimentation, *in situ* production of sediment relieves us from this consideration. Rising water will simply create more living space between the bottom and surface of the water and will tend to improve circulation of normal seawater to the shelf interior. At the same time, any problems created by clastic influx to the carbonate environment will be alleviated as clastic sediments are trapped in the estuaries and alluvial environments. Whereas rising water has a strong tendency to disrupt clastic sedimentation, it may actually improve conditions in carbonate environments (Figure 13.21).

Given new living space and improved circulation, each carbonate facies may tend to accumulate upward at a rate equal to or less than its own upward growth capability. If the rate of sea-level rise is sufficiently slow, the sediment-water interface will maintain its relationship to sea level and the character of the sediment being deposited will not change. If the sea-level rise is too fast, the sediment water interface will be left behind in deeper and deeper water and the carbonate facies will change accordingly.

For example, life on a shelf-margin coral reef will not change simply because sea level is rising at a rate of 1 meter per 1000 years. The day-to-day processes of calcification and biological erosion go on as usual. Yet, 1000 years later, the net result of continuing reef sedimentation during a time of sea-level rise will be the accumulation of an additional meter of *in situ* reef framework above the original substrate. Indeed, if the production of forereef detritus has been sufficient, the reef may even prograde seaward during a time of rising water. On the other hand, if the sea-level rise exceeds the maximum upward-growth capability of the coral reef, the reef-crest community finds itself struggling to survive in deeper and deeper water. Eventually it will give way to the deeper-water communities.

Shelf-margin oolite sand belts are likewise controlled largely by the water surface. If water depth increases slowly enough, more and more oolite accumulates. The oolite shoals and channels near the water surface maintain their ever-changing patterns of migration and sedimentation; yet each time a channel migrates, a little more oolite is left below in its final resting place.

We have also seen that facies of the shelf interior have their own sedimentation rate. Note further, however, that the sedimentation rate is greatest in environments having the best water circulation over the shelf. Skeletal sands of the Belize Barrier Platform, for example, have accumulated at the rate of 150 centimeters per 1000 years, whereas the pelleted muds of the more restricted Bahama shelf interior have accumulated at a rate of only 10 centimeters per 1000 years.

Shelf interiors, therefore, present an additional complication during conditions of rising water. First and foremost, facies of the shelf interior do have a sedimentation rate. Furthermore, freshening of shelf-interior water-circulation patterns may produce changes in the sediment-producing benthonic communities, thus changing the facies and increasing the sedimentation rate in that geographic position. The freshening of shelf-interior circulation is undoubtedly dependent upon complicated interaction between rising sea level and responses of the shelf-margin facies.

Taking a closer look at the shelf-margin and shelf-interior environments, we must once again observe an important consequence of *in situ* production of carbonate sediment. Specifically, adjacent sedimentary facies may have their own self-contained sediment supply. For example, the sedimentation rate of the skeletal sand apron of Belize barrier reefs (Figure 13.5) is totally unrelated to the sedimentation rate of Barrier Platform, *in situ,* skeletal sands. Similarly, each of the hard-bottom communities, sand-bottom communities, and mud-bottom communities within the shelf interior possesses its own potential sedimentation rate.

During times of constant sea level or slightly rising sea level, these distinctions may be minimized. Sedimentation tends to smooth out the low spots and a mosaic sea-level rise, one facies may keep pace with rising water while nearby facies do not. Topographic relief ensues. Inasmuch as light intensity is an important factor in carbonate-sediment production rates, the situation tends to become even more accentuated. Facies that maintain themselves at or near sea level will continue to receive sufficient light for rapid carbonate fixation. Facies that do not accumulate sufficiently fast to keep up with rising water will generally receive less light as water deepens and will therefore accumulate even more slowly.

The Barrier Platform of southern Belize appears to exemplify this principle. Where the Pleistocene preexisting topography lies within 15 meters of the present sea level, the Barrier Platform is a mosaic of coalescing small patch reefs, clean-sand bottom, and *Thallasia*-covered sand bottom. To the south, the same geomorphic features exist within the Pleistocene preexisting topography, but here they are as much as 25 meters below the present sea level. The pattern of coalescing facies noted to the north has broken down into isolated patch reefs standing high above the thin veneer of sediment in surrounding areas. The system is now locked in. The shoal areas receive sufficient light, so they can continue to calcify at rapid rates and thereby continue to grow upward toward sea level. The intervening areas have become too deep to support significant benthonic plant life; therefore, they continue to accumulate sediment at a slower rate than the shoal areas. Note that a long period of constant sea level will be required for this situation to rectify itself. The shoal areas can build upward only to sea level. Once they reach sea level, upward accumulation ceases. Then the intervening deep-water areas can begin to fill up, thus lessening the topographic distinction between shoals and surrounding areas.

Response of Carbonate Sedimentation to Conditions of Emergence. As the sea level recedes, supratidal deposits, subtidal platform-interior deposits, and shallow-water bank-margin environments all come under stress. Supratidal flats may regress over former subtidal shelf-interior sediments as the water retreats. At the shelf margin, however, this lateral shift of facies must come to an end. With the emergence of the shelf interior, shelf-margin oolite production will probably cease because there is no longer extensive tidal flow onto and off of the shelf interior.

Shelf-margin coral reefs are also faced with serious problems as the sea level falls. There will be a tendency for shallow-water coral faunas to migrate outward over deep-water coral faunas. But note that such regression demands a firm substrate in the forereef area. So long as the coral reef is producing sufficient forereef detritus, regression can occur (see Figure 13.22). Otherwise, the reef will tend to build outward as an overhanging structure, supported only by the strength of its bonding to the older reef rock. Such structures will ultimately collapse and become large blocks of forereef detritus themselves. This phenomenon can result in a near vertical wall just seaward of the highstand bank-margin environment. Vertical walls are an inhospitable place for reef-crest coral communities. Thus, only biogenic shelf-margin sedimentation has any chance of continuing during conditions of significant emergence. This sedimentation can take place only if there is adequate low-lying forereef topography or sufficient production of forereef detritus. The cessation of

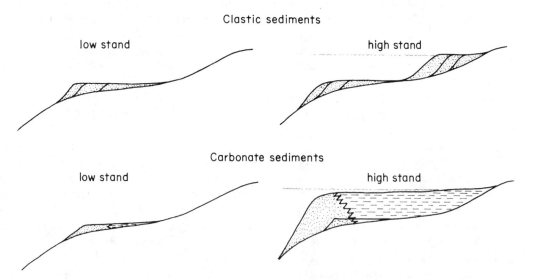

Figure 13.22 Comparison of clastic sedimentation with carbonate sedimentation during an event of submergence. Rising water tends to trap clastic sediments on the alluvial plain and in the newly drowned river-valley estuaries. Thus, sedimentation is disrupted by conditions of submergence and will be reestablished at the new shoreline during the new highstand. In contrast, carbonate sedimentation is invigorated by slowly rising water. First, rising water provides living space above the reef-crest community and thereby allows for vertical accumulation of reef-crest facies. Secondly, rising water tends to freshen marine circulation to the shelf interior, thereby increasing sedimentation rates. As a net result of this contrast between clastic sedimentation and carbonate sedimentation. Recent coastal clastic sediments are commonly found well back onto the modern continental shelf, whereas Recent carbonate sediments are typified by a shallow-water shelf-margin adjacent to deep water. Similar contrast would be expected in the stratigraphic record.

carbonate sedimentation by conditions of emergence is perhaps one of the most fascinating contrasts between clastic and carbonate sedimentation. No matter whether the sea level rises or falls, clastic sediments continue to come down the rivers and must be deposited somewhere. In contrast, when the sea level recedes from carbonate shelves, sedimentation ceases.

Concurrent with cessation of sedimentation on carbonate shelves, emergent conditions bring with them exposure of the sediments to meteoric water. It is well demonstrated that the unstable carbonate minerals, aragonite and high-magnesium calcite, recrystallize to low-magnesium calcite within a matter of a few thousand years when exposed to fresh water (Matthews, 1968, 1974, for example). Inasmuch as dolomitization may also result from the mineralogical stabilization process, this situation is discussed in greater detail in Chapter 14.

An Ancient Example: Pleistocene Coral Reefs of Barbados, the West Indies

Perhaps your immediate reaction is to question the use of the Pleistocene as an "Ancient" example. Geologically speaking, the Pleistocene happened only yesterday. Most of the species present in the Pleistocene sediments of Barbados are present in modern carbonate environments of the Caribbean.

Yet, in a very real sense, the Pleistocene is as much a part of the geologic record as any Paleozoic sequence. Even though Pleistocene sediments are very young, no one was there to see them deposited. By and large, in our approach to understanding these sediments, we must use the same sedimentological and stratigraphic inferences as we would apply to any Ancient sequence.

Regional Setting and Earlier Stratigraphic Work. Barbados is the easternmost island of the Lesser Antilles. Unlike its volcanic neighbors to the west, Barbados is composed entirely of sedimentary rocks. Oligocene and Miocene pelagic sediments now stand as much as 300 meters above sea level, attesting to the general tectonic uplift of the sea. Over approximately 85 percent of the island, the folded and faulted Tertiary strata are overlain by shallow-water Pleistocene carbonate deposits that average about 80 meters in thickness (Figure 13.23).

The surface of these Pleistocene shallow-water carbonates is distinctly terraced; the origin of these terraces has been the subject of some controversy since the 1890's. From the 1930's until fairly recent times, the views of Trechmann (1933, 1937) prevailed. He claimed that the entire terraced coral cap of Barbados had been deposited as a single, large, shallow, subtidal bank at the time when the rising mass of folded and faulted Tertiary strata began to approach sea level. Subsequent tectonic uplift of the bank produced faults and broad gentle folds that are the first-order topography of the island today. As these structures rose out of the water, there was tectonic quiescence from time to time. During quiescent times, pronounced wave-cut terraces were gouged out of the carbonate rocks by the erosional action of the surf.

Thus, to summarize Trechmann's views, the Pleistocene coral cap of Barbados

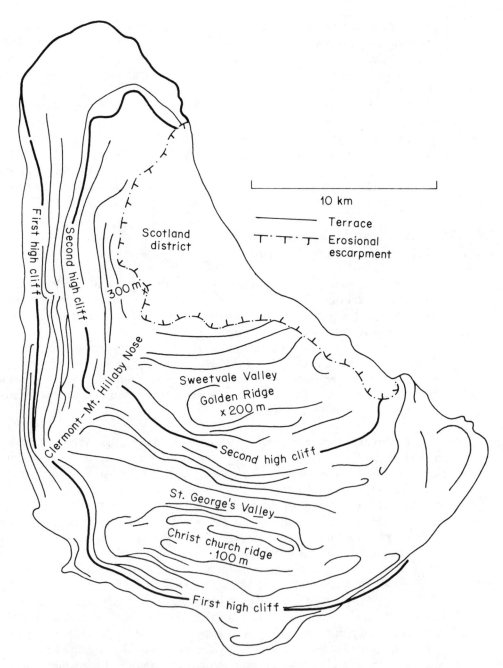

Figure 13.23 The terraced coral cap of Barbados, the West Indies. The first-order topography of the island consists of a series of large hills (Clermont–Mt. Hillaby Nose, Golden Ridge, and Christ Church Ridge) separated by broad valleys (Sweetvale Valley and St. George's Valley). Prominent terraces constitute the second-order topography of the island.

is isochronous. The terraces are of erosional origin, wave-cut during the intermittent uplifts that followed subtidal deposition of the coral cap.

Sedimentological Observations. By the time Mesolella *et al.* (1969, 1970) renewed geological investigation of the Pleistocene carbonates of Barbados, considerable progress had been made in the study of Recent carbonate sedimentation. Particularly important were the developing ideas concerning coral zonation from reef crest into forereef slope. Before the days of scuba diving, the geologist's knowledge of coral reefs was confined primarily to a faunal list. On this basis, for example, Trechmann and works before him recognized that the Pleistocene of Barbados did contain corals and was therefore a shallow-water deposit. With the advent of scuba diving, the coral-reef zonation from *A. palmata* reef crests to *A. cervicornis* foreslope to deep-water head zone has become well-documented in several areas.

Armed with this new knowledge of the Recent epoch, these workers have recognized that familiar corals occupy familiar positions (Figure 13.24). In particular, massive accumulations of *A. palmata* commonly occur at the forward edge of the terraces. On the foreslopes of terraces, *A. palmata* gives way downward to *A. cervicornis,* which in turn is replaced by a mixed coral assemblage predominated by large heads such as *Montastrea, Siderastrea,* and *Diploria sp.*

From the front edge back onto the terrace, *in situ A. palmata* usually passes into a rubble zone of transported *A. palmata* and skeletal sand. This sediment in turn gives way to clean skeletal sand with occasional coral heads or branching colonies. Further back on the terrace, there is skeletal sand with a lime-mud matrix. Finally, near the base of the next higher terrace, low-angle, cross-stratified, clean, skeletal sands are predominant.

On modern Caribbean reefs, *A. palmata* thrives in water depths ranging from zero to 6 meters. At some localities in the Pleistocene relicts of Barbados, *A. palmata* facies attain thicknesses of 30 meters. We must ask whether the ecological range of *A. palmata* has changed from Pleistocene to Recent, or whether this unusual thickness of *A. palmata* has changed from Pleistocene to Recent, or whether this unusual thickness of *A. palmata* records the dynamic interaction between upward reef growth and conditions of submergence.

Data relevant to this question can be obtained by studying the contact between *A. palmata* and *A. cervicornis* zones. When we measure the position of this contact on the foreslope surface, relative to the general level of the terrace (see Figure 13.24), we determine a lower limit to the Pleistocene *A. palmata* facies. This limit agrees with similar data for recent Caribbean reefs (Table 13.1). However, when we ex-

---

Early geologists considered that the entire coral cap had accumulated as a single subtidal carbonate bank, which was later folded to form the first-order highs and lows of the island. The second-order topography (the terraces) presumably resulted from coastal erosion during intermittent times of tectonic quiescence, which punctuated the general history of uplift of the island. Modern stratigraphic studies, however, have discounted this view.

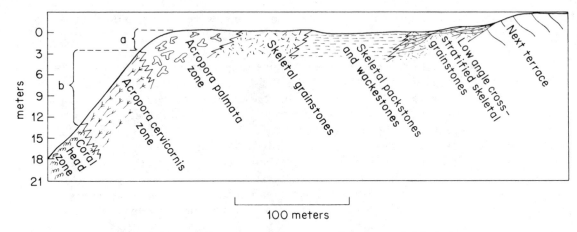

Figure 13.24 Sediment types encountered in a traverse through a typical Barbados terrace. Note that the facies contacts between *A. palmata* and *A. cervicornis* zones record two types of information. First, the depth a and (a + b) record the paleoecology of these coral zones. Secondly, the three-dimensional relationship between *A. palmata* and *A. cervicornis* should indicate whether reef growth occurred under conditions of submergence, constant sea level, or emergence. [After K. J. Mesolella, "Zonation of Uplifted Pleistocene Coral Reefs on Barbados, West Indies," *Science,* **156,** 638–640, Table 1 (5 May 1967), with modification.] Copyright 1967 by the American Association for the Advancement of Science.

amine this contact between *A. palmata* and *A. cervicornis* in the third-dimension exposure provided by road cuts and quarries, the contact between the two zones typically has a strong vertical component.

Sedimentological Interpretations. The data suggest that each Barbados terrace represents a discrete time of coral-reef development around an emerging island. Furthermore, observations concerning thick sections of *A. palmata* reef-crest community as well as the contact between the *A. palmata* community and the *A. cervicornis* foreslope community suggest that many of these reef-tract terraces were deposited under conditions of *submergence*. At first glance, this fact is rather surprising in view of the island's general history of *emergence* throughout the Pleistocene.

The key to reconciling this apparent contradiction comes from radiometric dating of the lower Barbados terraces by the thorium-230 growth method (Mesolella *et al.,* 1969). In particular, a 125,000-year-old terrace is well known from widely separated parts of the world. It is commonly encountered 2 to 6 meters above the present sea level. On Barbados, this terrace is encountered at elevations ranging from 20 to 60 meters above present sea level.

Thus, the elevation of this terrace on Barbados not only confirms the general history of tectonic uplift of the island but also provides a basis for estimating the rate of tectonic uplift. On the average, Barbados is undergoing tectonic emergence at the rate of 30 centimeters per 1000 years.

Immediately, we realize that 30 centimeters per 1000 years is an extremely slow

Table 13.1 Depth Range of Coral Zone below Reef Crest, Barbados Terrace Data
(n.e., not exposed; n.d., not developed)

Section	A. palmata zone (meters)	Buttress-zone (meters)	A. cervicornis zone (meters)	Coral-head zone (meters)
BD	0–6.8	n.d.	6.8–17.8[a]	n.e.
CZ	0–4.0	n.d.	4.0–12.0[a]	n.e.
BT	0–6.8	n.d.	6.8–18.2	18.2–27.7[a]
BS	0–2.8	n.d.	2.8–10.2[a]	n.e.
BE	0–3.7	n.d.	3.7–13.8	13.8–15.4[a]
AEO	0–4.0	4.0–9.5	9.5[b]–17.8[a]	n.e.
DM	0–3.1	n.d.	3.1–13.5	13.5–20.3[a]
BF	0–5.5	n.d.	5.5–13.5[a]	n.e.
BY	0–4.0	n.d.	4.0–13.2	13.2–21.2[a]
EH	0–9.2	n.d.	9.2–26.5	26.5–33.2[a]
EW	0–5.2	n.d.	5.2–23.7[a]	n.e.
AFM	0–6.2	n.d.	6.2–16.9	absent
D	0–4.6	4.6–12.3[a]	n.e.	n.e.
IF	0–4.9	n.d.	4.9–15.1	15.1–23.7[a]
AEU	0–3.7	n.d.	3.7–7.7[a]	n.e.
E	0–2.2	n.d.	2.2–7.1	7.1–20.3[a]
QU	0–4.6	4.6–10.2[a]	n.e.	n.e.
HG	0–5.5	n.d.	5.5–10.2[a]	n.e.
BV	0–3.7	n.d.	3.7–10.8	10.8–12.6[a]
Average	*0–4.8*	*4.4–9.5*	*4.8–15.0*	*14.8*
Recent Jamaican reefs (Goreau, 1959)	*0.5–6.0*	*1–10*	*7.0–15.0*	*15.0*

(a) Minimum figure; base not exposed.
(b) Lowered due to presence of buttress zone. After Mesolella, 1967.

rate of *emergence* when compared to rates of *submergence* associated with Pleistocene sea-level fluctuations. For example, the post-Wisconsin transgression involved eustatic sea-level rise at the rate of 8 meters per 1000 years (Figure 9.1).

Such considerations suggest that each Barbados coral-reef terrace records, not a tectonic event in the history of uplift of the island, but rather a eustatic highstand of the sea during the Pleistocene. Gradual tectonic uplift of the island has simply raised each terrace up out of the realm of the next eustatic highstand.

Thus, an understanding of Recent carbonate sedimentation and the dynamics of Pleistocene sea-level fluctuations has led to a major reevaluation of terraced coral caps. Prior to this work, for example, scientists interested in the history of Pleistocene sea-level fluctuations stayed away from tectonically active areas. How can we possibly recognize a *eustatic* terrace among all of those *tectonic* terraces?

The Barbados data cast serious doubt on the hypothesis of a strictly tectonic origin for terraced coral caps. Instead, it appears that tectonism is simply working for us as a natural strip-chart recorder. As the sea-level data are recorded by the formation of a coral-reef terrace, tectonic uplift moves that information along to where it can be preserved for us to read; it does not allow additional eustatic data to be superimposed in an unintelligible fashion. Indeed, the eustatic data contained in Pleistocene coral caps of tectonically emergent islands promise to be the foundation in formulating a proper understanding of Pleistocene history (Mesolella *et al.,* 1969; Veeh and Chappell, 1970), a point we shall return to in a later chapter.

Citations and Selected References

AKIN, R. H. JR., and R. W. GRAVES, JR. 1969. Reynolds oolite of southern Arkansas. Bull. Amer. Assoc. Petrol. Geol. 53: 1909–1922.
Subsurface study of the transition from shelf margin to subtidal shelf interior in the economically important Smackover formation Jurassic of the Gulf Coast.

BALL, M. M. 1967. Carbonate sand bodies of Florida and the Bahamas. J. Sed. Petrology 37: 556–591.
Dynamics and geometry of carbonate sand accumulation.

BATHURST, R. G. C. 1971. Carbonate sediment and their diagenesis. Development in Sedimentology 12. Elsevier, Amsterdam. 620 p.
A thoughtful synthesis.

BLOOM, A. L. 1974. Geomorphology of reef complexes, p. 1–8. *In* L. F. Laporte (ed.), Reefs in time and space: selected examples from the Recent and Ancient. Soc. Econ. Paleont. Mineral. Spec. Pub. 18.

BUBB, J. N., and W. G. HATLELID, 1978. Seismic stratigraphy and global changes of sea level, part 10: seismic recognition of carbonate buildups. Bull. Amer. Assoc. Petrol. Geol. 62 (5): 772–791.

BUCHANAN, H. 1970. Environmental stratigraphy of Holocene carbonate sediments near Frazer's Hog Cay, British West Indies. Ph.D. dissertation, Columbia University. 257 p.

CANN, R. S. 1962. Recent calcium carbonate facies of the north-central Campeche Bank, Yucatan, Mexico. Ph.D. dissertation, Columbia University. 225 p.

CLOUD, P. E. 1962. Environment of calcium carbonate deposition west of Andros Island, Bahamas. U. S. Geol. Survey Prof. Paper 350. 138 p.

DODD, J. R., and C. E. SIEMERS. 1971. Effect of late Pleistocene karst topography on Holocene sedimentation and biota, lower Florida keys. Bull. Geol. Soc. Amer. 82: 211–218.

ENOS, P., and R. D. PERKINS. 1979. Evolution of Florida Bay from island stratigraphy. Bull. Geol. Soc. Amer. 90: 59–83.

ENOS, P., and L. H. SAWATSKY. 1981. Pore networks in Holocene carbonate sediments. J. Sed. Petrology 51: 961–985.

ESTEBAN C. MATEO. 1976. Vadose pisolite and caliche. Bull. Amer. Assoc. Petrol. Geol. 60 (11, pt. I): 2048–2057.
A caliche expert takes on the classic interpretation of Permian Reef pisolites; extensive bibliography.

EVAMY, B. D., and D. J. SHEARMAN. 1965. The development of overgrowths from Echinoderm fragments. Sedimentology 5: 211–233.
Petrography of cementation in French Jurassic limestones.

FISHER, W. L., and P. U. RODDA. 1969. Edwards Formation (Lower Cretaceous), Texas: dolomitization in a carbonate platform system. Bull. Amer. Assoc. Petrol. Geol. 53: 55–72.
Paleogeography and sedimentology of an economically important carbonate unit.

FROST, STANLEY H., MALCOLM P. WEISS, and JOHN B. SAUNDERS (eds.). 1977. Reefs and related carbonates—ecology and sedimentology. Amer. Assoc. Petrol. Geol., Studies in Geology 4. 421 p.

GINSBURG, R. N. 1956. Grain size of Florida carbonate sediments. Bull. Amer. Assoc. Petrol. Geol. 40: 2384–2427.
Grain size and constituent-particle composition of sediments from the Florida keys eastward to the shelf margin.

———. 1964. South Florida carbonate sediments. Guidebook for field trip no. 1. Geol. Soc. Amer. annual convention, Miami Beach, Fla. 72 p.

———. 1969. Presidential address before the Society of Economic Paleontologists and Mineralogists.

———. 1974. Introduction to comparative sedimentology of carbonates. Bull. Amer. Assoc. Petrol. Geol. 58 (5): 781–786.
Introduces 80 pages of high-quality papers resulting from a "Comparative Sedimentology of Carbonates Symposium."

GOREAU, T. F. 1959. The ecology of Jamaican coral reefs, part 1: species composition and zonation. Ecology 40: 67–90.

GOREAU, T. F., and L. LAND. 1972. Forereef slope ecology and depositional processes in Jamaica. *In* L. F. Laporte (eds.), Reef complexes in time and space: their physical, chemical, and biological parameters. Soc. Econ. Paleont. Mineral. Spec. Pub. 19.

HINE, ALBERT C., and A. CONRAD NEUMANN. 1977. Shallow carbonate-bank-margin growth and structure, Little Bahama Bank, Bahamas. Bull. Amer. Assoc. Petrol. Geol. 61 (3): 376–406.

HOFFMEISTER, J. E., and H. G. MULTER. 1964. Growth-rate estimates of a Pleistocene coral reef of Florida. Bull. Geol. Soc. Amer. 75: 353–358.

HRISKEVICH, M. E. 1970. Middle Devonian reef production, Rainbow area, Alberta. Bull. Amer. Assoc. Petrol. Geol. 54: 2260–2281.

IMBRIE, J., and H. BUCHANAN. 1965. Sedimentary structures in modern carbonate sands of the Bahamas, p. 149–172. *In* J. V. Middleton (ed.), Primary sedimentary structures and their hydrodynamic interpretation. Soc. Econ. Paleont. Mineral. Spec. Pub. 12.

IMBRIE, J., and E. G. PURDY. 1962. Classification of modern Bahamian carbonate sediments, p. 253–272. *In* W. E. Hamm (ed.), Classification of carbonate rocks. Amer. Assoc. Petrol. Geol. Mem. 1.

JAMES, N. P. 1977. Facies models 7: shallowing-upward sequences in carbonates. Geoscience Canada 4: 126–136.

JAMES, N. P., and R. N. GINSBURG. 1979. The seaward margin off Belize barrier and atoll reefs. Internat. Assoc. Sedimentol. Spec. Pub. 3. 200 p.

JAMES, N. P., ROBERT N. GINSBURG, DONALD S. MARSZALEK, and PHILIP W. CHOQUETTE. 1976. Facies and fabric specificity of early subsea cements in shallow Belize (British Honduras) reefs. J. Sed. Petrology 46 (3): 523–544.

JONES, O., and R. ENDEAN (eds.). 1972. Biology and geology of coral reefs, vol. 1: geology. Academic Press, New York.

KENDALL, C. G. ST. C., and P. A. D. SKIPWITH. 1969. Geomorphology of a recent shallow-water carbonate province: Khor Al Bazam, Trucial Coast, southwest Persian Gulf. Bull. Geol. Soc. Amer. 80: 865–892.

KLOVAN, JOHN EDWARD. 1964. Facies analysis of the Redwater reef complex, Alberta, Canada. Bull. Can. Petrol. Geol. 12 (1): 1–100.

KONISHI, K., S. O. SCHLANGER, and A. OMURA. 1970. Neotectonic rates in the central Ryukyu islands derived from thorium-230 coral ages. Marine Geol. 9: 225–240.

LOGAN, B. W., J. L. HARDING, W. M. AHR, J. D. WILLIAMS, and R. G. SNEED. 1969. Carbonate sediments and reefs, Yucatan shelf, Mexico, p. 5–198. Amer. Assoc. Petrol. Geol. Mem. 1. Sec. 1.

LOGAN, B. W., *et al.* 1970. Carbonate sedimentation and environments, Shark Bay, Western Australia. Amer. Assoc. Petrol. Geol. Mem. 13. 223 p.

MACINTYRE, IAN G., RANDOLPH B. BURKE, and ROBERT STUCKENRATH. 1977. Thickest recorded Holocene reef section, Isla Perez core hole, Alacran Reef, Mexico. Geology 5 (12): 749–754.

MAIKLEM, W. R. 1968. The Capricorn reef complex, Great Barrier Reef, Australia. J. Sed. Petrology 38: 785–798.
Physiogeography and gross aspects of sedimentation from the southern portions of the Great Barrier Reef.

MATTHEWS, R. K. 1966. Genesis of Recent lime mud in southern British Honduras. J. Sed. Petrology 36: 428–454.
Proposes a predominantly skeletal origin for lime mud in southern British Honduras.

————. 1968. Carbonate diagenesis: equilibration of sedimentary mineralogy to the subaerial environment: coral cap of Barbados, West Indies. J. Sed. Petrology 38: 1110–1119.

————. 1969. The Coral Cap of Barbados: Pleistocene studies of possible significance to petroleum geology, p. 28–40. *In* J. G. Elam and S. Cuber (eds.), Cyclic sedimentation in the Permian Basin. West Texas Geol. Soc. Publ. 69–56.

————. 1974. A process approach to diagenesis of reefs and reef associated limestones, p. 234–356. *In* L. F. Laporte (ed.), Reefs in time and space: selected examples from the Recent and Ancient. Soc. Econ. Paleont. Mineral. Spec. Pub. 18.

MAXWELL, W. G. H., and J. P. SWINCHATT. 1970. Great Barrier Reef: regional variation in a terrigenous- carbonate province. Bull. Amer. Assoc. Geol. 81: 691–724.
A convenient reference to the general features of the Great Barrier Reef.

MESOLELLA, KENNETH J. 1967. Zonation of uplifted Pleistocene coral reefs on Barbados, West Indies. Science 156: 638–640.

———. 1978. Paleogeography of some Silurian and Devonian reef trends, central Appalachian Basin. Bull. Amer. Assoc. Petrol. Geol. 62: (9): 1607–1644.

MESOLELLA, KENNETH J., R. K. MATTHEWS, W. S. BROECKER, and D. L. TURNER. 1969. The astronomical theory of climatic change: Barbados data. J. Geology 77:250–274.

MESOLELLA, KENNETH J., H. A. SEALY, and R. K. MATTHEWS. 1970. Facies geometries within Pleistocene reefs of Barbados, West Indies. Bull. Amer. Assoc. Petrol. Geol. 54: 1899–1917.

MOUNTJOY, ERIC W., and ROBERT RIDING. 1981. Foreslope stromato-poroid-renalcid bioherm with evidence of early cementation, Devonian Ancient Wall reef complex, Rocky Mountains. Sedimentology 28 (3): 299–320.

NELSON, CAMPBELL S. 1978. Temperate shelf carbonate sediments in the Cenozoic of New Zealand. Sedimentology 25 (6): 737–772.

NEUMANN, A. CONRAD, and LYNTON S. LAND. 1975. Lime mud deposition and calcareous algae in the Bight of Abaco, Bahamas: a budget. J. Sed. Petrology 45: 763–786.

PURDY, EDWARD G. 1963. Recent calcium carbonate facies of the Great Bahama Bank. J. Geology 71: 334–355, 472–497.
The basic definition of Bahamian carbonate facies.

———. 1974a. Karst-determined facies patterns in British Honduras: Holocene carbonate sedimentation model. Bull. Amer. Assoc. Petrol. Geol. 58 (5): 825–855.

———. 1974b. Reef configurations: cause and effect, p. 9–76. *In* L. F. Laporte (ed.), Reefs in time and space: selected examples from the Recent and Ancient. Soc. Econ. Paleont. Mineral. Spec. Pub. 18.

PURDY, EDWARD G., and J. IMBRIE. 1964. Carbonate sediments, Great Bahama Bank. Guidebook for field trip no. 2. Geol. Soc. Amer. annual convention, Miami Beach, Fla. 66 p.

PURDY, EDWARD G., and R. K. MATTHEWS. 1964. Structural control of Recent calcium carbonate deposition in British Honduras [abstract], p. 157 Geol. Soc. Amer. annual convention, Program with abstracts.

ROSE, P. R. 1976. Mississippian carbonate shelf margins, western, United States. J. Res. U. S. Geol. Surv. 4: 449–466.

RÜTZLER, KLAUS, and IAN G. MACINTYRE. 1982. The Atlantic barrier reef ecosystem at Carrie Bow Cay, Belize, I: structure and communities. Smithsonian Institution Press, Contributions to the Marine Sciences 12, Washington, D. C. 539 p.

SCOTT, ROBERT W. 1979. Depositional model of early cretaceous coral-algal-rudist reefs, Arizona. Bull. Amer. Assoc. Petrol. Geol. 63: (7) 1108–1127.

SEARS, S. O., and F. J. LUCIA. 1979. Reef-growth model for Silurian pinnacle reefs, northern Michigan reef trend. Geology 7 (6): 299–302.

SHINN, E. A. 1969. Submarine lithification of Holocene carbonate sediments in the Persian Gulf. Sedimentology 12: 109–144.

STOCKMAN, K. W., R. N. GINSBURG, and E. A. SHINN. 1967. The production of lime mud by algae in south Florida. J. Sed. Petrology 37: 633–648.
Green algae appear to be capable of producing all the aragonite mud present in the Recent environment.

STODDART, D. R., and C. M. YONGE (eds.). 1971. Regional variation in Indian Ocean coral reefs. Zool. Soc. London, Symposia 28. Academic Press, New York. 608 p.

———. 1978a. The northern Great Barrier Reef, part A: Phil. Trans. Roy. Soc. London A 291: 1–297.

———. 1978b. The northern Great Barrier Reef, part B: Phil. Trans. Roy. Soc. London B 284: 1–163.

SWINCHATT, J. P. 1965. Significance of constituent composition, texture, and skeletal break-down in some Recent carbonate sediments. J. Sed. Petrology 35: 71–70.

TRECHMANN, C. T. 1933. The uplift of Barbados. Geol. Mag. (Great Britain) 70 (823): 19–47.

———. 1937. The base and top of the coral rock in Barbados. Geol. Mag. (Great Britain) 74 (878): 337–358.

VEEH, H. H., and J. CHAPPELL. 1970. Astronomical theory of climate change: support from New Guinea. Science 167: 862–865.

VINCELETTE, R. R., and R. A. SOEPARJADI. 1976. Oil-bearing reefs in Salwati Basin of Irian Jaya, Indonesia. Bull. Amer. Assoc. Petrol. Geol. 60: (9): 1448–1462.

WANLESS, H. R. 1970. Influence of preexisting bedrock topography on bars of "lime" mud and sand, Biscayne Bay, Florida [abstract]. Bull. Amer. Assoc. Petrol. Geol. 54: 875.

WANTLAND, K. F., and WALTER C. PUSEY III. 1971. The southern shelf of British Honduras. New Orleans Geol. Soc. New Orleans, La. 98 p. (plus appendices).
Well-illustrated guidebook containing much three-dimensional information.

———. (eds.) 1975. Belize shelf—carbonate sediments, clastic sediments, and ecology. Amer. Assoc. Petrol. Geol. Tulsa, Studies in Geology 2. 599 p.

WILSON, JAMES LEE. 1974. Characteristics of carbonate-platform margins. Bull. Amer. Assoc. Petrol. Geol. 58: (4): 810–824.

———. 1975. Carbonate facies in geologic history. Springer-Verlag, New York. 471 p.

WINLAND, H. D., and R. K. MATTHEWS. 1974. Origin of Recent grapestone grains, Bahama Islands. J. Sed. Petrology 44: 921–927.

Carbonates and Evaporites of the Intertidal and Supratidal Shelf

14

This chapter is the logical continuation of the subject matter in Chapter 13. In the previous chapter, we broke off our discussion simply to consolidate some of the observations concerning the shelf margin and subtidal shelf interior. Now we return to our consideration of the restricted environments of the shelf interior.

Recent Intertidal and Supratidal Environments

We have noted that subtidal shelf interiors tend to become restricted environments, dominated by conditions other than those of the normal marine environment. On the Great Bahama Bank and in the Persian Gulf, increased salinity resulting from poor circulation generally excludes many marine organisms from the subtidal shelf interior. In Florida Bay, a similar reduction in fauna has been caused by poor circulation and the additional complication of a variable freshwater input from the north.

As we move into the intertidal and supratidal environments, we see that the fauna must face the further hazard of intermittent exposure above sea level. The result is even further restriction of the fauna. This situation characterizes intertidal-supratidal environments.

The Importance of Subtidal to Supratidal Transitions. A volumetrically significant amount of carbonate and evaporite sedimentation occurs in the supratidal environment. For example, much of the dolomite to be found in Recent environments occurs in association with supratidal sediments. Inasmuch as dolomite is an abundant carbonate mineral in Ancient sequences, Recent supratidal dolomite is of considerable interest to us.

All Recent sedimentation of evaporite minerals essentially occurs on supratidal flats marginal to subtidal-marine environments. Furthermore, there is reasonably good evidence, and strong geologic implication, that pore water from supratidal evaporite deposits may flow back through underlying carbonate sediments and dolomitize them.

In addition, the intertidal-supratidal environment appears to be a major site of stromatolite deposition and preservation. By recognizing these deposits in Ancient rocks, we may have an excellent environmental datum from which to make estimates on the water depth of associated environments.

Finally, the subtidal-supratidal couplet, in which supratidal sediments are closely superimposed upon subtidal sediments, is one of the primary tipoffs of progradation with carbonate sequences. We shall try to identify these transitions for precisely the same reasons that we sought to identify the coarsening-upward cycle in coastal clastic sedimentation.

Let us now look at the transition from subtidal to supratidal environment in six examples where the water of the supratidal environment ranges from fresh to very highly saline. In the transition from the subtidal Florida Bay into the Everglades, there really is no transition from subtidal-marine to supratidal sedimentation. Subtidal marine simply becomes subtidal freshwater as we examine the coastal swamps. In portions of the Bahamas, supratidal carbonate flats exhibit well-developed stromatolites and there is supratidal dolomitization without associated evaporite minerals. In the sabkha environment of the Persian Gulf, gypsum and anhydrite precipitation in the supratidal environment is added to the Bahamas model.

The Florida Bay Facies Mosaic. In the northwest portion of Florida Bay, the transition from marine deposition in the bay to freshwater deposition in the Florida Everglades has been studied in considerable detail in connection with problems of the Holocene sea-level rise (Scholl and Stuiver, 1967). Only in the broadest sense does this transition go from subtidal-marine sedimentation to supratidal sedimentation. We would give a better description by saying that this is the area where fresh water and saline water meet. Water is everywhere, and the entire sediment-water interface is at or very near sea level. Just such features as natural levees on tidal channels are persistent topographic highs.

Traversing from Florida Bay into the Everglades, we find that Florida Bay lime mud gives way to a coastal mangrove swamp, which in turn gives way to the freshwater swamps of the Everglades. In vertical profile through the marine deposits, the sequence is repeated and records the way the Recent sea-level rise transgressed onto this very low lying, carbonate erosion surface (Figure 14.1).

Progradation in this area has a very interesting style. Just as we did not find bay deposits prograding seaward over the Chenier Plain (Figures 12.14 and 12.15), we would not expect to find mangrove or freshwater swamp deposits prograding out over such coastal marine deposits as the Cape Sable beach sands. In both cases, there is a problem of space so long as sea level is constant. Note, however, that the

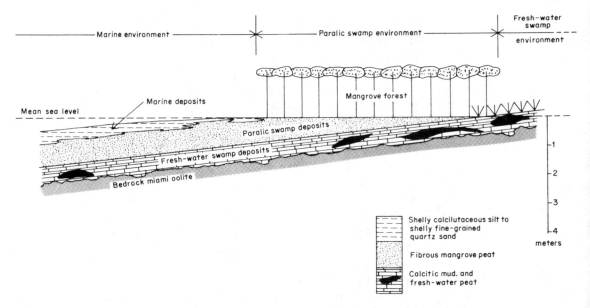

Figure 14.1 Idealized cross section of coastal swamps, south Florida. [After D. W. Scholl and M. Stuiver, "Recent Submergence of Southern Florida: A Comparison with the Adjacent Coasts and Other Eustatic Data," *Bull. Geol. Soc. America*, **78**, 437–454, (1967).]

marine mudbanks of Florida Bay (Figures 13.2, 13.13, and 13.15) tend to cut off isolated ponds among them. Similar features have been documented beneath the thin veneer of freshwater swamps east of Cape Sable (Gebelein, 1977). Here marine mudbanks cut off ponds that developed progressively fresher water as sedimentation continued. Thus, a complex pattern of marine mudbank deposits with intervening freshwater lake deposits is generated. To the southeast, in the more open-water environments of Florida Bay, Enos and Perkins (1979) consider a similar situation to be unfolding with regards to the whole of the bay. Throughout the history of the bay, mangrove swamps have from time to time colonized the tops of mudbanks. The mangrove swamp then creates a sheltered environment, both within its roots and to the leeward of the mangrove island as a whole. Supratidal sediments build larger islands, lakes become more and more isolated and take on a salinity history of their own. In a very real sense, the net result of these processes is progradation. In a general sense, there is migration of the "shoreline" seaward by sedimentation processes. However, "the position of the shoreline" at any particular time during this process presents an extremely complicated geometry. It may be useful to back off from such strict definition of progradation and recognize the concept of sub-tidal-supratidal or subtidal-freshwater facies mosaic. The culmination of the progradational sequence here is a thin veneer of freshwater swamp. The vertical facies succession leading to that culmination may be quite varied from locality to locality, depending on small-scale variation in the paleogeography.

Andros Island: An Introduction to Supratidal Dolomite. In the northwest portion of Andros Island, the Bahamas, the transition from subtidal to supratidal takes place (1) without the influence of large amounts of fresh water, such as occurs in southern Florida, and (2) with some fair amount of topographic relief (Figure 14.2). The topography in this area is apparently generated by subaerial accumulation of pelleted mud, presumably trapped on the palm hammocks by the dense vegetation during times of unusually high water (during storms, for instance). Subtidal sediments here are likewise pelleted carbonate muds. Between these two sediment types, laminated algal mats and supratidal dolomites record the intertidal to supratidal sedimentation today.

Filamentous algae thrive wherever the sediment surface is wet. They are ubiquitous in subtidal, intertidal, and low-lying supratidal environments. In all of these regions, filamentous algae are capable of binding sediments together (Gebelein, 1969). In the subtidal and intertidal environments, however, the browsing marine benthos (principally gastropods) crop the algae as fast as they grow; burrowing activity also tends to disrupt any algal structure that may be formed. Therefore, it is quite common in modern environments to find laminated algal structures preserved only in the supratidal environment where stress conditions place severe limitations on browsing faunas. (See Logan *et al.*, 1964; and Shinn, 1968, for details of lithology and petrography of stromatolites.)

The Bahamas is our first example of the formation of Recent dolomite under supratidal conditions. Scientists do not completely understand the details of this

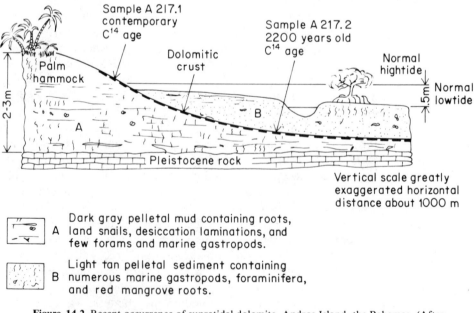

Figure 14.2 text labels:
Sample A 217.1 contemporary C^{14} age
Sample A 217.2 2200 years old C^{14} age
Dolomitic crust
Normal hightide
Normal lowtide
.5 m
Palm hammock
2–3 m
A
B
Pleistocene rock
Vertical scale greatly exaggerated horizontal distance about 1000 m

A — Dark gray pelletal mud containing roots, land snails, desiccation laminations, and few forams and marine gastropods.

B — Light tan pelletal sediment containing numerous marine gastropods, foraminifera, and red mangrove roots.

Figure 14.2 Recent occurrence of supratidal dolomite, Andros Island, the Bahamas. (After Shinn *et al.*, 1965.)

dolomite occurrence. However, two generalities have been made. First, most Recent supratidal dolomite appears to form by the replacement of aragonite rather than by direct precipitation of dolomite. Secondly, studies of pore-water chemistry indicate that the dolomitizing fluid has a magnesium-calcium ratio much greater than that of normal marine seawater.

Two mechanisms have been proposed for the generation of a high magnesium-calcium ratio in the pore water of supratidal sediments. If seawater is evaporated to the point where calcium sulfate is precipitated, a significant portion of seawater calcium is lost from the fluid phase, thus generating a higher magnesium-calcium ratio in the remaining brine (see Butler, 1969, for example). Alternatively, algae tend to concentrate magnesium in their vital fluids. Upon the death of the algae, these fluids are presumably released to the pore water and impart a high magnesium-calcium ratio (Gebelein and Hoffman, 1973). Clearly, the former mechanism requires evaporating conditions, whereas the latter mechanism does not. In the Bahaman example of supratidal dolomite, evidence of significant precipitation of calcium sulfate is lacking. It can be argued that calcium sulfate was never precipitated in this environment or, alternately, that calcium sulfate is precipitated during the dry season and dissolved away by rainwater during the wet season. This problem is important, so we shall dwell on it at some length as we attempt to formulate a general model for dolomite within stratigraphic sequences.

The Sabkhas of Eastern Arabia: Introduction to Supratidal Evaporites. Around the Persian Gulf, the Bahamas model of algal mats and supratidal dolomite is carried a step further. Here there is large-scale precipitation of calcium sulfate and thus the formation of extensive supratidal evaporite deposits (Figures 13.11, 14.3, 14.4, and 14.5).

The southwest coastline of the Persian Gulf is an area of virtually no rainfall and exceedingly high evaporation. For these reasons, the salinity of the Gulf as a whole is considerably higher than normal marine. Pore fluids within the low-lying supratidal flats adjacent to the Persian Gulf are even more concentrated by evaporation. The supratidal flats (sabkhas) are deflation surfaces, controlled by the level of the water table, which is in turn controlled by sea level. Sediment in proximity to the water table remains wet and is not blown away during windstorms. On the other hand, sediment well above the water table may dry out completely and is highly subject to wind transportation.

Because the water table is close to the deflation surface, evaporation from the water table is significant. Water lost to the atmosphere by evaporation must be replaced somehow. Replacement comes through (1) the landward flow of marine water into the water table beneath the sabkha and (2) the seaward flow of continental ground water from the interior of Arabia.

Continuing evaporation and continuing replenishment of the water lead to a steady-state condition of pore-water chemistry, such as is portrayed in Figure 14.5. Inland from the subtidal lagoons bordering the Gulf, pore water becomes more and more concentrated by evaporation. At some point in this process, there is an abrupt

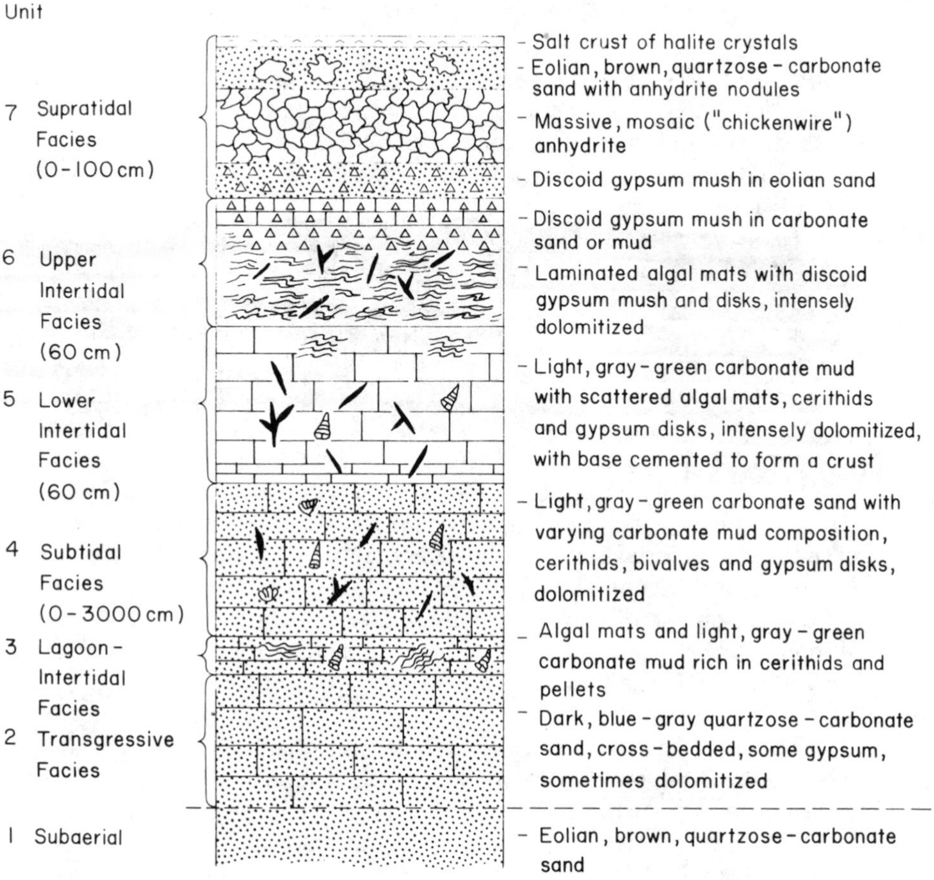

Unit
7 Supratidal Facies (0-100 cm)
6 Upper Intertidal Facies (60 cm)
5 Lower Intertidal Facies (60 cm)
4 Subtidal Facies (0-3000 cm)
3 Lagoon-Intertidal Facies
2 Transgressive Facies
1 Subaerial
- Salt crust of halite crystals
- Eolian, brown, quartzose-carbonate sand with anhydrite nodules
- Massive, mosaic ("chickenwire") anhydrite
- Discoid gypsum mush in eolian sand
- Discoid gypsum mush in carbonate sand or mud
- Laminated algal mats with discoid gypsum mush and disks, intensely dolomitized
- Light, gray-green carbonate mud with scattered algal mats, cerithids and gypsum disks, intensely dolomitized, with base cemented to form a crust
- Light, gray-green carbonate sand with varying carbonate mud composition, cerithids, bivalves and gypsum disks, dolomitized
- Algal mats and light, gray-green carbonate mud rich in cerithids and pellets
- Dark, blue-gray quartzose-carbonate sand, cross-bedded, some gypsum, sometimes dolomitized
- Eolian, brown, quartzose-carbonate sand

Figure 14.3 Idealized vertical profile of a Persian Gulf sabkha transgressive-regressive (progradational) sedimentary sequence. Note that dolomitization tends to be especially intense in intertidal facies (5) and (6). (After McKenzie *et al.,* 1980.)

Sabkha
13 km
Lagoon
Flood water
Algal mats
Gypsum
subtidal muds
Anhydrite
Algal mats
Gypsum cemented layer
Water saturated pleistocene sands
Miocene evaporitic rocks
3 meters
North
South

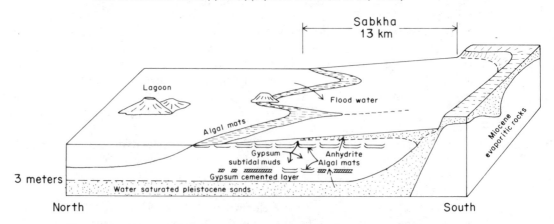

Figure 14.4 Schematic block diagram of facies relationships in the sabkha environment, southern Persian Gulf. (After Butler, 1969.)

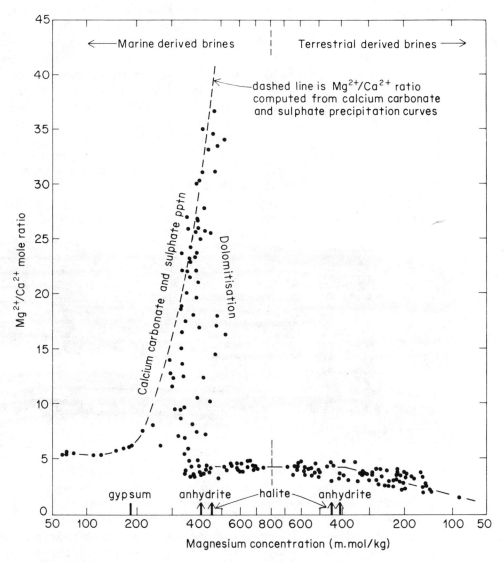

Figure 14.5 Variations in the magnesium-to-calcium ratio, with magnesium concentration, across the sabkha. From the left, sea water evaporates as it comes onto the sabkha. Then precipitation of calcium carbonate and calcium sulfate drives the magnesium-to-calcite ratio in the water very high. Apparently, dolomitization of preexisting calcium carbonate brings the magnesium-to-calcium ratio down again, even as further evaporation of the water continues. From the right side of the diagram, continental groundwaters enter the sabkha and evaporate to halite deposition without the production of a high magnesium-to-calcium ratio. (After Butler, 1969.)

change in the magnesium-calcium ratio of the pore fluid. This change reflects precipitation of calcium sulfate minerals at or near the water table. The calcium sulfate may be gypsum or anhydrite. For our purposes, the distinction is unimportant. Evaporation continues until the water is almost saturated with sodium chloride.

The lithologic consequences of this process are depicted in Figures 14.3 and 14.4. Traversing from the intertidal zone across the sabkha surface, we first encounter an extensive development of algal flat deposits. This environment is only slightly above high-tide level. The sediment surface is kept generally moist by close proximity to the water table and occasional tidal flooding. Dessication polygons are typically well developed. Farther inland, algal deposits give way to carbonate and quartz sands of the sabkha deflation surface. Crystals of both gypsum and anhydrite become volumetrically important in these sediments.

Calcium sulfate crystals commonly grow by displacive precipitation; that is, the mineral does not cement nearby sedimentary particles together but instead pushes them aside as the calcium sulfate crystal grows.

Two striking textures result from this displacive precipitation process. Calcium sulfate beds on the order of centimeters to tens of centimeters thick tend to grow near the water table. As the displacive growth continues, these horizontal layers of calcium sulfate undergo intense deformation. With continuous expansion of the layer by new precipitated material, folding of the calcium sulfate layer ensues. We see then the first texture: highly crenulated, laminated beds of anhydrite. Individual laminae range from millimeters to centimeters in thickness. The entire bed can be as much as 50 centimeters thick.

As the process continues, the second texture develops. Contortion and expansion can obliterate the laminations and produce a more or less massive bed of "chicken-wire anhydrite," in which large aggregates of calcium sulfate are separated from one another by thin irregular strips of the sediment that the calcium sulfate is displacing.

Dolomitization in the sabkha environment is a particularly elusive problem, although all of the elements are here for any possible model. Dolomitization is indeed locally important, but its occurrence is spotty. Therefore, we must be cautious about making generalizations on the origin and importance of dolomite formation in association with supratidal calcium sulfate precipitation.

Coring beneath the anhydrite deposits of the sabkha deflation surface yields a vivid picture of the dynamics of tidal-flat progradation following the post- Wisconsin rise in sea level (Figure 14.3). The underlying subaerial erosion surface is closely overlain by subtidal carbonate mud. The vertical transition from subtidal carbonates to algal stromatolites to anhydrite deposits of the sabkha flat is an excellent vertical record of precisely the horizontal facies variations that are observed today in going from the subtidal lagoon across the algal mats onto the sabkha deflation surface.

Pore Fluids in Pleistocene Carbonates: Introduction to Carbonate Diagenesis. Most carbonate sediments in Recent environments are composed of relatively unstable minerals such as aragonite, high-magnesium calcite, and protodolomite; whereas, Ancient carbonate rocks are low-magnesium calcite and well-ordered dolomite. Studies of Pleistocene carbonates indicate that this transition can occur relatively soon after sediment deposition through a process of dissolution-reprecipitation.

Figure 14.6 presents a schematic cross section through a carbonate coastline that includes late Pleistocene coral-reef terraces. When rainwater falls on the Pleistocene terraces, these carbonate sediments are exposed to fresh water in the *vadose* (above the water table) and *phreatic* (below the water table) environments. In the presence of fresh water, the original sedimentary mineralogy of aragonite and high-magnesium calcite recrystallizes to low-magnesium calcite within 10^3 to 10^5 years (Matthews, 1968, 1974; Steinen and Matthews, 1973; for example).

The situation becomes even more exciting in the mixing zone that separates the freshwater phreatic lens from salt water. Calculations indicate that the mixed water should be undersaturated with respect to calcite and supersaturated with respect to dolomite (Hanshaw *et al.*, 1971). Thus, the mixing zone between freshwater phreatic and marine water is a likely environment in which to precipitate dolomite. Indeed, Land (1973) reports partially dolomitized Pleistocene carbonates that may reflect this process. However, completely dolomitized sediments in this environment have not been documented with regard to modern phreatic lenses. Nevertheless, there are numerous Ancient examples for which this model appears to be valid—a point to which we shall return later.

Thus we see, both with respect to recrystallization and dolomitization, recognition of subaerial exposure surfaces in carbonate sequences is extremely important. If a sequence of carbonate sediments has been exposed to fresh water soon after deposition, then the diagenetic overprints discussed above are to be anticipated. Further, if it can be demonstrated that subaerial exposure resulted from sea-level fluctuation, then that sea-level fluctuation must likewise have affected all sedimentary facies within the regional paleogeography. Documentation of amplitude and

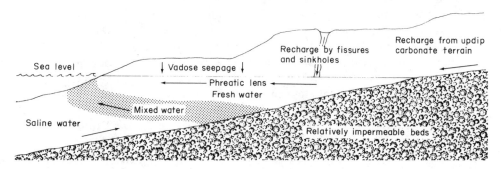

Figure 14.6 Schematic cross section depicting the major diagenetic environments associated with freshwater flow through Pleistocene bank-margin carbonates. The term "vadose" refers to the diagenetic environment above the water table and the term "phreatic" refers to the diagenetic environment below the water table. (After Matthews, 1974.)

frequency of sea-level fluctuations is, therefore, an important goal, and carbonate sequences offer good opportunity for gaining this information.

Subaerial exposure horizons can sometimes be recognized as soil zones or as characteristic caliche crusts that form immediately beneath soil zones (Harrison and Steinen, 1978, for example). In the absence of straightforward lithologic evidence of subaerial exposure horizons, there is a carbon and oxygen isotope technology which commonly picks up these discontinuities (Allan and Matthews, 1977, 1982). Highly negative $\delta^{13}C$ values are locked into carbonate rocks that have recrystallized immediately beneath a soil zone. The highly negative $\delta^{13}C$ values result from incorporation of soil gas derived from rotting organic matter. Oxygen isotope data may also show important variations across subaerial exposure surfaces.

Continuing Development of a Basic Model for Carbonate and Evaporite Sedimentation

The following discussion concerning the generalities of intertidal and supratidal sedimentation runs parallel to similar discussions in Chapter 13. Note well that these remarks will be simply additions to models already under development in the previous chapter.

Aerial Distribution of Contemporaneous Sedimentary Facies. We have previously observed that sediments in the subtidal shelf-interior environment are strongly influenced by the conditions of seawater circulation and the climate. The transition from subtidal to supratidal is subject to even greater variation as a function of these variables. If carbonate sedimentation is marginal to a clastic land mass, such as in Belize, then the transition from subtidal to supratidal along the mainland coast will occur in coastal clastic facies rather than in carbonate facies. Without a large clastic input, the transition from intertidal to supratidal is primarily dependent on the presence or absence of fresh water. Sediment types anticipated in this geographic position range from freshwater swamps (the Everglades, southern Florida) to mildly evaporating, dolomitic, stromatolitic tidal flats (the Bahamas) to highly evaporatic sabkha environments with crenulated anhydrite, chicken-wire anhydrite, or halite (the Persian Gulf). The stromatolitic, lower supratidal environment would be expected to provide a good environmental datum, with the one qualification that the absence of a subtidal browsing and burrowing fauna may allow stromatolite development in subtidal environments.

Progradation in Intertidal-Supratidal Carbonates and Evaporites. Certainly the concept of progradation has a straightforward application in examples like the sabkha flats of the Persian Gulf. Clearly, relationships between subtidal lagoonal lime mud, high intertidal to supratidal stromatolites, and sabkha evaporite flats are quite analogous to offshore silts, beaches, and topset marsh deposits of the Chenier Plain, for example.

Note that progradation of supratidal algal mat and sabkha environment over subtidal lagoonal carbonates tends to produce a sedimentary sequence in which mineralogy as well as sedimentary facies varies systematically. A sedimentary couplet of subtidal limestone below and supratidal dolomite above will be a fundamental building block of many Ancient carbonate sequences.

The transition from northern Florida Bay into the freshwater swamps of the Everglades offers us an interesting example of just how complicated a prograding biogenic sequence can become. This very complicated facies mosaic is the direct result of very local production of carbonate and bituminous sediments.

As in our discussion of topset accumulations in coastal clastic environments, note that there is a serious space problem that precludes the accumulation of thick, topset depositions under conditions of constant sea level.

Invasion of Subtidal Carbonates by Freshwater Phreatic Lenses. The invasion of aragonite and high-magnesium calcite sediments by fresh water may result in recrystallization in the freshwater diagenetic environment or dolomitization in the mixed-water diagenetic environment. While most of our examples of these processes are taken from phreatic lenses presently situated in Pleistocene carbonates, one can imagine similar processes occurring wherever progradation allows freshwater runoff to invade former subtidal sediments. On the outcrop or well-site scale of observation, it is important to note that phreatic lens processes tend to generate limestone-dolomite couplets in which the dolomite (mixing zone) is overlain by limestone (freshwater phreatic lens), just the reverse of the dolomite-limestone couplets produced by progradation of supratidal facies (dolomite) over subtidal facies (limestone).

Response of Supratidal Environments to Conditions of Submergence. We note once more the parallel between supratidal carbonates and the topset beds of coastal clastic sedimentation. In both cases, submergence is required if there is to be sufficient room to accommodate a thick section of these facies.

The rate of sea-level rise will also affect the geometric relationships of supratidal facies and related sediment. In the coastal swamp environment, rising water may accommodate vertical accumulation of subtidal to intertidal marine-carbonate deposits, brackish-water swamp deposits, and freshwater swamp deposits, more or less as they appear in map view. On the sabkha flats, rising sea level means a rising water table (deflation surface) and therefore opportunity for vertical accumulation of supratidal deposits. As long as net deposition of eolian detritus and precipitated evaporite minerals can keep this area above sea level, vertical contacts will be generated among evaporite facies, algal stromatolite facies, and subtidal lagoonal carbonate-mud facies. If sediment accumulation on the sabkha flat cannot keep pace with rising water, transgression will occur. Algal stromatolite facies may migrate landward over evaporite facies in a continuous fashion, or the entire sabkha flat may develop subtidal conditions by undergoing discontinuous floods.

Response of Supratidal Environments to Conditions of Emergence. Supratidal environments associated with carbonate deposition are closely related to the water table, which is in turn closely related to sea level. As the water table falls during conditions of emergence, former supratidal deposits are left high and dry. Freshwater swamps may become dry land and thus cease to produce carbonate mud and peat. Sabkhas will very likely undergo eolian deflation.

An Ancient Example: The Helderberg Group, Devonian of New York

Let us now attempt to apply our general model to an Ancient stratigraphic sequence. Does the model explain what we see in Ancient rocks? Does the model help us to understand better the precise earth history recorded by these rocks?

Regional Setting and Earlier Stratigraphic Work. The Helderberg Group of western New York State has long been recognized as containing some of the oldest Devonian fossils in North America. The Group has been designated as the type section for the Helderbergian stage, the lowermost Devonian stage of North American biostratigraphy. Because of the biostratigraphic significance attached to the Helderberg Group, an understanding of its sedimentology and precise depositional history is fundamental to the understanding of the Silurian-Devonian contact. For example, is there or is there not a major worldwide transgressive episode separating the Silurian from the Devonian? If a major transgressive event can be demonstrated, then transgressive basal Devonian deposits may be synchronous around the world. If, on the other hand, no major transgressive event exists, then we must face the possibility that the biostratigraphically defined transition from Silurian to Devonian may be diachronous on a worldwide scale.

For many years, geologists believed that the Helderberg Group was a layer-cake arrangement of lithologically defined formations (Figure 14.7). Each lithology was thought to represent a discrete "time." The Rondout and Manlius formations were considered Upper Silurian, and the Coeymans was the base of the Lower Devonian Helderbergian stage.

The carbonate sequences under consideration rest with an angular unconformity upon Middle Ordovician shales and sandstones. The Rondout and Manlius formations are poorly fossiliferous. The Manlius is primarily dark grey to black, fine-grained, carbonate rock, with occasional bituminous laminated horizons. The fine grain size, lack of fossils, and bituminous nature of the Manlius led early workers to think that this environment had been relatively deep (below wave base) stagnant water.

Coeymans rocks, with their diverse fauna and high abundance of skeletal sands, were taken to indicate "a time" of shallower water and better circulation. Muddy Kalkberg rocks suggest a deepening of the water and the first significant influx of clastic detritus into this area from the east. In "New Scotland time," marine, fine-grained, clastic sedimentation dominated the area.

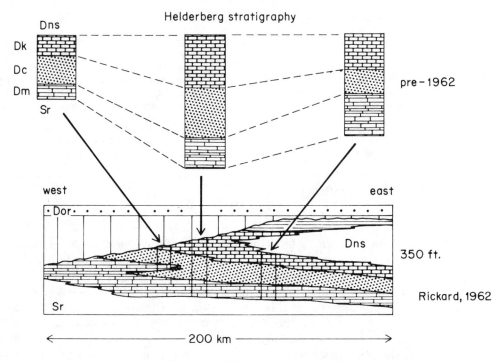

Figure 14.7 Stratigraphic diagram of different interpretations of Helderberg stratigraphy in New York. Whereas the Helderberg Group had peviously been considered layer-cake stratigraphy in which each rock unit represented a different period of time, Rickard (1962) demonstrated intertonguing relationships among these units. (After Laporte, 1969.)

Rickard (1962) demonstrated intertonguing relationships between Manlius and Coeymans (Figure 14.7). This demonstration of diachronous relationships required that geologists rethink the layer-cake version of earth history. If the Manlius formation does not represent "a time of...," then what does it represent?

Sedimentological Observations within the Manlius Formation. Recognizing the major questions of sedimentological history that are posed by Rickard's stratigraphic observations, Laporte (1967, 1969) undertook detailed sedimentological and paleoecological studies within the Helderberg Group. In the following discussion, we shall emphasize his work in the Manlius formation, because understanding the intertonguing relationship between Manlius and Coeymans provides the key to understanding the history recorded by the Helderberg Group.

Laporte recognized three distinct facies within the Manlius formation. Their characteristics are summarized in Table 14.1, where the facies are labeled *A, B,* and *C.* Let us work along with Laporte in attempting to assess their paleogeographic significance.

Consider first facies *A*. The fauna contains a bit of everything. The lithology is generally carbonate mudstone with only rare occurrences of moderately clean

Table 14.1 Lithologic and Paleontologic Attributes of Manlius Facies

Facies	Lithology	Paleontology
A	Pelletal, carbonate mudstones and reefy biostromes. Medium to massively bedded; in places "reefoid."	Biota relatively abundant and diverse. Stromatoporoids, rugose corals, brachiopods, ostracods, snails, and codiacean algea.
B	Interbedded, pelletal, carbonate mudstone and skeletal calcarenite; a few limestone-pebble conglomerates and mud cracks.	Fossil types few but individuals abundant. Ostracods, tentaculitids, brachiopods, oncolites, bituminous laminations.
C	Dolomitic, laminated mudstone. Disruption of laminations common.	Fossils scarce; bituminous laminations, ostracods, and burrows.

After Laporte, 1967, with modification.

sands. This lithologic and paleontologic description sounds reasonable for a subtidal platform-interior environment. The presence of green algae and oncolite structures suggests only moderate water depth, but the precise depth is undetermined. Perhaps we can learn something about that feature from the other facies.

Next consider facies *B* in comparison with facies *A.* Green algae, corals, and echinoderms, which are present in facies *A,* are all absent from facies *B.* Well-preserved laminations, lacking in facies *A,* are quite common in facies *B.* Furthermore, the abundance of snails is reduced and spar-cemented carbonate sands are common. Indeed, occasional beds within facies *B* are coquinas of mollusc and possibly brachiopod debris.

In facies *C,* the fauna is still further reduced; spar-cemented sediments are absent; dolomite, detrital quartz, and clay are more abundant than elsewhere; and bituminous laminations are common.

These bituminous laminations are the features that led early geologists to conclude that the Manlius environment had been deep, stagnant water. Closer examination proves interesting. Laminae and groups of laminae are commonly disrupted (Figures 14.8 and 14.9). Typically, disrupted segments of laminae are somewhat curled; on occasion, curled chips have apparently been overturned and transported. In still other cases, disrupted laminations appear to have formed in a convex configuration, whereas intervening laminations appear to have formed in a concave configuration.

Sedimentological Interpretation. The interpretation placed on these observations by Laporte (1967) is summarized in Table 14.2 and Figure 14.10. Facies *A* is taken to represent moderately restricted, platform-interior sedimentation. Facies *B,* with restricted fauna and biopelsparites, is the intertidal environment. The bituminous laminations of facies *C* are interpreted as algal stromatolites, complete with dessication chips and mud cracks, as pictured in Figures 14.8 and 14.9. Facies *C* is identified as a supratidal environment.

These facies interpretations are exciting pieces of paleogeographic information.

Figure 14.8 Negative print of a peel of facies *C,* Manlius, showing the discontinuity of laminations. Observe the concave-up tendency in disrupted laminae at the upper right. (Peel courtesy of L. F. Laporte.)

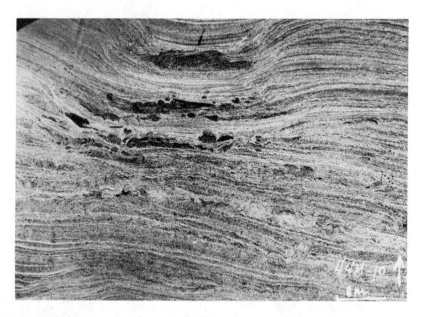

Figure 14.9 Negative print of a peel of facies *C,* Manlius, showing the concave-down arrangement of laminae. Whereas Figure 14.8 depicts laminae that appear to have been broken, many of these laminae actually seem to have grown in a discontinuous concave-down configuration. (Peel courtesy of L. F. Laporte.)

Table 14.2 Lithologic and Paleontologic Attributes of Manlius Facies with Interpretation

Facies	Lithology	Paleontology	Interpretation
C	Dolomitic, laminated mudstone; mud cracks; "birds's-eye."	Fossils scarce; algal laminae, ostracods, and burrows.	Supratidal
B	Interbedded, pelletal, carbonate mudstone and skeletal calcarenite; a few limestone-pebble conglomerates and mud cracks.	Fossil types few but individuals abundant. Ostracods, tentaculitids, brachiopods, algal stromatolites, and oncolites.	Intertidal
A	Pelletal, carbonate mudstones and reefy biostromes. Medium to massively bedded; in places "reefoid."	Biota relatively abundant and diverse. Stromatoporoids, rugose corals, brachiopods, ostracods, snails, and codiacean algae.	Subtidal

After Laporte, 1967, with modification.

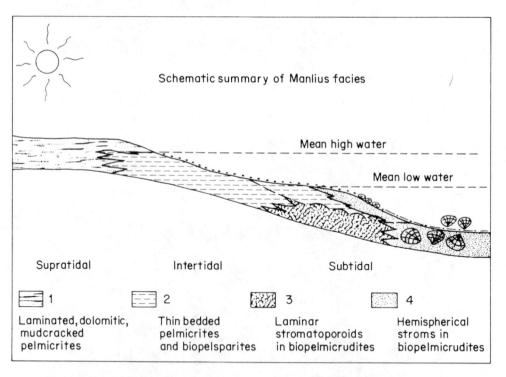

Figure 14.10 Schematic cross section showing interpreted relationships among four Manlius facies. Sediment types are as follows: (1) supratidal laminated, dolomitic, mud-cracked pelmicrites; (2) intertidal thin-bedded pelmicrites and biopelsparites, with stromatolites, and mud cracks; (3) subtidal channel-floor deposits, predominantly biopelmicrudites with laminar stromatoporoids; (4) subtidal biopelmicrudites with hemispherical stromatoporoids and oncolites. In the intertidal to subtidal portion of the diagram, symbols below the dashed line represent tidal-channel deposits, whereas symbols above the dashed line represent tidal-flat and normal subtidal deposits. (After Laporte, 1967, with modification.)

If we examine the Coeymans biosparites with diverse fauna, they look like shelf-margin facies. Kalkberg biomicrites become foreshelf facies. The fact that Kalkberg overlies Coeymans and Manlius to the southeast indicates that submergence occurred within a rather lowp-lying paleotopography. These views are put forward in detail by Laporte (1969).

Let us, therefore, acknowledge three major environments within the Manlius formation and recognize the paleogeographic relationships within the Helderberg Group. What can we then say about the history of the Helderberg transgression? Figure 14.11 summarizes facies relationships within the Manlius.

The Thachter member of the Manlius formation was apparently deposited on a broad, shallow shelf interior. Persistent occurrences of supratidal facies within predominantly subtidal Thachter facies confirm rather shallow-water conditions.

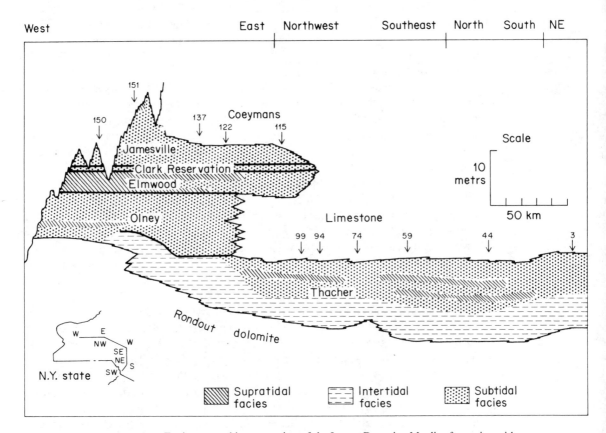

Figure 14.11 Environmental interpretation of the Lower Devonian Manlius formation with its five members. Lateral migration of facies depicted in Figure 14.10 resulted in the complex facies mosaic that we see in the Manlius today. (After Laporte, 1967, with modification.)

With continuing submergence, the shelf margin apparently succumbed to rising water and reestablished itself in the vicinity of localities (99) to (115). Again, the generally shallow character of the shelf interior during this time is confirmed by the occurrence of supratidal facies in among predominantly subtidal facies of the Olney member. Upward accumulation of sediment was in close balance with the rate of submergence, thus generating the relatively restricted zone of interfingering between the Manlius and Coeymans lithologies throughout the deposition of more than 10 meters of sediment.

Note at locality (115) the occurrence of subtidal Manlius over Coeymans, a "regressive" relationship. Regression at this time is further confirmed by the supratidal Elmwood over the subtidal Olney member. But what kind of regression allows subtidal shelf-interior sediment to be deposited overlying shelf-margin biosparite facies?

According to our general model of carbonate sedimentation, this sequence of lithologies records progradation under conditions of net submergence. The Coeymans shelf-margin facies was capable of producing more sediment that was required to keep pace with rising water. Therefore, this facies prograded seaward. The Elmwood supratidal flats had no difficulty keeping up with rising water and so began to prograde out across the shallow subtidal environments. The subtidal Manlius in this interval was simply the shallow water that was left *behind* the prograding Coeymans shelf margin and *in front of* the prograding Elmwood supratidal flats.

Moving up the section, we see that the rate of submergence once more exceeded sedimentation rate. Jamesville subtidal Manlius overlies Elmwood supratidal Manlius, and Coeymans shelf-margin facies once again migrates westward.

Let us return to the problem of the Silurian-Devonian boundary, which we mentioned briefly in the introduction to the Helderberg Group. We are familiar with the age-old problem of where to place boundaries between systems and periods. If there is a major event, such as a worldwide transgression, we use it. Rocks above it can be separated from rocks below it, and the basal transgressive rocks are more or less synchronous around the world. Helderberg rocks are traditionally regarded as the lowest Devonian rocks of North America. Is the transgression that brought these rocks into existence likely to be of worldwide significance?

Manlius and Coeymans taken together never exceed 50 meters in thickness. Contrary to earlier beliefs that the Manlius represented relatively deep water (and therefore a major transgression), it now appears that the Manlius was deposited at or near sea level during conditions of submergence. Taking isostasy into account, we observe that a relative sea-level rise of less than 20 meters is probably sufficient to account for the measured thickness of Manlius and Coeymans sediments. Such a scale of sea-level fluctuation will hardly prove useful for worldwide correlation purposes. Thus, the Helderberg Group is probably just a little vignette in earth history. It does not record a worldwide event of profound importance; instead, another small piece of low-lying coastline happened to get a little water over it about 350 million years ago.

An Ancient Example:
The Permian of West Texas

Permian strata of west Texas not only provide an excellent opportunity to test our carbonate sedimentation models, but they also exemplify the economic importance of carbonate rocks. In this particular example, the discovery of oil preceded systematic study of the carbonate rocks. However, let us review that facts so that this exploration province can serve as a model in which stratigraphic geology may lead to the discovery of petroleum, whether it be in a new field in the Permian of west Texas or the first discovery in some new and previously unexplored basin.

Regional Setting and Older Stratigraphic Work. The discovery of major petroleum accumulations in the Permian of west Texas occurred in the 1920's. The productive formations were at relatively shallow depths (700 to 1500 meters). As was the custom of the day, wildcatters punched down holes at random, wherever they could put together a large block of land and get sufficient backers to pay for the drilling of the hole. This process was rather unscientific, but it found a lot of oil.

As time went by, more and more data accumulated, and more scientific methods were applied to petroleum exploration in this region. Early drilling documented the existence of the Delaware basin and the Midland basin (Figure 14.12). Both of these basins are Permian topographic features that have no expression in the Cretaceous rocks now covering the area. Wells drilled in the eastern shelf area or on the central basin platform encountered Permian carbonate rocks at relatively shallow depth, and a significant number of such wells found oil. Wells drilled in the basin areas encountered thick sections of anhydrite and seldom found Permian oil.

As this paleogeography became apparent, geologists realized that rocks of equivalent age to those in the central basin platform and eastern shelf producing horizons cropped out in the Guadalupe Mountains. Clearly, detailed knowledge of the surface exposures of these economically important units could be valuable in further exploration of Permian strata in the subsurface of west Texas.

In the Guadalupe and Delaware mountains of west Texas and New Mexico, Permian paleogeography can be easily reconstructed. Fine-grained sandstones and occasional black limestones of the Delaware mountain group interfinger with limestones and dolomites of the Guadalupe Mountains (Figure 14.13). Furthermore, the high angle of dip and the abundant evidence of transportation led early workers to recognize the limestones of the eastern front of the Guadalupe Mountains as talus deposits marginal to shallow-water, carbonate-shelf deposits. Indeed, on the basis of field examination and stratigraphic relationships, it appeared quite reasonable to consider the massive gray to white limestone updip from these talus deposits as the "reef" from which the talus was derived. In this manner, the "Permian reefs" of western Texas became legendary.

Note well that from the very beginning of stratigraphic studies in the Guadalupe

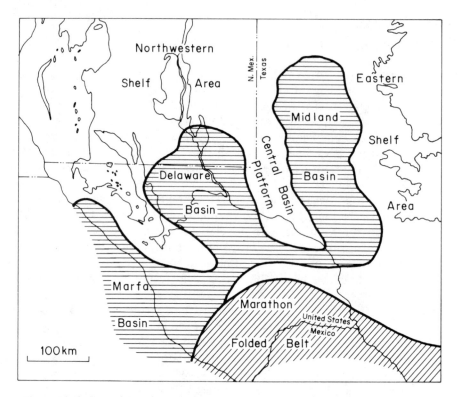

Figure 14.12 General Permian paleogeography in west Texas and adjoining New Mexico.

Mountains, the basic platform and basin paleogeography was recognized, and attempts were made to work with the stratigraphy in the context of the paleogeography. This starting point is tremendously elevated, compared to early work in the Helderberg Group, for example.

Yet, even with this early recognition of basic paleogeography, lithologic descriptions continued to emphasize gross aspects of the rock unit rather than sedimentology of the unit. Thus, Figure 14.13 is only partially analogous to studies of Recent sediment. With the stratigraphy under control, it was only a matter of time until geologists would want to know more about the details of sedimentation in the "Permian reefs."

Sedimentological Observations in the Guadalupe Mountains. Figure 14.14 presents brief lithologic descriptions in modern sedimentological terms. Consider these descriptions from the basin, up onto the shelf, and back across the shelf.

The lower portion of the "forereef talus" is wackestone, which is carbonate sand with an abundant carbonate-mud matrix. Proceeding updip, we see that the carbonate sands contain less mud. Skeletal grainstones predominate at the top of the eastern part of the present-day Guadalupe Mountains. In terms of paleogeog-

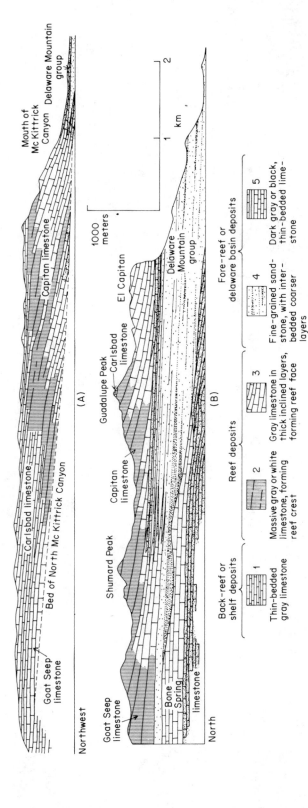

Figure 14.13 Two cross sections through the Permian rocks of the Guadalupe Mountains, west Texas. Section A is stratigraphically higher than section B. Rock types are as follows: (1) thin-bedded gray limestone, (2) massive gray or white limestone, (3) gray limestone in thick inclined layers, (4) fine-grained sandstone with interbedded coarser layers, (5) dark-gray or black thin-bedded limestone. (After King, 1948.)

337

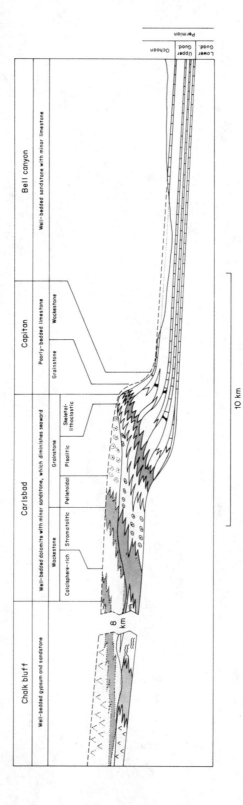

Figure 14.14 Sedimentary facies of Upper Guadalupian (Permian) carbonate rocks in the Guadalupe Mountains and surrounding area. (After Dunham, 1972.)

338

raphy, this is the area that separates flat-lying, shallow-water carbonates to the west from gently dipping, "forereef" carbonates to the east. Interior from the skeletal grainstone facies is a pisolitic facies; interior to the pisolitic facies is pelletoidal grainstone; and so on further back onto the shelf, through stromatolitic wackestone, into interbedded evaporite and sandstone.

Sedimentological Interpretation. After you examine Figures 14.13 and 14.14, it will come as no surprise that the Permian of west Texas has become a classic example of regressive shelf-margin facies with a restricted, shallow, subtidal to supratidal shelf interior. That the basinward progradation of the shelf margin occurred during a time of net submergence is amply confirmed by the thick stromatolitic and evaporite sequences of the shelf interior.

Note that Figure 14.4 makes no mention of "reef." Simply stated, the massive *in situ* skeletal structures are not an important part of the shelf-margin carbonate complex in the Permian of west Texas. Terminology which seemed quite appropriate in the light of megascopic field examination (that is, massive gray or white limestone = a reef crest) has simply not withstood detailed petrologic investigation.

Citations and Selected References

ALLAN, J. R., and R. K. MATTHEWS. 1977. Carbon and oxygen isotopes as diagenetic and stratigraphic tools: surface and subsurface data, Barbados, West Indies. Geology 5: 16–20.

――――. 1982. Isotope signatures associated with early meteoric diagenesis. Sedimentology 29: 797–817.

BADIOZAMANI, K. 1973. The Dorag dolomitization model—application to the Middle Ordovician of Wisconsin. J. Sed. Petrology 43: 965–984.
See also Discussion in Sed. Petrology 46(1): 254–258.

BORCHERT, H. 1969. Principles of oceanic salt deposition and metamorphism. Bull. Geol. Soc. Amer. 80: 821–864.
A summary paper in English by a deep-water evaporite specialist who usually publishes in German.

BUTLER, G. P. 1969. Modern evaporite deposition and geochemistry of coexisting brines, the sabkha, Trucial Coast, Arabian Gulf. J. Sed. Petrology 39: 70–89.
General stratigraphy and pore-water chemistry of dolomitization and evaporite deposition on supratidal flats.

――――. 1970. Holocene gypsum and anhydrite of the Abu Dhabi Sabkha, Trucial Coast: an alternative explanation of origin, p. 120–152. *In* J. L. Rau and L. F. Dellwig (eds.), Third symposium on salt. Northern Ohio Geol. Soc.
An extensive and well-illustrated discussion.

BUZZALINI, A. D., F. J. ALDER, and R. L. JODRY (eds.). 1969. Evaporites and petroleum. Bull. Amer. Assoc. Petrol. Geol. 53: 775–1011.
Entire issue of the bulletin devoted to evaporite sedimentation.

DEFFEYS, K. S., F. J. LUCIA, and P. K. WEYL. 1965. Dolomitization of Recent and Plio-Pleistocene sediments by marine evaporite waters on Bonaire, Netherlands Antilles, p. 71–88. *In* L. C. Pray and R. C. Murray (eds.), Dolomitization and limestone diagenesis. Soc. Econ. Paleont. Mineral. Spec. Pub. 13.

DUNHAM, R. J. 1972. Capitan Reef, New Mexico and Texas: facts and questions to aid interpretation and group discussion. Soc. Econ. Paleont. Mineral., Permian Basin Section (Midland, Texas), Pub. 72–14. 213 p.

ENOS, PAUL, and RONALD D. PERKINS. 1979. Evolution of Florida Bay from island stratigraphy. Bull. Geol. Soc. Amer. 90(1): 59–83.

GEBELEIN, C. D. 1969. Distribution, morphology, and accretion rate of Recent subtidal algal stromatolites, Bermuda. J. Sed. Petrology 39: 49–69.
A "breakthrough" study concerning the details of sediment entrapment by filamentous algae.

———. 1977. Dynamics of recent carbonate sedimentation and ecology: Cape Sable, Florida. E. J. Brill, Leiden. 120 p.

GEBELEIN, C. D., and P. HOFFMAN. 1973. Algal origin of dolomite laminations in stromatolitic limestone. J. Sed. Petrology 43: 603–613.

HANSHAW, B. B., W. BACK, and R. G. DEIKE. 1971. A geochemical hypothesis for dolomitization by groundwater. Eco. Geol. 66: 710–724.

HARDIE, LAWRENCE A., and HANS P. EUGSTER. 1980. Evaporation of seawater: calculated mineral sequences. Science 208 (4443): 498–501.

HARRISON, R. S., and R. P. STEINEN. 1978. Subaerial crusts, caliche profiles and breccia horizons: comparison of some Holocene and Mississippian exposure surfaces, Barbados and Kentucky. Bull. Geol. Soc. Amer. 89: 385–396.

HSÜ, K. J., W. B. F. RYAN, and M. B. CITA. 1973. Late Miocene desiccation of the Mediterranean. Nature 242: 240–244.
The "modern way" to make "deep-water evaporites": Just evaporate all of the water out of the basin.

KENDALL, A. C. 1978. Facies models 11: continental and supratidal evaporites. Geoscience Canada 5: 66–78.

KENDALL, C. G. ST. C., and P. A. D. SKIPWITH. 1969. Geomorphology of a Recent shallow-water carbonate province: Khor Al Bazam, Trucial Coast, southwest Persian Gulf. Bull. Geol. Soc. Amer. 80: 865–892.
Especially noteworthy for its large foldout map (plate 6).

KING, P. B. 1948. Geology of southern Guadelape Mountains, Texas. U. S. Geol. Survey Prof. Paper 215, 183 p.

KIRKLAND, D. W., and R. Y. ANDERSON. 1970. Micro-folding in the Castile and Todilto evaporites, Texas and New Mexico. Bull. Geol. Soc. Amer. 81: 3259–3282.

LAND, LYNTON S. 1973. Holocene meteoric dolomitization of Pleistocene limestone, North Jamaica. Sedimentology 20: 411–424.

————. 1980. The isotopic and trace element geochemistry of dolomite: the state of the art. p. 87–110. *In* D. H. Zenger, J. B. Dunham, and R. L. Ethington (eds.), Concepts and models of dolomitization. Soc. Econ. Paleont. Mineral. Spec. Pub. 28.
Note that chemical data on Ancient dolomites does not bear great similarity to chemical data on Holocene supratidal dolomite. Either Holocene supratidal dolomites are not as good a model for Ancient dolomites as previously proposed, or dolomites tend to undergo recrystallization in much the same way as aragonite and high-magnesium calcite recrystallized to low-magnesium calcite.

LAPORTE, L. F. 1967. Carbonate deposition near mean sea level and resultant facies mosaic: Manlius formation (Lower Devonian) of New York State. Bull. Amer. Assoc. Petrol. Geol. 51: 73–101.
Modern reinterpretation of classical sequences that were grossly misunderstood by early workers.

————. 1969. Recognition of a transgressive carbonate sequence within an epeiric sea: Helderberg Group (Lower Devonian) of New York State, p. 98–118. *In* G. M. Friedman (ed.), Depositional environments in carbonate rocks. Soc. Econ. Paleont. Mineral. Spec. Pub. 14.

————. 1971. Paleozoic carbonate facies of the central Appalachian shelf. J. Sed. Petrology 41:724–740.

LOGAN, B. W., R. REZAK, and R. N. GINSBURG. 1964. Classification and environmental significance of stromatolites. J. Geology 72: 68–83.

LUCIA, F. J. 1968. Recent sediments and diagenesis of south Bonaire, Netherlands Antilles. J. Sed. Petrology 38: 845–858.

MATTHEWS, R. K. 1968. Carbonate diagenesis: equilibration of sedimentary mineralogy to the subaerial environment: Coral Cap of Barbados, West Indies. J. Sed. Petrology 38: 1110–1119.

————. 1974. A process approach to diagenesis of reefs and reef-associated limestones, p. 234–256. *In* L. F. Laporte, Reefs in time and space. Soc. Econ. Paleont. Mineral. Spec. Pub. 18.

MCKENZIE, JUDITH ANN, KENNETH J. HSÜ, and JEAN F. SCHNEIDER. 1980. Movement of subsurface waters under the sabkha, Abu Dhabi, UAE, and its relation to evaporative dolomite genesis. p. 11–30. *In* D. H. Zenger, J. B. Dunham, and R. L. Ethington (eds.), Concepts and models of dolomitization. Soc. Econ. Paleont. Mineral. Spec. Pub. 28.

MEYERS, W. J. 1978. Carbonate cements: their regional distribution and interpretation in Mississippian limestones of southwestern New Mexico. Sedimentology 25 (3): 371–399.

MURRAY, R. C., and F. J. LUCIA. 1967. Cause and control of dolomite distribution by rock selectivity. Bull. Geol. Soc. Amer. 78: 21–35.
Selective dolomitization was apparently controlled by the flow path of early diagenetic fluids.

————. 1969. Hydrology of south Bonaire, Netherlands Antilles: a rock selective dolomitization model. J. Sed. Petrology 39: 1007–1013.

PATTERSON, R. J., and DAVID J. J. KINSMAN, 1977. Marine and continental groundwater sources in a Persian Gulf coastal sabkha, p. 381–397. Studies in Geology 4. Amer. Assoc. Petrol. Geol.

PINGITORE, NICHOLS E., JR. 1976. Vadose and phreatic diagenesis: processes, products and their recognition in corals. J. Sed. Petrology 46 (4): 985–1006.

PRAY, L. C., and R. C. MURRAY (eds.). 1965. Dolomitization and limestone diagenesis. Soc. Econ. Paleont. Mineral. Spec. Pub. 13. 180 p.

PURSER, B. H. (ed.). 1973. The Persian Gulf. Springer-Verlag, New York. 471 p.
A collection of papers concerning Holocene carbonate sedimentation and diagenesis.

RAUPE, O. B. 1970. Brine mixing: an additional mechanism for formation of basin evaporites. Bull. Amer. Assoc. Petrol. Geol. 54: 2246–2259.

RICKARD, L. V. 1962. Late Cayugan (Upper Silurian) and Helderbergian (Lower Devonian) stratigraphy in New York. New York State Museum and Science Service Bull. 386. 157 p.

SCHOLL, D. W., and M. STUIVER. 1967. Recent submergence of southern Florida: a comparison with adjacent coasts and other eustatic data. Bull. Geol. Soc. Amer. 78: 437–454.

SEARS, S. O., and F. J. LUCIA. 1980. Dolomitization of northern Michigan Niagara reefs by brine refluxion and freshwater/seawater mixing. p. 215–235. *In* D. H. Zenger, J. B. Dunham, and R. L. Ethington (eds.), Concepts and models of dolomitization. Soc. Econ. Paleont. Mineral. Spec. Pub. 28.

SHEARMAN, D. J., and J. G. FULLER. 1969. Anhydrite diagenesis, calcitization, and organic laminites, Winnipegosis formation, Middle Devonian, Saskatchewan. Bull. Can. Petrol. Geol. 17: 496–525.
Argues a shallow-water depositional environment for basin-center evaporites.

SHINN, E. A. 1968. Practical significance of birdseye structures in carbonate rocks. J. Sed. Petrology 38: 215–233.

SHINN, E. A., R. N. GINSBURG, and R. M. LLOYD. 1965. Recent supratidal dolomite from Andros Island, Bahamas, p. 112–123. *In* L. C. Pray and R. C. Murray (eds.), Dolomitization and limestone diagenesis. Soc. Econ. Paleont. Mineral. Spec. Pub. 13.

STEINEN, R. P., and R K. MATTHEWS. 1973. Phreatic vs. vadose diagenesis: stratigraphy and mineralogy of a cored borehole on Barbados, W.I. J. Sed. Petrology 43: 1012–1020.

THOMAS, G. E., and R. P. GLAISTER. 1960. Facies and porosity relationships in some Mississippian carbonate cycles of western Canada basin. Bull. Amer. Assoc. Petrol. Geol. 44: 569–588.

VON DER BORCH, C. C., and D. LOCK. 1979. Geological significance of Coorong dolomites. Sedimentology 26 (6): 813–824.

WAGNER, P. D., and R. K. MATTHEWS. 1982. Porosity preservation in the upper Smackover (Jurassic) carbonate grainstone, Walker Creek field, Arkansas: response to paleophreatic lenses to burial processes. J. Sed. Petrology 52: 3–18.

WALTER, M. R. (ed.). 1976. Stromatolites. Developments in sedimentology 20. Elsevier, Amsterdam. xi + 790 p. $99.95.
See also book reviews in J. Sed. Petrol. 47 (3): 1402.

WIGLEY, T. M. L., and L. N. PLUMMER. 1976. Mixing of carbonate waters. Geochimica et Cosmochimica Acta. 40: p. 989–995.

WILKINSON, BRUCE H. 1979. Biomineralization, paleoceanography, and the evolution of calcareous marine organisms. Geology 7 (11): 524–527.
Argues that aragonite may have been a much less abundant carbonate sediment type during portions of the geologic record. See also Discussion and reply in Geology 8 (6): 265–267.

WILLIAMS, R. E. 1970. Ground water flow systems and accumulation of the evaporite minerals. Bull. Amer. Assoc. Petrol. Geol. 54: 1290–1295.

ZENGER, DONALD H., JOHN B. DUNHAM, and RAYMOND L. ETHINGTON (eds.). 1980. Concepts and models of dolomitization. Soc. Econ. Paleont. Spec. Pub. 28. 320 p.
Symposium volume, sequel to Pray and Murray (1965); lots of mixing zone dolomite.

Shelf-to-Basin Transition
at Continental Margins

15

Gravity is a predominant driving force in sedimentary processes. Alluvial environments are characterized by water and sediment moving downhill. At the shoreline, gravity-driven processes are masked by tidal processes and by the action of wind-driven waves and currents. At some distance from the shore, however, in perhaps 50 to 100 meters of water, these shallow-water processes yield once again to the dominance of gravity.

Persistent ocean currents may play a role in the ultimate distribution and fabric in deeper-water environments, but the sediment is supplied chiefly by gravitational mechanisms. Clay particles suspended in water near the bottom may make the water sufficiently dense so that it flows slowly basinward as a low-velocity density current. Newly sedimented silt and clay may undergo plastic deformation and creep basinward. As creep becomes severe, slumping may occur. As slump blocks move faster, the sediment may become resuspended and flow basinward as a high-velocity turbidity current.

Recent Sedimentation along Continental Margins

The generalizations of shelf-to-basin sediment transport just outlined seem quite plausible; indeed, for many years, they have constituted *ad hoc* models for sedimentary sequences observed in the stratigraphic record. In recent years, however, scuba-diving equipment, research submarines, and improved oceanographic techniques for subbottom profiling, bottom photography, and sampling have allowed geologists to investigate these processes as they occur today along the continental margins. The following two examples are particularly instructive.

Submarine Canyons and Basins of the Southern California Transform Margin. As we noted earlier (Figure 10.13), continuing tectonic activity has produced pronounced topographic relief in southern California. On land, this tectonic topography takes the form of valleys and mountain ranges. In the offshore area, tectonism has resulted in relatively deep basins adjacent to the modern coastline. Obviously, we would expect to find that sediments derived from the continent may have been deposited in this relatively deep water.

Scientists have made detailed studies of the shore-to-basin transition in the vicinity of the La Jolla canyon and San Diego trough. The general bathymetry of the area is indicated in Figure 15.1. Observe the extremely narrow continental shelf here; the 100-meter contour lies within 4 kilometers of the shoreline. Outward from the 100-meter contour, the bottom slopes seaward at greater than 50 meters per kilometer. Below 800 meters, the topography begins to flatten out and the general

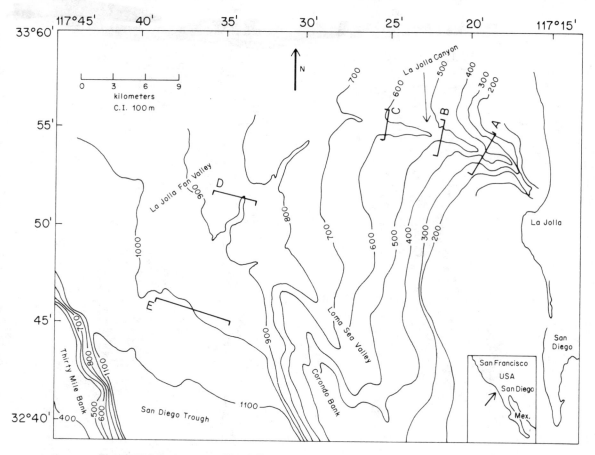

Figure 15.1 Bathymetry of La Jolla canyon, La Jolla fan, and San Diego trough.

slope of the sediment surface turns southward into the San Diego trough. Note that the La Jolla canyon and its seaward extension remain essentially perpendicular to the regional strike of the sea floor from the shelf to the San Diego trough. Bathymetric profiles across the La Jolla canyon and its seaward extension are indicated in Figure 15.2. The shallow end of the canyon is deeply incised into the shelf and slope topography, whereas the seaward extension of the canyon is simply a pronounced low-relief channel. We can see from its topography alone that the La Jolla canyon must be an important element in regional sedimentation.

The story of clastic sedimentation in this area begins with the longshore transport of sediment that the rivers have delivered to the coast. Sorting of the sediment accompanies longshore transport. Sands continue to be moved southward along the shore, whereas silt and clay tend to be winnowed out and moved seaward (Figure 15.3).

Bottom observations indicate that at least some of the clay-sized material is transported seaward by slow-moving density currents. A turbid layer, commonly called the *nepheloid layer*, is observed to occupy the first meter or so of the water

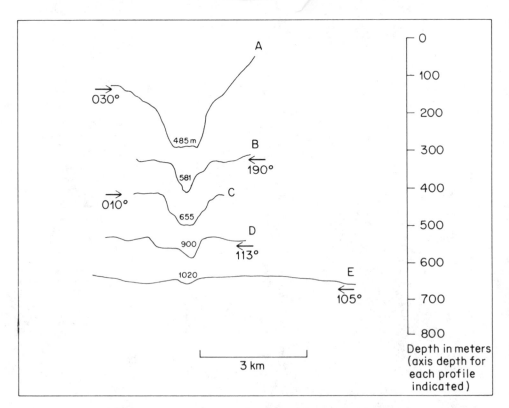

Figure 15.2 Bathymetric profiles of La Jolla canyon and La Jolla fan. (From the data of Shepard *et al.,* 1969.)

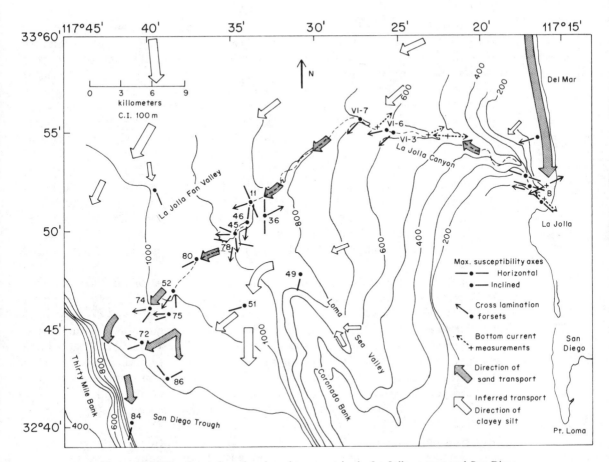

Figure 15.3 Directions of sand and mud transport in the La Jolla canyon and San Diego trough. Sand tends to be transported southward along the coast until it is funneled down La Jolla canyon. In contrast, mud travels in low-velocity turbid flows near the bottom across much of the shelf. [After U. von Rad, "Comparison of Sedimentation in the Bavarian Flysch (Cretaceons) and Recent San Diego Trough (California)," *Jour. Sed. Petrology,* **38**, 1120–1154 (1968).] Courtesy of the Society of Economic Palentologists and Mineralogists.

column above the shelf sediments. It is presumed that the clay minerals in this turbid suspension provide sufficient density so that the turbid water tends to flow slowly downslope. Although part of the silt and clay is deposited at the shelf margin and slope, some clay in the slow-moving density currents is also eventually channeled into the deep basins by submarine canyons. Apparently, the layers of turbid water move obliquely across the continental shelf until they are either funneled into a submarine canyon or spill over the continental slope. When he made seismic studies of unconsolidated fill in outer basins of the southern California continental borderline, Moore (1970) demonstrated striking differences in the amount of fill in adjacent basins. He found that these differences are related to the way the system of basins is connnected to submarine canyons. Where an outer basin is ultimately connected to basins with active submarine canyons, the basin receives abundant clay fill. An adjacent basin, topographically isolated from landward basins, receives very little clay fill. Prior to these observations, it had been assumed that the clay fill of the outer basins settled out of the entire water column. Moore's observations strongly suggest that low-velocity density currents, consisting of turbid water such as is observed on the outer continental shelf, are the transporting mechanism for outer-basin clay sedimentation. The observations further emphasize the importance of submarine canyons as funnels for these fine-grained sediments.

Let us return to the shore and pick up the story of sand transportation and deposition. Longshore transport of the sand-sized fraction ends abruptly at the La Jolla headland. Here the coastline presents a topography that is unfavorable to continued longshore transport. Sand-sized sediment, therefore, accumulates at the head of La Jolla canyon. This sediment will ultimately find its way by gravitational processes to the La Jolla fan, 20 to 40 kilometers seaward and beneath 800 to 1200 meters of water.

The precise mechanism of sand transport down the La Jolla canyon has been the subject of considerable study and discussion. Clearly, some high-energy currents must come down the valley from time to time; the canyon in places dissects rocks of Cretaceous and Eocene age, leaving vertical and overhanging walls. Boulders and cobbles of this material, as well as mud lumps of semiconsolidated shelf and slope sediment, accumulate in the central channel of the sediment fan that forms in front of the erosional portion of the La Jolla canyon. However, the strong currents that these sediment accumulations imply have not been observed firsthand.

Studies carried out by scuba divers have demonstrated a slow, downslope, sediment creep in the head of the canyon. It has also been noted that the sediment surface around the head of the canyon may undergo periodic deepening on the order of several meters. Presumably, such deepening is related to rather large-scale slumping, but details of the process are not known.

Thus, the study of La Jolla canyon reveals certain inconsistencies concerning the mode of sand transport from near shore into deep water. The sand does get transported; the La Jolla fan and San Diego trough contain large amounts of sand of demonstrable shallow-water origin. The morphology of the erosional portion of La Jolla canyon and the size of cobbles and boulders in the channel of the La Jolla fan certainly imply high-energy current systems. Yet observations of sediment transport today indicate primarily slow movement of sediment by creep or slump. Indeed, much of the floor of La Jolla canyon, La Jolla fan, and San Diego trough are now covered with a layer of silt and clay, as though the processes responsible for the movement of sand, cobbles, and boulders are dormant.

Primary structures revealed in box cores from the La Jolla fan and the San Diego trough (von Rad, 1968; Shepard *et al.,* 1969) provide a further complication concerning the generalities of sand sedimentation in this area. The primary structures are typical of traction transportation. Structureless graded beds, which experimentation and theory suggest should be typical of high-velocity turbidity- current deposition, are uncommon in the La Jolla fan. As shown in Figure 15.3, the data indicate that traction transport is southward down the axis of the San Diego trough. Whether this transport was the result of persistent marine currents of the result of currents associated with the La Jolla canyon is not certain. In either case, the process does not appear to be active today, because the entire region is covered with a layer of silt and clay.

Thus, the area from the head of La Jolla canyon, across La Jolla fan, and down the San Diego trough presents an interesting dichotomy. At some time in the near past, high-energy currents accomplished significant erosion within the walls of La Jolla canyon and delivered large volumes of sand, cobbles, and even boulders out onto the La Jolla fan. Bottom currents further reworked this material into traction-transported sand deposits. Yet, present-day large-scale sand transport in the La Jolla canyon is restricted to creep and slumping at the head of the canyon. Indeed, the lower reaches of the canyon, the La Jolla fan, and the San Diego trough are now covered with a thin accumulation of silty clay. Shepard, *et al.* (1969) suggest that this contrast of modern conditions with more energetic conditions in the past may be associated with the possibility of a greater sediment supply from the land at a time when higher rainfall brought more sand to coastal areas. On the other hand, studies by Hand and Emery (1964) and Moore (1970) note that the eustatic lowering of sea level during the Pleistocene would have brought longshore sediment transport into the steeper reaches of existing submarine canyons, thus producing a gravitational instability more pronounced than exists today.

Gorsline and Emery (1959) propose another way that southern California basins may have been filled more easily at a time of eustatic sea-level lowering. Figure 15.4 depicts the development of deep-water sand fans where sediment supply comes from submarine canyons, a generalization more or less compatible with La Jolla fan sedimentation described earlier. This figure indicates the possibility of deriving basin fill when silt and clay slumps from the outer-shelf and slope environment. Whether the outer-shelf, clayey silt creeps, slumps, or flows into the adjacent basin may be considered a point of secondary importance; the important point is that outer-slope sediments deposited during a eustatic highstand provide a convenient source of relatively shallow-water, fine-grained sediment, which may be remobilized for deposition in the basin. We would expect this process to occur especially with the increasing energy conditions that are attendant to the eustatic lowering of sea level.

Estimates of sedimentation rate in the deep-water basins are generally in the range of 5 to 15 centimeters per 1000 years. All of these estimates are based on the carbon-14 dating of sediments 10,000 years old and younger. As previously outlined, there is good reason to suspect that modern sedimentation is occurring at a slower rate than may have prevailed during Pleistocene eustatic lowstands of the sea or under wetter climatic conditions.

The Mature Passive Margin, East Coast of the United States and Canada. The transition from the eastern coastline of the United States to the abyssal plain of

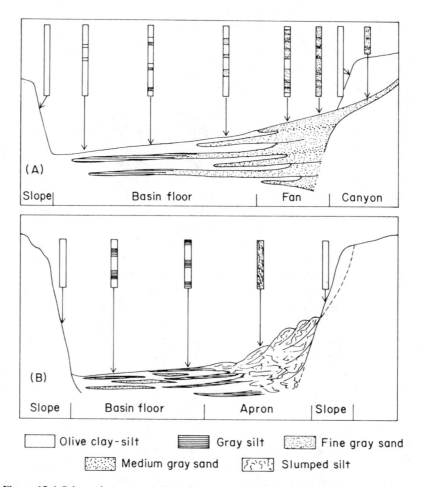

Figure 15.4 Schematic representation of two possible modes of basin filling in southern California. Where submarine canyons enter the basin, (A), sand transport down the canyon would be anticipated. Away from the canyon or in basins that do not attach to canyons, (B), slumping would occur as an important filling process. (After Gorsline and Emery, 1959.)

northwestern Atlantic provides an important contrast to sedimentation in the continental borderland basins of southern California. First and foremost, the East Coast of the United States is not a tectonically active area. Whereas tectonism of southern California chops up the area into small units that must be considered individually, the lack of tectonism on the East Coast allows us to apply broad generalizations to larger areas. The East Coast, therefore, displays contrast in terms of both tectonic setting and scale.

Figure 15.5 indicates the general bathymetry of a portion of the western Atlantic. Consider the area from latitude 34° northward. There is (1) a broad continental shelf, 60 to 300 kilometers in width, which exhibits slopes on the order of 0.6 to 3

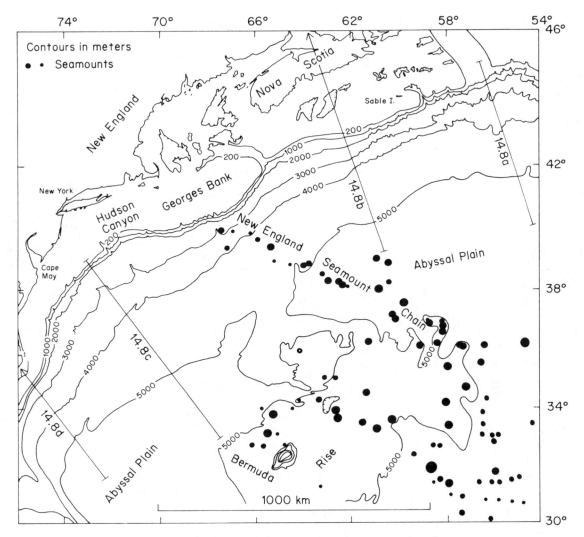

Figure 15.5 Bathymetry of a portion of the northwest Atlantic.

meters per kilometer; (2) a continental slope, usually ranging in the depth from 200 to 2000 meters and in width from 20 to 60 kilometers, with slopes averaging 40 meters per kilometer; (3) a continental rise, generally between the 2000 to 4500-meter bathymetric contour, 200 to 300 kilometers wide, and having an average slope of 10 meters per kilometer; and (4) an abyssal plain that is very nearly flat. Observe the sharp contrast between this bathymetry and scale and that of southern California (Figure 15.1). Here the continental shelf is extremely broad. Any basins that may have existed are now smoothed over by sedimentation. The continental slope and the continental rise have exceedingly similar profiles for more than 1000 kilometers along strike. If we consider the distance from continental shelf to abyssal plain as the measure of the shelf-to-basin transition, the transition zone here is hundreds of kilometers wide.

It is convenient to begin our discussion of the distribution of sediment types in the northwest Atlantic by referring to the characteristic of their reflections in echo-sounding profiles. Figure 15.6 presents a summary of such data. In general, a strongly reflecting bottom indicates the presence of firm substrate composed largely of sand. A weakly reflecting bottom indicates the predominance of clay and silt-

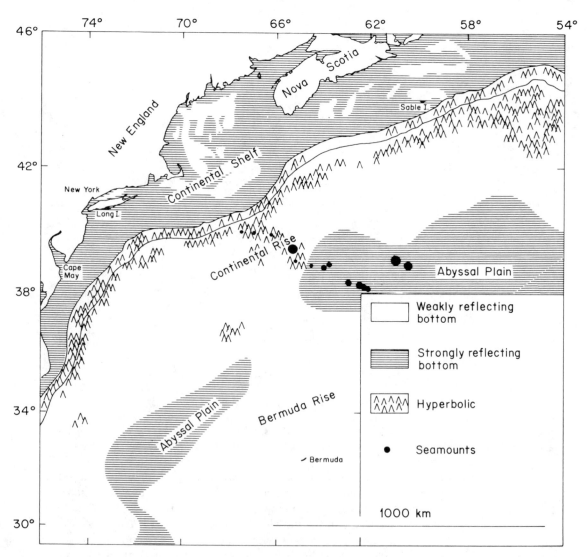

Figure 15.6 Physiography and sediment reflection characteristics of a portion of the northwest Atlantic. (After Emery *et al.,* 1970, with modification.)

sized material. Hyperbolic reflections occur where local topographic relief allows high points to act as point-source reflectors as the survey ship passes by. Note that the northwest Atlantic can be divided into four physiographic-sedimentologic provinces on the basis of these data. The continental shelf is typically an area of strongly reflecting bottom; the continental slope, an area of weakly reflecting bottom with abundant hyperbolic reflections; the continental rise, a smooth area of weakly reflecting bottom; and the abyssal plain, a region with a stronger reflecting bottom. Sedimentologic studies allow us to translate these observations directly to sediment type and sedimentation processes.

Sediments of the continental shelf are generally relict sand deposits or coarse glacial sediment accumulated during Pleistocene eustatic lowstands of the sea (Emery, 1968). The sediment load of rivers discharging into this area has been insufficient to keep pace with Holocene sea-level rise. River valleys along this coast are drowned estuaries. Modern deltas exist only at the heads of bays. Even the sand of modern beaches is derived from the relict deposits of the continental shelf.

Sedimentation on the continental rise runs strongly parallel to "hemipelagic" sedimentation in the outer basins of the southern California borderland. By relating variations in sedimentation rate in the outer basins to the presence or absence of a landward connection to submarine canyons, Moore (1970) inferred that "hemipelagic sediment" is not pelagic at all. Rather than settling down through the water column from the surface, these clays are deposited by low-velocity, turbid-density currents that follow bottom topography.

Similarly, a nepheloid layer exists above the continental rise of the eastern United States. However, the vast scale of this low-relief topography has allowed a very different current system to develop.

Ewing and Thorndike (1965) report on the observation of slightly turbid bottom water over the western Atlantic continental rise. *In situ* measurements made by optical-photometric methods indicate the lower 200 to 900 meters of the water column to be somewhat cloudy. Processing of a large volume of water samples suggests that bottom water contains on the order of 2 milligrams per liter of clay minerals not observed elsewhere in the water column.

Heezen *et al.* (1966) account for continental-rise sedimentation by means of a southward-flowing western boundary undercurrent, which follows the contours of the continental rise from Labrador to the Bahamas. The western boundary undercurrent is a geostrophic current that has its origin in the deep thermohaline circulation of the Atlantic. Near the western margin of the North Atlantic, the Coriolis force (to the right) is balanced by the gravitational force associated with the general slope of the continental rise (to the left). As the net result of the balance between these two forces, the current is guided along the contours of the continental rise for extremely long distances.

Heezen *et al.* (1966) suggest that this current transports the nepheloid layer observed by Ewing and Thorndike (1965) and calculate that deposition of 10% of the total sediment load would account for the total thickness of Recent continental-rise clay accumulation observed in piston cores.

The generalizations of Heezen *et al.* (1966) concerning the genetic relationships between continental shelf, continental rise, and abyssal plain are summarized in Figure 15.7. According to their generalizations, the continental shelf owes its origin largely to shallow-marine sedimentation processes, like those discussed in Chapters 12 to 14. The continental rise originated largely from clay sedimentation left by southward-flowing contour currents. They view the continental slope as the zone of discontinuity between continental-shelf processes and continental-rise processes. Clastic transport to the abyssal plain is limited to downslope turbidity currents that follow submarine canyons across the continental rise.

Studies of the continental slope in front of the Hudson canyon tend to bear out the generalities of Figure 15.7 concerning the genetic relationship between canyons and deep-sea sands (Ericson, Ewing, and Heezen, 1952, for example). Sediment cores taken in the topographic extension of the Hudson canyon 50 to 100 kilometers basinward from the continental slope and in 3000 meters of water contain graded beds of muddy gravel with pebbles of igneous, metamorphic, and sedimentary rocks up to 2 centimeters in diameter. Cores taken a few tens of kilometers to either side of this locality contain only the clay typical of the general continental-rise province.

Sediments of the abyssal plain are quartz sand and silt interbedded with deep-sea clays and pelagic ooze. Fauna contained in these sands indicate that the shallow water of the continental margin is the immediate source of the quartz sand and silt. Abrupt lower contact and a generally graded nature of many of these sand layers

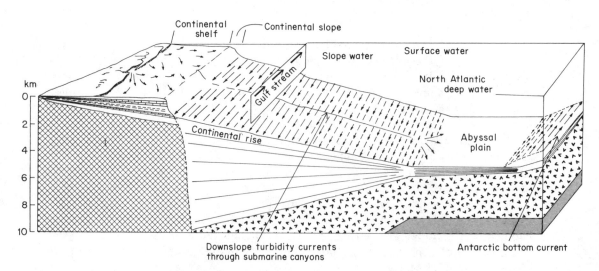

Figure 15.7 Diagram showing how the continental rise is shaped by geostrophic contour currents. [After B. C. Heezen *et al.*, "Shaping of the Continental Rise by Deep Geostrophic Contour Currents," *Science*, **152**, 502–508 (22 April 1966).] Copyright 1966 by the American Association for the Advancement of Science.

are considered evidence of deposition from turbidity currents. The occurrence of such layers as much as 1000 kilometers from any apparent source area seems to indicate that the turbidity currents were of considerable size and generated considerable momentum.

With this synoptic sketch of sediment types from the continental shelf, the continental rise, and the abyssal plain, it is important that we return to Figure 15.6 and consider once again the immense scale of these sedimentological provinces.

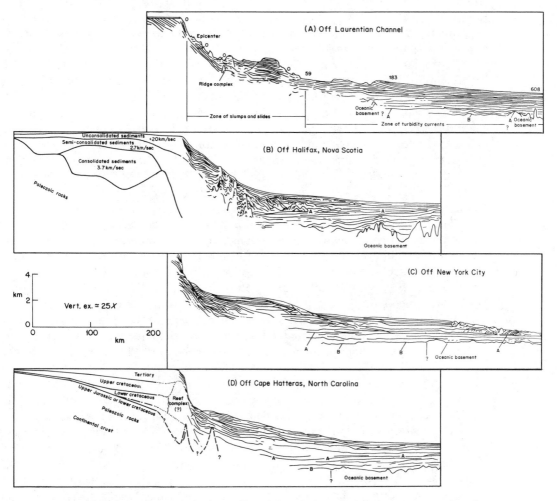

Figure 15.8 Continuous seismic profiles of the continental rise of southeastern Canada and eastern United States. Horizon (A) is a prominent reflecting horizon believed to be of Middle Eocene to Upper Cretaceous age. Horizon (B) is the reflection from the oceanic basement. (After Emery *et al.,* 1970.)

The continuous seismic profiling of Emery *et al.* (1970) provides additional insight on sedimentation and contemporaneous deformation on the continental rise. Figure 15.8 presents four profiles from the continental shelf to the abyssal plain. In each profile, horizon (A) is a prominent reflecting horizon, taken to be Middle Eocene to Upper Cretaceous in age; horizon (B) is the contact between sediment and oceanic basement. Observe carefully the vertical and horizontal scale on these diagrams. We are discussing truly large-scale sedimentation features. Of particular interest to us are the numerous evidences of large-scale slump structure. Note especially the upper kilometer of sediment in Figure 15.8 (C) between 200 and 500 kilometers off the continental shelf. An enormous slice of continental-rise sediment appears to have slid along planes more or less parallel to bedding. This situation is truly gravity tectonics on a grand scale.

A similar example of large-scale sediment transport from continental slope to continental rise is apparent in Figure 15.8(A). Here we have the additional benefit of data from the 1929 trans-Atlantic cable break to add vivid color to these large-scale sediment-transport processes. During 1929, an earthquake occurred near the Grand Banks. The approximate position of the epicenter is indicated in Figure 15.8(A). Following the earthquake, trans-Atlantic cables began to give way. A number of cables failed at the time of the earthquake; others broke a considerable time after the earthquake. The positions and times of some cable failures are given in Figure 15.8(A). Figure 15.9 gives a summary of time-distance relationships for cable failures.

The failure of cables at or near the time of the earthquake appears to have been caused by either the shock of the earthquake or local slumping of continental-slope sediment attendant to the earthquake. As geologists, we would not consider these failures spectacular or surprising. In retrospect, it was just a bit of poor geological engineering. If you lay a cable along the side of a slope, you are simply asking for small movements on the slope to stretch your cable and break it. However, about ten hours later, something rather spectacular did begin to happen. As indicated in Figure 15.9, seven cables, located 400 to 600 kilometers away from the epicenter, failed in a very systematic fashion. The first to go were the ones closest to the epicenter, and the last to go were the ones farthest away.

Inasmuch as these cable breaks occurred half a century ago and in 6000 meters of water, we shall probably never know what happened. However, a plausible and widely accepted explanation for the delayed breaks is that slumping and sliding on the continental slope and upper continental rise generated turbidity currents that flowed along the bottom at velocities in the range of 30 to 50 kilometers per hour. This interpretation would suggest that turbidity currents still contain considerable energy even after traveling 500 to perhaps 1000 kilometers outward from their source. In addition, it should be noted that the outer cable breaks occurred over an area several hundred kilometers in width, measured along the strike of the continental rise. Thus, the proposed turbidity currents were not only fast and far-reaching, they were also of wide lateral extent.

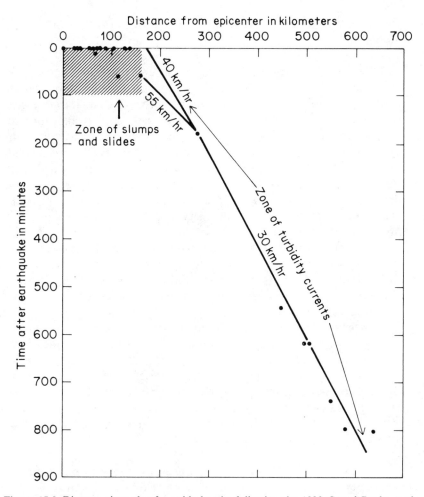

Figure 15.9 Distance-time plot for cable breaks following the 1929 Grand Banks earthquake. (After Emery *et al.,* 1970.)

Other Late Tertiary Examples
of Continental Margin Sedimentation

Studies of convergent margins and deep-sea fans provide additional information concerning additional sedimentation within the shelf-to-basin transition. However, the degree of complexity and the scale of observation set these studies apart from studies cited above. It is, therefore, useful to think of these studies collectively as intermediate between studies of Recent sedimentation and Ancient examples. We shall not use them specifically to formulate a "Recent sedimentation model," but we shall use them as an intermediate proving ground in which to gain experience with regard to diverse recombinations of various portions of our Recent sediments model.

Convergent Margins. The major concepts of sedimentation at convergent margins remain the same as the Recent sediment examples discussed above. There is a tendency for coarse-grained sediments to be separated from fine-grained sediments; there is a tendency for the coarse-grained sediments to be captured at the head of submarine canyons and passed down the canyons whereas mud transport is by low-velocity, turbid-density currents. Both of these types of sediment should be expected to fill any closed basin to sill depth. Once the local basin is filled to sill depth, gravity transport continues to the next lower basin, and ultimately to the abyssal plain or deep-sea trench. The new insight that convergent margins provide with regard to this overall picture lies in the diversity of basin geometries that can occur within the arc-trench system.

Seaward from the continental margin, one encounters first the forearc basin, then the slope basins, and finally the deep-sea trench (see Figure 10.11 and related discussion). Both the forearc basin and the slope basins may be expected to fill to sill depth with coarse clastics or nepheloid muds, depending upon configuration of submarine canyons feeding the area. The Sunda Arc southwest of Sumatra provides a good example of a forearc basin filled with coarse clastic turbidites and shales (Karig *et al.,* 1979). Offshore Oregon provides a good example of slope basins which are filling with silt turbidites and nepheloid muds (Kulm and Scheidegger, 1979). The northern portion of the Middle America Trench (south of southern Mexico) provides a good example of submarine canyons that transect the forearc basin and slope basins, and deliver coarse clastic sediments directly to the deep-sea trench (Ross, 1971; Underwood, Bachman, and Schweller, 1980; Underwood and Karig, 1980).

Deep-Sea Fans. The literature concerning deep-sea fans (sometimes called deep-sea cones) provides a fascinating example of the difficulty in synthesizing observations of divergent scale and accessibility. The students of modern fans (Normark, 1970; Damuth and Kumar, 1975; Damuth and Embley, 1981; for example) have the morphology of the modern fan laid out before them, but it is beneath 1000 to 4000 meters of water. There principal tools for study of these features are reflection seismic and side-scan sonar. When they want to sample the fan, they can barely prick the surface with conventional piston cores. On the other hand, students of late Tertiary examples of submarine fans now exposed in stratigraphic sequences on the continents (Mutti and Ricci Lucchi, 1972; Ricci Lucchi and Valmori, 1980; Walker, 1978; for example) have spectacular stratigraphic exposures available to them, but morphology and paleogeography of the fan must be inferred. Add to this the likelihood that these features probably really are complex and diverse, and you have a literature that will be full of controversy and disagreement for many years to come (Nilsen *et al.,* 1980, for example).

Controversy notwithstanding, there are some generalities concerning deep-sea fans that are worthy of our attention. The fan story generally begins where a submarine canyon incised into the continental slope breaks out of its incisement and spews its sediment load out on to the surrounding sea floor. It is convenient to think of upper fan, mid-fan, and lower fan. The upper fan and mid-fan are typically covered with braided channels. The outer reaches of the mid-fan region are characterized by *suprafan lobes*. These suprafan lobes appear as convex upward departures from the large-scale fan bathymetry.

In sedimentological terms, it is convenient to think of the deep-sea fan as somewhat analogous to a braided-stream environment surrounded by a delta-front environment. In the terrestrial stream environment, the density contrast between water and air is much greater than the density contrast of turbid water to clear water in the submarine fan environment. Nevertheless, in both systems it is higher density that keeps the turbid water constrained to channels. With regard to the delta-front analogy, the suprafan lobes mark the end of channel sedimentation just as the bar-finger sands of the delta environment mark the end of terrestrial stream sedimentation. Whereas the bar-finger sands of the delta environment are constrained by the air-water interface, no such constraint exists with regard to the level obtained by suprafan lobes.

The upper-fan region is the site of dumping of the coarsest material present. Suprafan lobes tend to be better-sorted sands and commonly exhibit tendency to graded bedding, especially apparent as concentration of pebbles low in each sandstone bed. Behind the suprafan lobes, abandoned channels may fill with shales deposited from nepheloid layers. In front of the suprafan lobes, complete classical A–E turbidite beds occur, with proximal turbidites in the suprafan lobe and distal turbidites well out on to the lower fan. While each individual bed constitutes a fining-upward cycle, an overall relationship of coarsening upward sequence may occur as suprafan lobes prograde out over lower-fan turbidites.

A Basic Model for Shelf-to-Basin Transition at the Continental Margin

The preceding discussions have been area-oriented. Let us now begin to extract the generalities and combine them into a flexible working model.

Sand Sedimentation. In both examples just cited, the most striking generality concerning sand sedimentation is the intimate relationship between deep-water sands and submarine canyons. Shallow-marine processes, such as discussed in Chapter 12, deliver sand to the heads of submarine canyons. As the sand accumulates, gravitational instability ensues and the stage is set for some kind of downslope transport.

Modern observations confirm that creep and slump move sand down submarine canyons. However, the topography of canyons and the existence of pebbles, cobbles, and boulders of soft sediment derived from canyon walls strongly suggest that any rapid movement of sand down the canyons occurs either periodically or under conditions different from those that may have existed for the last 5000 years. Such currents may have been genuine turbidity currents, in which the greater density of the turbid water mass provided the impetus for rapid downhill transport. Alternatively, high-velocity water currents generated by tides and storms may have moved large volumes of sediment down the canyons by traction transport. The choice between these two mechanisms remains a matter for speculation; neither process has been observed to operate in the Recent epoch on the scale that is implied by deep-water sand deposits. In any case, we must realize that the erosion accompanying this large-scale transportation is extremely important in maintaining submarine canyons.

Silt and Clay Sedimentation. Whereas material of sand size and larger is most commonly transported by traction, clay and, to some extent, silt is transported in suspension by low-velocity turbid flows. The existence of widespread turbid bottom water on the continental shelf, slope, and rise is well documented. This turbid water may undergo gravity flow because of the slight density contrast provided by the suspended mineral matter, or the turbid water may be moved along by persistent bottom currents of other origin.

Lack of Relationship between Sand and Clay Sedimentation in the Recent Epoch. As indicated in Chapter 12, shallow-marine processes tend to separate sand and clay. If poorly sorted material undergoes significant longshore transport, we anticipate that the mud fraction will be winnowed out of the sand. The sand will continue its movement along the shoreline, and the mud will be carried seaward.

In the previously cited examples of continental-margin sedimentation, deep-water sand and mud have been shown to accumulate by more or less independent processes. Therefore, a sedimentation model based strictly upon Recent studies must actually consist of two independent models: one for sand sedimentation and another for clay sedimentation.

To a large extent, the separation of sand and clay sedimentation in the Recent epoch is the direct result of the existence of continental shelves. Such a relatively broad, gently sloping platform allows sand to accumulate in a position of relative gravitational stability. Mud, however, tends to be carried to the edge of the shelf or off the shelf into the basin. Clearly, if there were no continental shelf, then shallow-water sand would accumulate in a position of much greater gravitational instability. Thus, a "no-shelf" model for basin-margin sedimentation would bring sand sedimentation and mud sedimentation into much closer relationship than is observed today.

Relationship of Basin-Margin Sedimentation to Sea-Level Fluctuations. In Chapter 12, we suggested that fluctuating sea level tends to modulate sediment supply to coastal environments. When the sea level falls, stream gradients usually increase, resulting in downcutting and more abundant sediment supply to the coastline. When the sea level rises, the sediment supply is trapped as alluvial valleys build upward to the new equilibrium profile and as deltaic sedimentation builds outward into newly formed estuaries. Obviously, these same considerations apply to the modulation of clastic sediment supply from the land to deep water.

Falling sea level has the additional effect of bringing the shoreline closer to the shelf margin. Sediment transported by longshore currents will then find its way to submarine canyons at a point where canyon topography is more pronounced. In Figure 15.1, for example, note that the present head of La Jolla canyon is a rather small area. In sharp contrast, a sea level 100 meters lower will intersect La Jolla canyon in a position where the canyon is 4 kilometers wide. Furthermore, there are numerous canyons that cut the outer shelf and slope but do not extend to the modern shoreline. With lowered sea level, longshore transport will tend to intersect the heads of many small canyons that are today bypassed by the coarse fraction as it moves along the present highstand shoreline.

Finally, falling sea level undoubtedly causes higher-energy conditions to impinge upon the silt and clay deposits of the outer shelf and slope. Presumably, these sediments will be remobilized under these new conditions, be it by increased rate of erosion, creep, slumping, or turbid flow.

Thus, numerous lines of reasoning suggest that shelf-to-basin transport of clastic sediments would be more rapid during lowstands of the sea than during highstands. Furthermore, the relatively broad width of continental shelves at times of high sea-level stand would afford greater opportunity for clean separation of sand-sedimentation processes from mud-sedimentation processes. During lowstands of the sea, sediment is introduced at a shoreline much closer to gravitational instability. Therefore, we would suspect that there is less opportunity for separation of sand from mud prior to downslope transport.

Relationship to Tectonic Setting. Tectonic activity tends to create local basins and highs that may behave as almost independent sedimentation systems during their early history. Juvenile basins that are fed by large submarine canyons usually fill rapidly, whereas nearby basins may receive little or no sediment. As the juvenile basins fill to sill depth, a graded profile develops. Downslope transportation bypasses the filled basin and begins to fill the next basin downslope (see, for example, Gorsline and Emery, 1959; and Moore, 1970). Given tectonic quiescence and continuing sedimentation, the basic sedimentation model takes on larger and larger proportions.

With reference to Recent examples cited earlier, this model is the transition from a San Diego trough model to a northwestern Atlantic model. Indeed, Emery *et al.* (1970) suggest that this transition occurred along the eastern coast of North America throughout the Mesozoic and into the early Tertiary. Seismic evidence on the continental shelves and slopes indicates local basin filling throughout the Mesozoic with little contribution of turbidities to the abyssal plain. During the mid-Eocene, a relatively smooth continental shelf and slope grew up. Then the continental rise developed, and turbidity currents began depositing material on the abyssal plain.

An Ancient Example: The Taconic Sequence, Cambro-Ordovician, Eastern New York State

It has long been recognized that mountain chains are added to continents as "mobile belts" more or less parallel to the shape of the old continent (or *craton*). The western margin of the Americas provides a Mesozoic-Cenozoic example of this association; the Appalachians of the eastern United States and Canada provide and equally spectacular Paleozoic example.

Yet for many years, geologists did not rely heavily upon the analogy between modern continental margins and the sedimentary sequences that preceded mountain building. We referred to those Ancient continental margins as *geosynclines*, and we were not certain that there was anything like them on the face of the earth today. We were not sure why geosynclines subsided and filled with sediment, and we were not sure why they came up again to form mountain ranges.

With the rapid development of "the new global tectonics" during the 1960's, geologists began to unravel many of the apparent discrepancies between geosynclines and continental margins. As a broad generalization, for instance, Appalachian tectonism records the mid-Paleozoic closing of the Atlantic. Perhaps, therefore, early Paleozoic sediments of the Appalachians are a better analogy to the modern East Coast continental margin than we had previously believed. With this possibility in mind, let us examine briefly one of the classical areas of Appalachian geology.

General Lithostratigraphy of Eastern New York and Vermont. Figure 15.10 is a generalized geological sketch map of the Taconic region. Billion-year-old Grenville rocks constitute the basement of the area and outcrop in the Adirondacks, Berkshires, and Green Mountains. The basement is overlain by Cambrian orthoquartzite sandstone, which gives way upsection to Cambro-Ordovician carbonate rocks. This section is exposed along the eastern margin of the Adirondacks and along the western margin of the Green Mountains. The area between the Adirondacks and the Green Mountains is largely occupied by Cambro-Ordovician shales, slates, and minor sandstones. This region is the Taconic facies. To the east of the Green Mountains, shale, sandstones, and volcanic rocks have undergone considerable metamorphism. Here chlorite schist predominates. To the west of the Taconic facies and southeast of the Adirondacks, there is a large area of shales and sandstones, deposited at the time of, and in intimate association with, major regional tectonism. Posttectonic sediments of Siluro-Devonian age occur in the southwestern and eastern portion of the map area.

Cambro-Ordovician Biostratigraphic Correlation. Figure 15.11 presents general facies information and biostratigraphic correlation among lithologies (2) through (4) of the map in Figure 15.10. The general location of these composite stratigraphic sections is indicated in the upper portion of Figure 15.10. Observe especially that

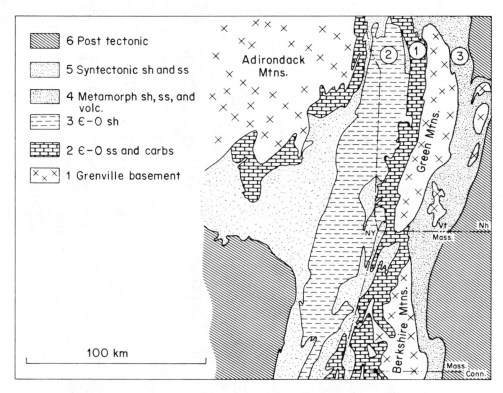

Figure 15.10 Geologic sketch map of the Taconic region. Major lithologic terranes of the area are as follows: (1) billion-year-old Grenville basement rocks; (2) Cambro-Ordovician sequence beginning with sandstones and giving way upward to carbonate rocks; (3) Cambro-Ordovician sequence predominantly of shale; (4) Cambro-Ordovician shale, sandstone, and volcanics now metamorphosed; (5) shales and sandstones deposited during major regional tectonism; (6) posttectonic sediment, including the Helderberg and younger sequences of eastern New York and the Triassic red beds of the Connecticut Valley.

lithostratigraphic units (2), (3), and (4) comprise the same biostratigraphic interval, namely Cambrian through Lower Ordovician. Geologists have long realized that sediments of the Taconic sequence must have accumulated somewhere to the east of their present position and must have been thrust westward prior to the metamorphism of their equivalents in eastern Vermont. These Taconic thrust sheets comprise an area some 20 kilometers wide and 200 kilometers long and appear to involve westward thrusting at least on the order of 30 to 50 kilometers.

Structure and Stratigraphy within the Taconic Thrust Sheets. Figures 15.12 and 15.13 summarize the results of detailed structural and stratigraphic analysis of the Taconic thrust sheets. The physical limits of each thrust sheet are worked out on the basis of field relationships. Within each thrust sheet, the biostratigraphic sequence has been determined. It is interesting to note that the first thrust sheets to be emplaced contain the most complete representation of Cambrian–Lower Ordov-

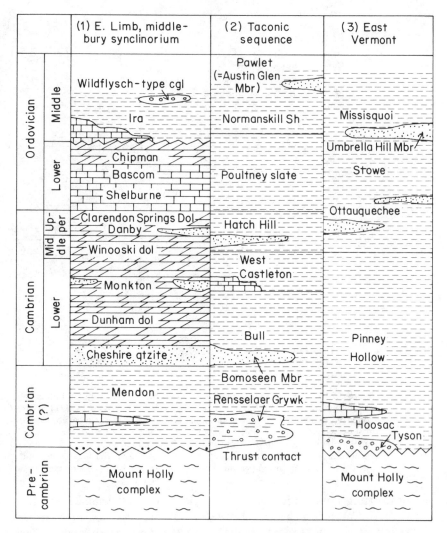

Figure 15.11 Biostratigraphic correlation of principal sequences within the Taconic region. See Figure 15.10 for location of composite sections. (After Zen, 1968.)

ician stratigraphic units and that the youngest thrust sheets contain only lowermost Cambrian-Precambrian rocks. The emplacement of these thrust sheets is interpreted as a gravity-tectonic unroofing of the source area to the east.

Time-Space Synthesis. At least two hypotheses can be formulated within the context of plate tectonics to explain the major relationships observed in the Taconics. One involves an analogy to the modern eastern North American passive margin continental-rise sedimentary prism. The other involves an analogy to the accretionary prism of modern subduction zones.

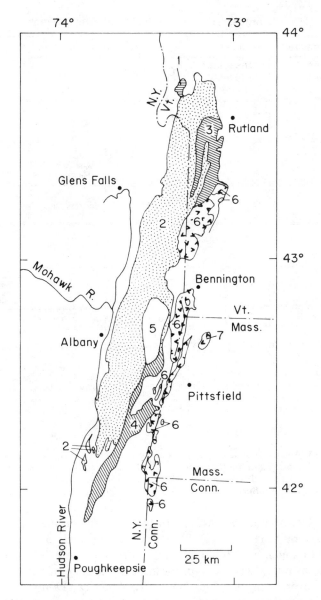

Figure 15.12 Sketch map depicting the extent of major thrust sheets within the Taconics. The thrust sheets are numbered more or less in order of their tectonic emplacement. See Figure 15.13 for additional details. [After E-An Zen, "Time and Space Relationships of the Taconic Allochthon and Authochthon," *Geol. Soc. America* Spec. Paper No. 97, 107 p. (1967).]

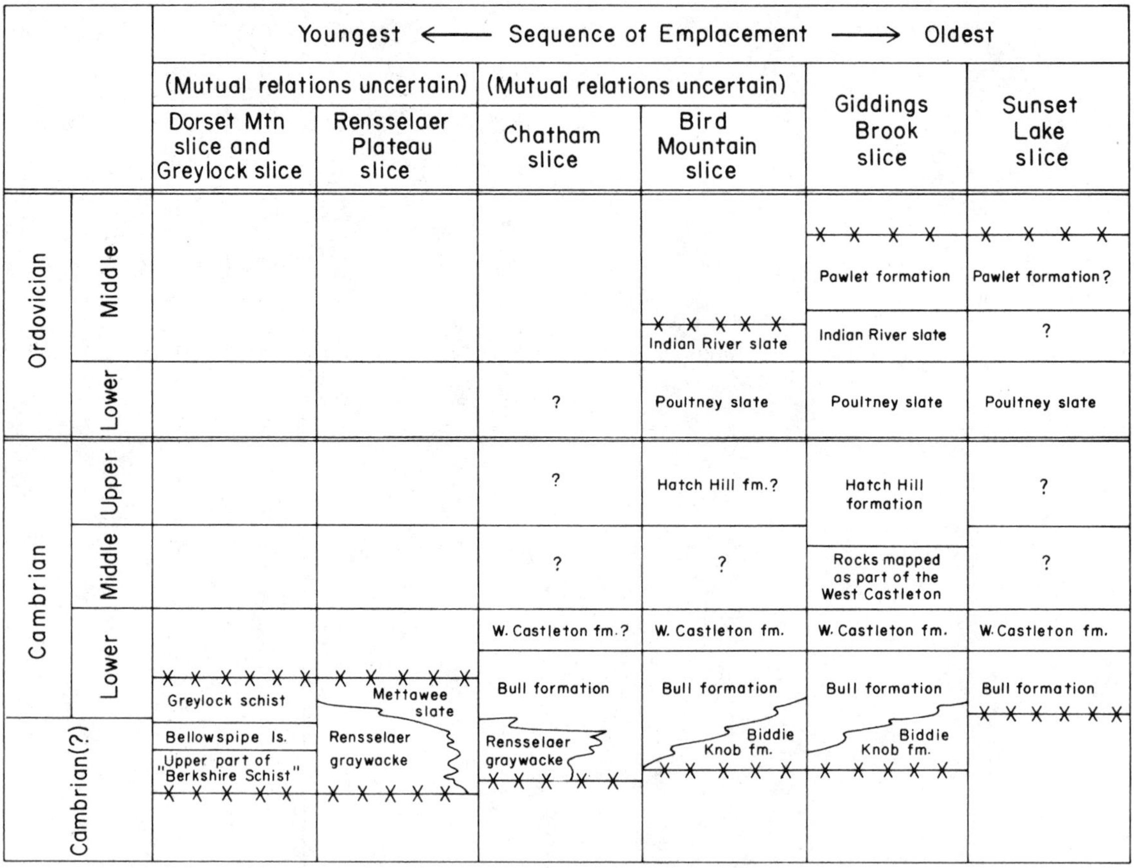

		Dorset Mtn slice and Greylock slice	Rensselaer Plateau slice	Chatham slice	Bird Mountain slice	Giddings Brook slice	Sunset Lake slice
		Youngest ← Sequence of Emplacement → Oldest					
		(Mutual relations uncertain)		(Mutual relations uncertain)			
Ordovician	Middle					x x x x Pawlet formation	x x x x Pawlet formation?
					x x x x x Indian River slate	Indian River slate	?
	Lower			?	Poultney slate	Poultney slate	Poultney slate
Cambrian	Upper			?	Hatch Hill fm.?	Hatch Hill formation	?
	Middle			?	?	Rocks mapped as part of the West Castleton	?
	Lower			W. Castleton fm.?	W. Castleton fm.	W. Castleton fm.	W. Castleton fm.
		x x x x x x Greylock schist	x x x x x x Mettawee slate	Bull formation	Bull formation	Bull formation	Bull formation x x x x x x
		Bellowspipe ls.	Rensselaer graywacke	Rensselaer graywacke	Biddie Knob fm.	Biddie Knob fm.	
Cambrian(?)		"Upper part of Berkshire Schist" x x x x x	x x x x x	x x x x x	x x x x x x	x x x x x	

Figure 15.13 Stratigraphic ranges of rocks within the various thrust slices of the Taconics. Lines with crosses indicate the beginnings and ends of sections within these slices. [After E-An Zen, "Time and Space Relationships of the Taconic Allochthon and Authochthon," *Geol. Soc. America* Spec. Paper No. 97, 107 p. (1967).]

Recognizing the vastness of the modern continental-rise sedimentary prism off of eastern North America (Figures 15.5 through 15.8), it is interesting to ask what might happen if such a sedimentary province were delivered to an active subduction zone, as might well be hypothesized at the time of the closing of a North Atlantic by convergence of Africa on North America. Figure 15.14 presents a time-space schematic diagram that might explain the geology of the Taconic region. Cambrian–Lower Ordovician sandstones and carbonates [lithology (2) of Figure 15.10] appear analogous to the modern continental shelf. Cambro-Ordovician fine-grained sediments of lithologies (3) and (4) of Figure 15.10 are considered analogous to the modern continental-rise deposits. With the closing of the Atlantic, commencing in Middle Ordovician time, the continental-rise sedimentary prism underwent collapse and subsequent metamorphism of the Taconic thrust sheets were replaced by gravity

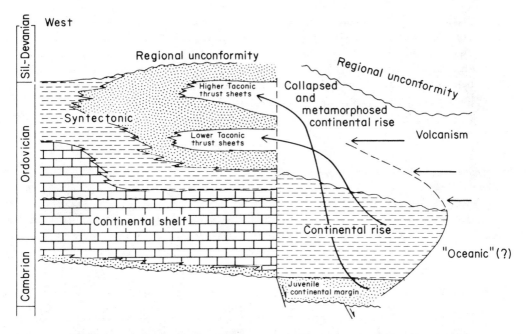

Figure 15.14 Schematic diagram showing one possible time-space relationship to explain the geology of the Taconic region. During Cambro-Ordovician time, the deposition of continental-shelf and continental-rise sequences took place. Middle Ordovician closing of the paleo-Atlantic Ocean resulted in collapse of the continental-rise sedimentary prism and emplacement of the various Taconic thrust sheets from the east by gravity tectonics. (After Bird and Dewey, 1970, with extreme simplification.)

tectonics early in this process, thus escaping the metamorphism. Age relationships within the Taconic thrust sequence (Figure 15.13) are explained as progressive unroofing of the collapsed continental rise sedimentary prism. Continued compression formed the large uplift that shed clastic debris to the west to form the Catskill clastic wedge (Figure 10.14). Whereas the original authors of this concept propose that a paleo-Atlantic plate was consumed beneath the North American plate, this scenario could also be conveniently scripted with the geometry of subduction reversed.

An alternative scenario (Rowley and Kidd, 1981, for example) is that the Taconic thrust sequence is actually the remnant of an accretionary prism thrust onto the North American continent during aborted subduction such as is depicted in Figure 10.17. By this hypothesis, the age relationships among "thrust sheets" depicted in Figure 15.13 were developed while the entire Taconic sequence was an active accretionary prism. Then, the accretionary prism was thrust over Cambro-Ordovician continental-shelf deposits at the time of aborted subduction of the North American continent.

In summary, the Taconics provide a marvelous example of complex interrelationships between field observations and "the big picture" of plate tectonics. Everyone would agree that we now recognize a Cambro-Ordovician passive margin that predates the emplacement of the Taconic sequence. This is a major step forward in our conceptual understanding of Appalachian geology. But was the Taconic sedimentary prism a passive-margin continental-rise prism or an active-margin accretionary prism? Was the paleo-Atlantic plate subducted beneath North America, beneath Africa, or beneath an oceanic subduction zone? There are and will remain large questions concerning the geology of the Taconics; we shall only find the answers by returning to the outcrops with competing plate tectonics hypotheses firmly in mind.

Citations and Selected References

AUDLEY-CHARLES, M. G., A. J., BARBER, D. J. CARTER, and ANGELO CROSTELLA. 1979. Geosynclines and plate tectonics in Banda Arcs, Eastern Indonesia: discussion and reply. Bull. Amer. Assoc. Petrol. Geol. 63 (2): 249–252.

BARKER, P. F., and I. A. HILL. 1980. Asymmetric spreading in back-arc basins. Nature 285 (5767): 652–654.

BEN-AVRAHAM, ZVI, and AMOS NUR. 1980. The elevation of volcanoes and their edifice heights and subduction zones (Paper 80B0616). J. Geophys. Res. 85 (B8): 4325–4335.

BIRD, J., and J. F. DEWEY. 1970. Lithosphere plate-continental margin tectonics and the evolution of the Appalachian orogen. Bull. Geol. Soc. Amer. 81: 1031—1060.

BOUMA, A. H., and A. BROUWER. 1964. Turbidites. Elsevier, Amsterdam. 264 p.
A collection of papers. Extensive bibliography.

BRIGGS, G., and L. M. KLINE. 1967. Paleocurrents and source areas of late Paleozoic sediments of the Ouachita Mountains, southeastern Oklahoma. J. Sed. Petrology 37: 985–1000.
Turbidity current deposition is dominated by current movement down the axis of the geosyncline throughout the deposition of 4000 meters of sediment.

BUFFINGTON, E. C., D. G. MOORE, R. F. DILL, and J. W. VERNON. 1967. From shore to abyss: near-shore transport, slope deposition and erosion, canyon transport, and deep-basin sedimentation [abstract]. Bull. Amer. Assoc. Petrol. Geol. 51: 456–457.
One of the most informative abstracts ever written.

DAMUTH, J. E., and R. W. EMBLEY. 1981. Mass-transport processes on the Amazon Cone: western equatorial Atlantic. Bull. Amer. Assoc. Petrol. Geol. 38: 629–634.

DAMUTH, J. E., and N. KUMAR. 1975. Amazon Cone: morphology, sediments, age, and growth pattern. Bull. Geol. Soc. Amer. 86: 863–878.

DICKINSON, W. R., and D. R. SEELY. 1979. Structure and stratigraphy of forearc regions. Bull. Amer. Assoc. Petrol. Geol. 63 (1): 2–31.

DIETZ, R. S. 1963. Collapsing continental rises: an actualistic concept of geosynclines and mountain building. J. Geology 71: 314–333.

DIETZ, R. S., and J. C. HOLDEN. 1966. Miogeoclines in space and time. J. Geology 74: 566–583.

DOTT, R. H., JR. 1963. Dynamics of subaqueous gravity depositional processes. Bull. Amer. Assoc. Petrol. Geol. 47: 104–128.

A readable summary of the various transport processes that deliver clastic sediments to deep-water environments.

EMERY, K. O. 1968. Relict sediments on continental shelves of the world. Bull. Amer. Assoc. Petrol. Geol. 52: 445–464.

EMERY, K. O., and J. D. MILLIMAN. 1978. Suspended matter in surface waters: influence of river discharge and of upwelling. Sedimentology 25 (1): 125–140.

EMERY, K. L., E. UCHUPI, J. D. PHILLIPS, C. O. BOWIN, E. T. BUNCE, and S. T. KNOTT. 1970. Continental rise off eastern North America. Bull. Amer. Assoc. Petrol. Geol. 54: 44–108.

Reports extensive geophysical surveys concerning structure, sedimentation, and history of the northwestern Atlantic. This entire issue of the bulletin is devoted to the geology of continental margins.

ENOS, P. 1969. Anatomy of a flysch. J. Sed. Petrology 39: 680–723.

ERICSON, D. B., M. EWING, and B. C. HEEZEN. 1952. Turbidity currents and sediments in the north Atlantic. Bull. Amer. Assoc. Petrol. Geol. 36: 489–511.

EWING, M., and E. M. THORNDIKE. 1965. Suspended matter in deep ocean water. Science 147: 1291–1294.

GEALEY, W. K. 1977. Ophiolite obduction and geologic evolution of the Oman Mountains and adjacent areas. Bull. Geol. Soc. Amer. 88 (8): 1183–1191.

GIBBS, RONALD J. 1981. Sites of river-derived sedimentation in the ocean. Geology 9 (2): 77–80.

GORSLINE, E. S., and K. O. EMERY. 1959. Turbidity current deposits in San Pedro and Santa Monica basins of southern California. Bull. Amer. Assoc. Petrol. Geol. 70: 279–290.

HAND, B. M., and K. O. EMERY. 1964. Turbidites and topography of the north end of San Diego trough, California. J. Geology 72: 526–542.

HEEZEN, B. C., and C. D. HOLLISTER. 1964. Deep sea current evidence of abyssal sediments. Marine Geol. 1: 141–174.

Bottom photographs provide widespread evidence of current transport in the deep sea.

HEEZEN, B. C., C. D. HOLLISTER, and C. L. DRAKE. 1964. Grand banks slump. Bull. Amer. Assoc. Petrol. Geol. 48: 221–224.

Slump and turbidity currents presumably triggered by an earthquake.

HEEZEN, B. C., C. D. HOLLISTER, and W. F. RUDDIMAN. 1966. Shaping of the continental rise by deep geostrophic contour currents. Science 152: 502–508.

HERSEY, J. B., and M. EWING. 1949. Seismic reflections from beneath the ocean floor. Amer. Geophys. Union, Trans. 30: 5–14.

HUBERT, J. F. 1964. Textural evidence for deposition of many western North Atlantic deep-sea sands by ocean-bottom currents rather than turbidity currents. J. Geology 72: 747–785.

IBRIHAM, ABOU-BAKR K., GARY V., LATHAM, and JOHN LADD. 1979. Seismic refraction and reflection measurements in the middle America trench offshore Guatemala. J. Geophys. Res. 84 (B10): 5643–5650.

KARIG, D. E., S. SUPARKA, G. F. MOORE, and P. E. HEHANUSSA. 1979. Structure and Cenozoic evolution of the Sunda Arc in the Central Sumatra Region, p. 223–238. *In* Joel S. Watkins, Lucien Montadert, and Patricia Wood Dickerson (eds.), Geological and geophysical investigations of continental margins. Amer. Assoc. Petrol. Geol. Memoir 29.

KEITH, BRIAN D., and GERALD M. FRIEDMAN. 1977. A slope-fan-basin-plain model, Taconic Sequence, New York and Vermont. J. Sed. Petrology 47 (3): 1220–1241.

KLEIN, G. DEV. 1966. Dispersal and petrology of sandstones of Stanley-Jackfork, boundary, Ouachita foldbelt, Arkansas and Oklahoma. Bull. Amer. Assoc. Petrol. Geol. 50: 308–326.

KUENEN, P. H. 1967. Implacement of flysch-type sand beds. Sedimentology 9: 203–243.
A veteran geologist takes strong exception to those who question the turbidity current hypothesis for implacement of deep-water sands.

KULM, L. D., and K. F. SCHEIDEGGER. 1979. Quaternary sedimentation on the tectonically active Oregon continental slope, p. 247–263. *In* L. J. Doyle and O. H. Pilkey, Jr. (eds.), The geology of continental margins. Soc. Econ. Paleont. Mineral. Spec. Pub. 27.

MCBRIDE, EARLE F., and ROBERT L. FOLK. 1977. The Caballos Novaculite revisited, part 2: chert and shale members and synthesis. J. Sed. Petrology 47 (3): 1261–1286.
The blind men and the elephant revisited: Geologists see what they are trained to see. They are also trained to argue convincingly. Lots of fun!

MCCAVE, N. (ed.). 1976. The benthic boundary layer. Proceedings of a conference, Les Arcs, France, Nov. 1974., vol. 1. Plenum Press, New York. 324 p.
Reviewed by W. H. Berger in Science 194: 417.

MCLEAN, H. 1981. Reservoir properties of submarine-fan facies: Great Valley sequence, California. J. Sed. Petrology 51 (3): 865–872.

MOORE, D. G. 1970. Reflection profiling studies of the California continental borderland: structure and Quaternary turbidite basins. Geol. Soc. Amer. Spec. Pap. 107. 142 p.

MOORE, J. C. 1975. Selective subduction. Geology 3: 530–532.

MUTTI, E., and F. RICCI LUCCHI. 1972. Le torbiditi dell' Appennino settentrionale: Introduzionne all' analisi di facies. Soc. Geol. Ital. Mem. 11: 161–199.
A classic paper. Also available in translation as "Turbidites of the northern Appennines: introduction to facies analysis," 1977, T. H. Nilsen (trans.), Internat. Geol. Rev. 20 (2): 125–166.

NILSEN, T. H., R. G. WALKER, and W. R. NORMARK. 1980. Modern and Ancient submarine fans: discussion and replies. Bull. Amer. Assoc. Petrol. Geol. (7): 1094–1112.

NORMARK, W. R. 1970. Growth patterns of deep-sea fans. Bull. Amer. Assoc. Petrol. Geol. 54: 2170–2195.

NORMACK, W. R., D. J. W. PIPER, and G. R. HESS. 1979. Distributary channels, sand lobes, and mesotopography of Navy Submarine Fan, California Borderland, with applications to ancient fan sediments. Sedimentology 26 (6): 749–774.

PAK, H., J. R. V. ZANEVELD, and J. KITCHEN. 1980. Intermediate nepheloid layers observed off Oregon and Washington, J. Geophys. Res. 85 (C11): 6697–6708.

RICCI LUCCHI, F., and E. VALMORI. 1980. Basin-wide turbidites in a Miocene, over-supplied deep-sea plain: a geometrical analysis. Sedimentology 27: 241–270.

ROBERTS, D. G. 1981. Continental margin processes. Nature 290 (5809): 733–734.

RODGERS, J. 1968. The eastern edge of the North American continent during the Cambrian and early Ordovician, p. 141–150. *In* E. Zen *et al.* (eds.), Studies of Appalachian geology: northern and maritime. Wiley-Interscience, New York.

———. 1970. The Taconics of the Appalachians. Wiley-Interscience, New York. 271 p.

ROSS, D. A. 1971. Sediments of the northern Middle America Trench. Bull. Geol. Soc. Amer. 82: 303–322.

ROWLEY, DAVID B., and W. S. F. KIDD. 1981. Stratigraphic relationships and detrital composition of the medial Ordovician flysch of western New England: implications for the tectonic evolution of the Taconic Orogeny. J. Geology 89: 199–218.

SANDERS, J. E. 1965. Primary sedimentary structures formed by turbidity currents and related resedimentation mechanisms, p. 192–219. *In* G. V. Middleton (ed.), Primary sedimentary structures and their hydrodynamic interpretation. Soc. Econ. Paleont. Mineral. Spec. Pub. 12.
Proposes that many so-called "turbidites" are not deposited from turbulent suspensions and suggests criteria for distinction among various deep-water resedimentation mechanisms.

SHEPARD, F. P., R. F. DILL, and U. VON RAD. 1969. Physiography and sedimentary processes of La Jolla submarine fan and fan-valley, California. Bull. Amer. Assoc. Petrol. Geol. 53: 390–420.
Excellent summary of physiography, sediment types, and sedimentary processes.

STAUFFER, P. H. 1967. Grain-flow deposits and their implications, Santa Ynez Mountains, California. J. Sed. Petrology 37: 487–508.
Discusses 4000 meters of conformable lower Tertiary sequence from deep-water turbidites through nonmarine beds.

UNDERWOOD, M. B., S. B. BACHMAN, and W. J. SCHWELLER. 1980. Sedimentary processes and facies associations within trench and trench-slope settings. Pacific Coast Paleography Symposium 4, p. 211–229.

UNDERWOOD, M. B., and D. E. KARIG. 1980. Role of submarine canyons in trench and trench-slope sedimentation. Geology 8 (9) 432–436.

UYEDA, S. and H. KANAMORE. 1979. Back-arc opening and the mode of subduction. J. Geophys. Res. 84 (B3): 1049–1062.

VALLONI, R., and J. B. MAYNARD. 1981. Detrital modes of recent deep-sea sands and their relation to tectonic setting: a first approximation. Sedimentology 28: 75–83.

VON RAD, U. 1968. Comparison of sedimentation in the Bavarian flysch (Cretaceous) and recent San Diego trough (California). J. Sed. Petrology 38: 1120–1154.
A convenient and well-illustrated discussion concerning the choice between turbidity current or normal bottom-current origin for deep-water sediment accumulation. Extensive bibliography.

WALKER, R. G. 1973. Mopping up the turbidite mess, p. 1–37. *In* Robert N. Ginsburg, (ed.), Evolving concepts in sedimentology. Johns Hopkins Univ., Studies in Geology, 21, Baltimore.
A particularly useful paper in that it is written in the context of the history of a scientific revolution.

———. 1978. Deep-water sandstone facies and Ancient submarine fans: models for exploration for stratigraphic traps. Bull. Amer. Assoc. Petrol. Geol. (6): 932–966.
See also Nilsen, Walker, and Normark 1980.

ZEN, E. 1967. Time and space relationships of the Taconic allochthon and authochthon. Geol. Soc. Amer. Spec. Pap. 97. 107 p.

———. 1968. Nature of the Ordovician orogeny in the Taconic area, p. 129–140. *In* Zen *et al.* (eds.), Studies of Appalachian geology: northern and maritime. Wiley-Interscience, New York.

———. 1972. Some revisions in the interpretation of the Taconic Allochthon in west-central Vermont: Bull. Geol. Soc. Amer. 83: 2573–2588.

ZEN, E., W. S. WHITE, J. B. HADLEY, and J. B. THOMPSON, JR. (eds.), 1968. Studies of Appalachian geology: northern and maritime. Wiley-Interscience, New York. 475 p.

Dynamics
in the Context of Time

V

In this section, we shall attempt to derive a general overview concerning the history of motion of the continents, the history of sea level, and resultant transgressions and regressions throughout the Phanerozoic. This is potentially a huge task. We must first be very clear about how we shall go about this, or we shall become bogged down in the huge volume of geological literature that is available for our consideration.

Among the sciences, geology is especially cursed with its humble beginnings, its popular "natural curiosity," and its practical applications. Pioneering stratigraphers did not map strata to understand the history of the earth; they mapped strata in order to build canals more efficiently. Fossils were not collected in order to understand the history of the earth; rather, it was popular to display them in museums. Even today, the petroleum companies spend billions of dollars on drilling, not to learn about the history of the earth; but to find oil. For these rather unscientific reasons, the science of geology has become the heir to huge amounts of data.

In the same way, individual geologists have a certain amount of unscientific training in their background. For example, geology students are still taught to map things because that is "what geologists do." They are set down in an unknown terrain, the blindfold is removed, and they are told to "map it." If they are halfway into the project and don't understand what they are looking at, they are confidently told that the answer to their problem lies in doing the job in even greater detail. Here again, we see geology as a science in which vast amounts of data precede real understanding of the scientific question involved.

This history of the science and of the geologist carry forward to the modern practice of geology. Geologists tend to try to integrate every fragment of data available to them. When they attempt, however, to integrate every fragment of data available concerning the whole earth for the entirety of the last 570 million years, the task becomes impossible. Attempts to do a detailed summary of everything that is known lead to impossible labyrinths of conflicting *ad hoc* scenarios. One common way of trying to avoid this to "get away from the data" is by building general summaries from more specific summaries. This rapidly degenerates into regurgitation of "who said what." No one looks at the original data any more (too much of it anyway!); the "common wisdom" *must* be true because so many people are saying the same thing. Further embellishment of the common wisdom becomes the goal of new data generation, a phenomena referred to as "reinforcement syndrome."

Physicists have avoided most of the curses outlined above. By and large, understanding of scientific principle has preceded application. Whereas the geologist has learned to drill for oil but still does not understand its formation and accumulation, physicists did not just stumble onto the atomic bomb and then go on to describe the structure of the atom. Physicists generally proceed systematically. They begin with simple models based on a small amount of fundamental data. When a theory comes to a possible branching point, experiments are designed to generate new data that will allow choice among competing hypotheses.

Within the earth sciences, the geophysicists are the descendants of the physicists described above. Their roots are in the science of physics, not in the building of canals. The approach that we shall follow in this section is that of the geophysicist-stratigrapher. We shall build simple models for various time intervals within the last 570 million years. We shall work only with fundamental, physically based data. We shall reach into the enormous storehouse of traditional geological data only to pull out a result that can be restructured to be the result of a physicist's experiment designed to choose among competing hypotheses.

To the geophysicists-stratigrapher there are four really fundamental data sets concerning physical description of earth history. These are (1) radiometric dates, (2) magnetic data concerning age of oceanic crust and apparent polar wandering, (3) $\delta^{18}O$ data concerning continental ice volume and/or local temperature, and (4) the seismic stratigraphy record of on-lap–off-lap relationships within relatively undeformed sediments. In the following pages, we shall attempt to derive a general model for various time intervals from these basic input data. Where questions arise that cannot be resolved within these data, we shall seek to define concise questions that may have already been answered by random data gathering.

The Quaternary
as the Key to the Past

16

The principle of uniformitarianism is often stated, "The present is the key to the past." In a dynamical sense, the concept of uniformitarianism can be restated as a set of nested models, each dealing with a large period of time. Just as the principle of uniformitarianism has long told us that things we observe in the present should not be disregarded as we look at Ancient rocks, we shall now consider that dynamics of the Quaternary should not be disregarded as we look at the Tertiary, dynamics of the Tertiary should not be disregarded as we look at the Mesozoic, and so on back through time.

The Quaternary offers another advantage as a dynamic model. There are more relevant quantitative data points per unit time for the Pleistocene than for any other time frame in the geologic record.

The 18,000-Year-Ago Glacial World

We are fortunate in stratigraphy that the last major Pleistocene glaciation occurred within a time interval that is well within the reach of carbon-14 dating. In the following pages, we shall use carbon-14 time control to demonstrate the systematic relationships among (1) the last great advance of ice sheets on North America and Europe, (2) sea-level history, (3) sea-surface temperature, and (4) $\delta^{18}O_{water}$ variation in the world ocean.

The Continental Record of Ice. For a great many years, it has been recognized that geomorphic features of central North America and northern Europe were the result of the previous existence of large continental ice sheets (Bowen, 1978; Bloom, 1978). Expressed in sedimentological terms that can be carried forward into the study of Ancient strata, we should note especially the occurrence of glacial striations on bedrock, the extremely poor sorting of tillite, and the occurrence of dropstones or other ice-rafted detritus in otherwise fine-grained marine or lake sediments. The late Pleistocene of the continents provides us an excellent look at these materials within the context of well-preserved geomorphologic features. As we go back into the geologic record, we shall have to rely more extensively upon the sedimentary criteria and less on geomorphology.

Figure 16.1 depicts the geometry of late Pleistocene Northern Hemisphere ice sheets. Moraines and tills of the Weichsel of northern Europe and the Wisconsin of North America record the last major advances of Northern Hemisphere continental ice sheets and are extensively dated by the carbon-14 method applied to wood and peat associated with tills and moraines. For purposes of this discussion, we shall consider maximum advance of these two ice sheets to be synchronous and to be approximately 17,000 to 21,000 years before present. By convention, we shall refer to this as "the 18,000 years before present world" (18,000 B.P.).

With extensive mapping of ice-related geomorphic features and extensive carbon-14 dating, the periphery of global continental ice sheets at 18,000 B.P. can be determined. Given documentation of the extent of the ice sheet and some general rules concerning stability of continental ice sheets, the volume of ice can be calculated. Such a calculation reveals that the volume of water removed from the ocean and stored on the land is indeed sufficient to cause significant changes in global sea level. Denton and Hughes (1981, p. 274) calculate that sea level in the 18,000 B.P. world stood somewhere between 91 and 163 meters below present sea level (depending upon assumptions concerning size of the ice sheet and rate of isostatic equilibration of the sea floor to changing load). Once again, we shall see that the wonders of carbon-14 dating allow us to check-out this calculation at will.

The Last 18,000 Years of Sea-Level History. If one is careful to choose materials from carefully defined stratigraphic context, a carbon-14 dated history of sea level rise throughout much of the last 18,000 years can be established. The best materials are obtained by coring. Favorite materials are coastal peat deposits overlying the Pleistocene subaerial exposure surfaces. Also popular are certain species of molluscs that occur only in shallow, low-salinity bays that existed during previous stands of the sea. Unfortunately, the literature also contains numerous carbon-14 dates on materials that were not collected within a useful stratigraphic context. Most common of these are "random shells washed up on a beach." In a similar vein, a dredge haul across a submerged beach ridge might also recover deeper-water molluscs that lived there long after sea level had submerged the ridge. Nevertheless, taking precautions to sample only good sea-level indicators and to sample them only in stratigraphic context, one does arrive at the observation that sea level has tended to rise since 18,000 B.P. Figure 9.1 presents such a sea-level rise curve. While data older than 12,000 B.P. are scant, a general tendency of sea-level rise from 17,000 B.P. to 6000 B.P. coincides with the carbon-14 dated continental record of glacial retreat. Thus, the phenomena of glacio-eustatic sea-level rise is well documented within the carbon-14 record of glacial retreat on land and sea-level rise in coastal

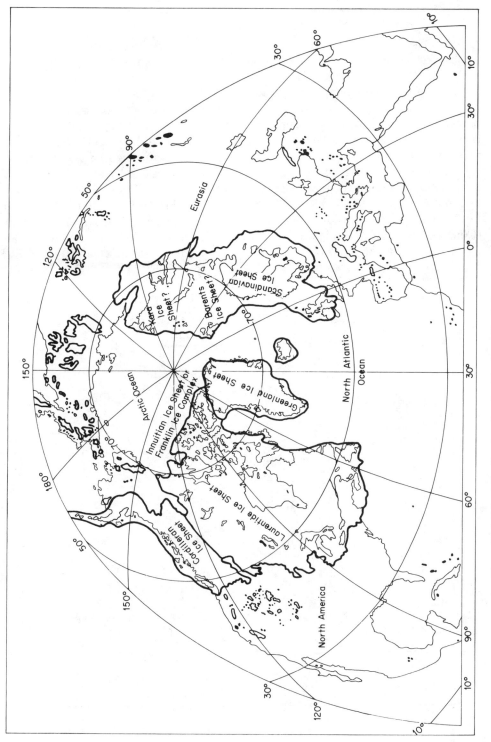

Figure 16.1 Schematic diagram showing Northern Hemisphere ice sheets approximately 18,000 B.P. Accumulation of ice on the continents is recorded by a stratigraphy of terminal moraines and tills that have been studied extensively in Europe and North America [After G. H. Denton and T. J. Hughes, *The Last Great Ice Sheets* (copyright 1981 John Wiley & Sons, Inc.), p. viii.]

377

regions. The absolute magnitude of the 18,000 B.P. lowstand of the sea remains a subject for which we shall seek additional confirmation later in this chapter.

18,000 B.P. Sea-Surface Temperature. As noted in Chapter 6, it is possible to derive quantitative estimates of sea-surface temperature from the assemblages of planktic organisms that have accumulated within deep-sea sediments. Quantitative data concerning abundance of various taxa in core-top samples can be used to write regression equations for sea-surface temperature of the modern ocean. These regression equations may then be applied to down-core samples to estimate environmental conditions at times in the past. The basic methodology is discussed in detail in Imbrie and Kipp (1971) and in Kipp (1976). The CLIMAP Project used this technique to construct global maps of sea-surface temperature at the 18,000 B.P. time interval (Cline and Hays, 1976). Figure 16.2 presents a comparison of core-top and 18,000 B.P. faunal data for the North Atlantic. Figure 16.3 presents a global picture of 18,000 B.P. sea-surface temperature for August, summarized as 18,000 B.P. depar-

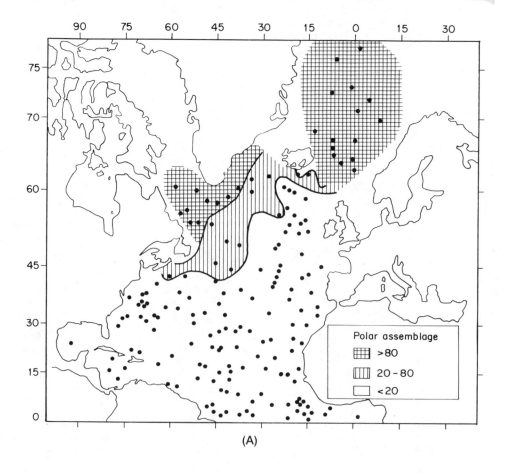

(A)

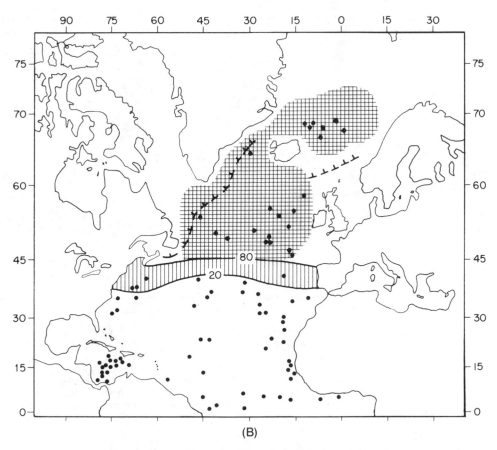

(B)

Figure 16.2 Distribution of polar assemblage of planktic foraminifers for core-top and 18,000 B.P. North Atlantic sediment samples. Whereas high abundance of this assemblage in core-top materials [Figure (A)] is restricted to a line running from Nova Scotia to northern Scotland, 18,000 B.P. distribution of this fauna [Figure (B)] extends southward to a line from Cape Cod to northern Spain. Thus, vast regions of the North Atlantic were much colder at 18,000 B.P. than they are today. Temperature regression equations based on core-top calibration of faunal and floral data allow quantitative temperature estimates to be made for the 18,000 B.P. level. Figure 16.3 presents a summary of such data. (After Kipp, 1976; and McIntyre *et al.,* 1976, with modification.)

ture from modern August conditions. Major decreases in sea-surface temperature are noted at high latitude. Additionally, eastern equatorial regions of the Atlantic and Pacific are cooler because of increased upwelling. Note, however, temperature estimates for large areas of the temperate and tropical ocean do not change much from 18,000 B.P. to the present. This observation is fundamental to interpretation of $\delta^{18}O$ data on planktic foraminifers.

18,000 B.P. $\delta^{18}O$ Values for Benthic and Tropical Planktic Foraminifers. Figure 16.4 presents a record of $\delta^{18}O$ variation in benthic foraminifers and shallow-dwelling

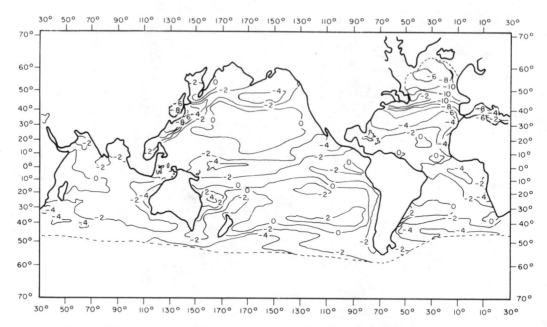

Figure 16.3 Temperature difference map comparing 18,000 B.P. August estimated sea-surface temperatures with modern sea-surface temperatures. Negative numbers mean 18,000 B.P. estimate is colder than modern; positive numbers mean 18,000 B.P. estimate is warmer than modern. Note exceedingly large temperature changes in North Atlantic and North Pacific. Note a tendency for eastern tropical Atlantic and tropical Pacific to be cooler at 18,000 B.P., reflecting increased upwelling. Note that the rest of the tropical and temperate 18,000 B.P. world ocean displays temperatures rather similar to modern. (After CLIMAP Project Members, 1981, with simplification.)

planktic foraminifers down a deep-sea core from the tropical Indian Ocean. Note that a change from isotopically light values to isotopically heavy values occurs within the upper meter of the core. This interval is likewise within the carbon-14 dating time interval and the calcium carbonate of planktic foraminifers is suitable material for carbon-14 dating. The isotopically heavy values that occur at around 100 centimeters in the core coincide with the 18,000 B.P. glacial maximum. Recall from previous discussion of the hydrologic cycle (Figure 6.5, for example) that this observation is consistent with storage of ice on the continents at the time of Wisconsin-Weichselian glaciation. However, recall that $\delta^{18}O$ values on foraminifers are a function not only of $\delta^{18}O_{water}$ variations but also of local temperature variations. Total faunal analysis temperature estimates such as depicted in Figure 16.3 can be used to constrain the temperature effect on shallow-dwelling planktic $\delta^{18}O$ values. Shackleton (1967) has argued convincingly that modern bottom water is sufficiently cold that the more positive $\delta^{18}O$ values given by 18,000 B.P. benthic foraminifers could not have resulted from yet colder conditions. Thus, both $\delta^{18}O$ records depicted in Figure 16.4 are taken to represent primarily fluctuations in $\delta^{18}O_{water}$ resulting from

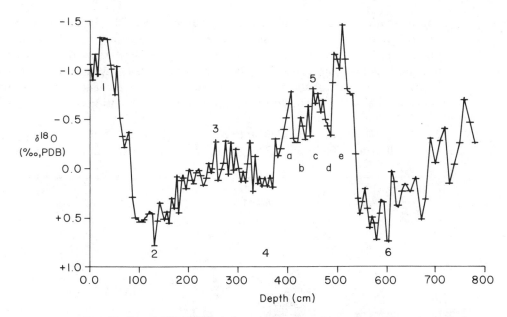

Figure 16.4 Shallow-dwelling planktic foraminifer oxygen isotope record from the tropical Indian Ocean (V34–88). Relatively light $\delta^{18}O$ values at zero to 50 centimeters and 480 to 520 centimeters record interglacial conditions. Relatively heavy $\delta^{18}O$ values at 100 to 150 centimeters and 550 to 600 centimeters record glacial conditions. (After Prell, 1983, with modifications.)

variations in the size of continental ice sheets. We shall seek additional confirmation of this point further on in this chapter.

The carbon-14 chronology in deep-sea cores does not extend beyond approximately 25,000 B.P. Note, however, in Figure 16.4 that the $\delta^{18}O$ signal in benthic and planktic foraminifers goes right on oscillating between interglacial (core-top) and glacial (100 centimeters) values throughout the length of the core. Having used the carbon-14 chronology to establish the upper portion of this record to be primarily a global ice-volume signal, it is tempting to accept that calibration and regard the $\delta^{18}O$ down-core signal as an ice-volume—glacio-eustatic sea-level signal with an amplitude of about 1.6%. However, this is a sufficiently important point that it deserves independent confirmation.

The $\delta^{18}O$ Signal as a Record of Continental Ice Volume. Carbon-14 dating of various stratigraphic records has provided us with an exciting picture of 18,000 B.P. glaciers on continental North America and northern Europe, 18,000 B.P. lower sea level, and 18,000 B.P. relatively enriched $\delta^{18}O$ values in foraminifers from deep-sea cores. Of these various records, only the $\delta^{18}O$ signal in deep-sea cores holds any prospect of yielding an intelligible continuous record within a chrono-stratigraphic framework beyond the range of carbon-14 dating. To a first approximation, many deep-sea cores can be regarded as having a constant sedimentation rate. To a first approximation, this rate can be estimated from carbon-14 dating of the upper portion of the core. Thus, for example, in Figure 16.4 one might project the sedimentation rate from core top to 100 centimeters (18,000 years ago) and estimate that the isotopically light values at 520 centimeters record events 94,000 B.P. In the following pages, we shall seek to develop additional documentation of the $\delta^{18}O$ signal as a reliable record of continental ice-volume fluctuations.

Chronology and $\delta^{18}O$ Calibration via Thorium-230 Dating of Coral-Reef Terraces. Thorium-230 dating of coral-reef terraces provides an important window of

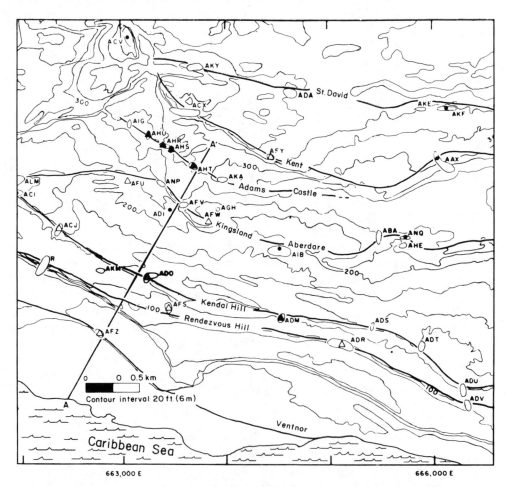

Figure 16.5 Map and cross section of late Pleistocene coral-reef terraces, Christ Church, Barbados, West Indies. Radiometric dating and $\delta^{18}O$ data on corals and molluscs allows construction of a $\delta^{18}O$ record that can be correlated to the deep-sea isotope record. Further, elevation relationships within the Barbados isotope record allow calibration of $\delta^{18}O$ as a function of glacio-eustatic sea-level lowering. (After Fairbanks and Matthews, 1978, with modification.)

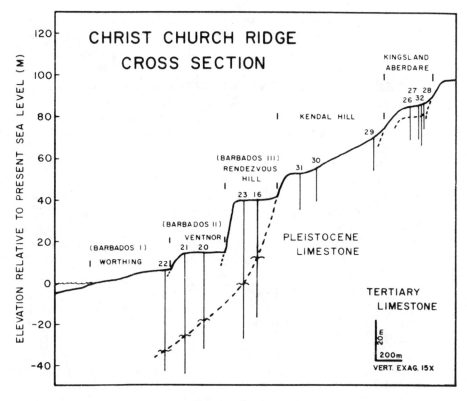

Figure 16.5 (*cont.*).

late Pleistocene chronological control that is fortuitously applicable to the last interglacial period. A coral-reef terrace somewhat above present sea level is well known from numerous tectonically stable regions (Neumann and Moore, 1975, for example). This terrace consistently dates around 125,000 B.P., conveniently in the mid-range of applicability of the thorium-230 dating method.

In tectonically active regions, such as Huan Peninsula of New Guinea, Ryukyu Islands of southern Japan, and the island of Barbados, Lesser Antilles, tectonic uplift allows access to several other terraces within the range of thorium-230 dating (see Bloom *et al.*, 1974, for example).

Figure 16.5 presents a map and cross section of several late Pleistocene coral-reef terraces on Barbados. Worthing, Ventnor, and Rendezvous Hill terraces are extensively dated at 82,000, 105,000, and 125,000 B.P. Kendal Hill terrace is only sparsely dated but probably records a highstand of the sea in the range of 180,000 to 200,000 B.P. It has long been recognized that Rendezvous Hill terrace is correlative with isotope stage (5e) (Broecker *et al.*, 1968; Mesolella *et al.*, 1969) and is likewise correlative with well-dated coral-reef terraces only slightly above present sea level in tectonically stable regions. Recognizing that the Rendezvous Hill terrace equivalent is only slightly above sea level in tectonically stable regions, we can calculate an average uplift rate for the island of Barbados over the last 125,000 years. Such estimates of uplift rate vary from place to place on Barbados but do not exceed 30 centimeters per 1000 years. Note that this rate of tectonic uplift is quite small compared to the rate of glacio-eustatic sea-level fluctuations (Chapter 9). Thus, we see the prospect that the major cause of elevation difference among the lower Barbados terraces is change in glacio-eustatic sea level. Assuming a constant uplift rate based upon the elevation of Rendezvous Hill terrace, it can be estimated that Ventnor and Worthing terraces were deposited at sea levels 20 or 25 meters below the sea level represented by Rendezvous Hill terrace (Matthews, 1973). Note that such a relationship is in good agreement with the relative magnitude of deep-sea oxygen isotope stages (5e) (Rendezvous Hill), (5c) (Ventnor), and (5a) (Worthing), thus providing additional evidence that deep-sea $\delta^{18}O$ variation reflects glacio-eustatic sea-level fluctuations resulting from variation in the amount of water stored in continental ice sheets.

Further confirmation of the relationship between $\delta^{18}O$ and glacio-eustatic sea-level fluctuation is provided by isotopic data on Barbados molluscs (Shackleton and Matthews, 1977) and corals (Fairbanks and Matthews, 1978). The coral data are especially noteworthy because they have been used to explore variation in $\delta^{18}O$ below and above the subaerial exposure surface that separates Kendal Hill forereef skeletal sands from skeletal sands transgressive to Rendezvous Hill terrace [see Figure 16.5(B)].

As indicated in Figure 16.6, there is a good agreement between relative differences in isotopic values in the Barbados terrace data and their correlative peaks and valleys in the generalized isotope curve for deep-sea foraminifers. Further, elevation differences among the Barbados samples [Figure 16.5(B)] may be taken to reflect primarily glacio-eustatic sea-level fluctuation. This affords direct calibration of the changing $\delta^{18}O$ value as a function of the changing elevation of sea level. This relationship is estimated at -0.011% per meter. Thus, the 1.6% amplitude of glacial-interglacial isotopic variation observed in deep-sea cores is taken to represent approximately 150 meters of sea-level fluctuation. This is within the realm of estimates derived from consideration of probable size of 18,000 B.P. continental ice sheets (Denton and Hughes, 1981, discussed above). Thus, late Pleistocene coral-reef terraces provide important chronology and calibration concerning the $\delta^{18}O$ record of fluctuations in continental ice volume. Further documentation of chronology can be obtained by extending the $\delta^{18}O$ record in deep-sea cores into the realm of magnetic stratigraphy.

Magnetic Stratigraphy Chronology and the Long Deep-Sea $\delta^{18}O$ Record. The work horse of Pleistocene deep-sea stratigraphy throughout the 1960's and 1970's was the conventional piston corer. Because they rely on gravity to ram the core barrel into the floor of the ocean, there are practical limits to the length of piston cores. Ten to 12 meters is a very respectable length for a piston core. If one wants to recover a relatively detailed record from a core, one tries to go to an area with

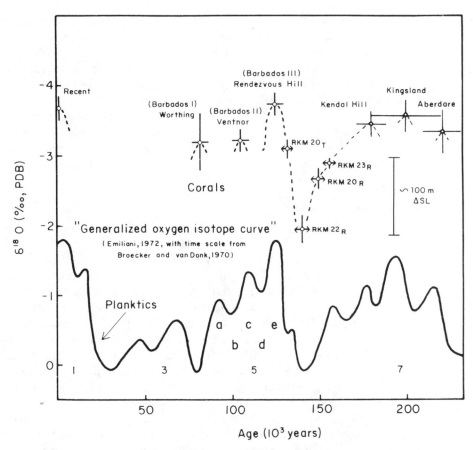

Figure 16.6 Comparison of Barbados coral $\delta^{18}O$ record with the generalized curve for tropical planktic foraminifers. The Barbados data provides a combination of $\delta^{18}O$ data and paleo-sea-level elevation data, thus allowing calibration of the $\delta^{18}O$ signal to glacio-eustatic sea-level fluctuations. (After Fairbanks and Matthews, 1978.)

a fairly high sedimentation rate, perhaps 3 centimeters per 1000 years. Thus, a 10- or 12-meter core from such a region would yield a sedimentary record of approximately the last 300,000 to 400,000 years of earth history. To achieve a longer record, one must sacrifice sedimentation rate and therewith detail. The activity of burrowing organisms at sea bottom tends to homogenize the upper 6 centimeters of the sediment. Therefore, the slower the sedimentation rate, the longer the amount of time represented by a single homogenized sample. For these reasons, a large amount of literature from the 1960's and 1970's emphasized earth history through isotope stage (7), that portion of the deep-sea record easily accessible with existing piston-core technology. The addition of hydraulic piston-coring technology to the Glomar Challenger in the early 1980's. (Prell *et al.,* 1980; Prell, 1982, for example) opened a whole new vista for the study of Lower Pleistocene and older deep-sea sediments. Virtually undisturbed long cores could now be extracted from high-sedimentation-rate areas for the first time.

Figure 16.7 presents planktic $\delta^{18}O$ data and magnetic stratigraphy for an hydraulic piston core from DSDP Site 502, western Caribbean. Penetration of the Brunhes-Matuyama magnetic boundary affords the next important control on chronology of late Pleistocene $\delta^{18}O$ stratigraphy and substantially confirms age models based on 125,000 B.P. age for isotope stage (5e) (see especially Shackleton and Opdyke, 1973). Note further in Figure 16.7 that the $\delta^{18}O$ signal shows high-frequency oscillation throughout its length, a point to which we shall return when we look at the Tertiary $\delta^{18}O$ record.

The Long Record
of Pleistocene Continental Ice Advances

Advance and retreat of continental ice sheets in northern Europe and North America are well known from geomorphology and till stratigraphy and consistently reoccupy the same general regions indicated in Figure 16.1 ($\pm$ a few hundred kilometers). For many years, vast compilations of Pleistocene continental stratigraphy within these two regions have stood as strictly physical stratigraphic representations of earth history, chronology beyond the range of carbon-14 dating being basically lacking. In recent years, some degree of success has been attained by placing the Brunhes-Matuyama boundary within the context of the classical physical stratigraphy. This job seems especially well done by Kukla (1977) with regards to correlation of classic European Pleistocene glacial geology with the deep-sea record.

Figure 16.8 summarizes the correlation of deep-sea $\delta^{18}O$ stages to classical Pleistocene geology of northern Europe. The key to this correlation is recognition of well-defined environmental cyclicity within the loess sequences of Czechoslovakia and recognition within those loess sequences of the Brunhes-Matuyama boundary. Whereas classical Pleistocene geology of northern Europe had recognized four advances and retreats of ice, the loess record of Czechoslovakia, downwind from the glaciated area, unmistakably records at least seventeen alterations of glacial advance and intervening interglacial conditions. In view of the fragmentary nature of the northern European glacial record, Kukla urgently recommends abandonment of the classical terminology in all interregional correlations. Rather, he would use the deep-sea $\delta^{18}O$ record as the worldwide standard.

Thus, we have learned from the study of continental glacial stratigraphy that the continental record is a very hard place to obtain an accurate chronology of the

---

Figure 16.7 (*opposite*) Magnetic stratigraphy and oxygen isotope stratigraphy for Pleistocene deep-sea hydraulic piston core 502. Penetration of the Brunhes-Matuyama boundary at around 16 meters provides an independent check on late Pleistocene chronology. Note that an important shift in average and amplitude of the $\delta^{18}O$ signal occurs between the upper and lower halves of this record. Such a shift presumably reflects a change in mode of ice-volume fluctuation. (After Prell, 1982, with modification.)

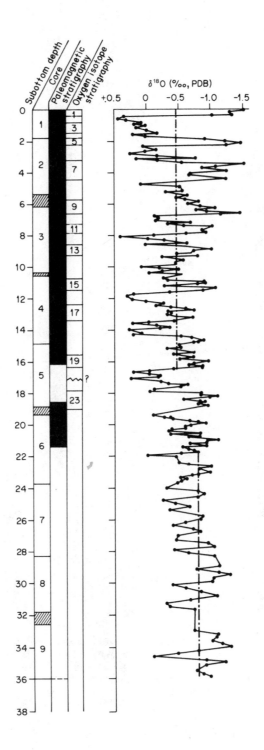

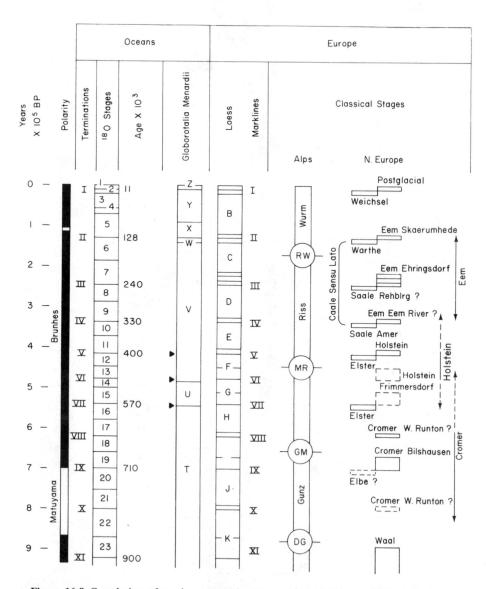

Figure 16.8 Correlation of marine oxygen isotope stages with European classical Pleistocene terminology. Kukla (1977) proposes that the classical names are of only local utility. [After G. J. Kukla, "Pleistocene Land-Sea Correlations, I. Europe," *Earth-Sci. Rev.* 13, 307–374] (with simplification). Copyright © 1977 Elsevier, Amsterdam.

history of continental ice. Nevertheless, these classic glacial deposits have served us well as a proving ground for the recognition of glacial sedimentary fabrics. As we proceed back in the geologic record, we shall look to the continents for ground truth concerning the existence of an ice sheet, but we shall look to marine sediments for the detailed record of glacio-eustatic sea-level fluctuations resultant from fluctuations in the size of continental glaciers.

Amplitude and Frequency
of Pleistocene Cyclicity

Numerous late Pleistocene $\delta^{18}O$ curves confirm a 1.6‰ amplitude for the late Pleistocene global ice-volume signal (Shackleton, 1977). Several long records confirm that this amplitude is representative throughout the Brunhes magnetic epoch. Further, $\delta^{18}O$ values appear to be normally distributed and the amplitude of the glacial interglacial cycle can be accurately represented as ± 2 standard deviations. Note in Figure 16.7 that a shift in the average and the amplitude of the $\delta^{18}O$ signal appears to occur slightly below the Brunhes-Matuyama boundary. The interpretation of this observation is that the earth tends to oscillate in a given mode for long periods of time and then tends to shift mode in response to some fundamental change in plate geometry or continental albedo or global ocean circulation.

It is apparent to even the most casual observor that there are tendencies to predominant frequencies within the $\delta^{18}O$ continental ice-volume signal (Mesolella *et al.*, 1969). This observation has been put on sound quantitative basis by application of power spectrum analysis techniques to long records of $\delta^{18}O$ and other data on deep-sea cores (Hays *et al.*, 1976, for example). The dominant periodicities within the late Pleistocene $\delta^{18}O$ signal are 100,000, 40,000 and 20,000 years.

Overview of the Pleistocene
as a Dynamic Model

It has often been said that the study of stratigraphy is the study of "one damn transgressing sea after another." The seas come in and the seas go out *ad nauseam*. Throughout the historical development of stratigraphy as a science, this fact has been handled in a substantially qualitative, descriptive fashion. The seas went out, but why and by how much? It has long been recognized that glacio-eustatic sea-level fluctuations were a possible causal mechanism, but were we confident that there was ice around at some particular time interval? Could we estimate the amount by which glacio-eustatic sea level was fluctuating?

Development of the deep-sea $\delta^{18}O$ record as a calibrated ice-volume indicator offers quantification of a concept that has been qualitatively familiar to stratigraphers for many years.

The Deep-Sea $\delta^{18}O$ Signal as a Monitor of Glacio-Eustatic Sea-Level Fluctuations. The deep-sea $\delta^{18}O$ record is highly correlated with glacio-eustatic sea-level fluctuations. Based primarily upon the direct calibration of Fairbanks and Matthews (1978), we shall take the late Pleistocene 1.6‰ amplitude to represent approximately 150 meters of sea-level fluctuation. This number is in reasonable agreement with ice-volume estimates of Denton and Hughes (1981). Further, the general tendency is clearly confirmed by the carbon-14 dated sea-level curve based upon submerged coastal peats and bay molluscs. Although the general relationship is indisputable, precise calibration is surely subject to some error. A large potential source of error arises from questions concerning thickness of Arctic floating ice pack during glacial times. The thickness of Arctic floating ice today is an inconsequential number (generally less than 7 meters). If floating ice attained considerable thickness, that ice could store water with a highly negative $\delta^{18}O$ value without affecting sea level. The magnitude of this effect is about 0.3‰ uncertainty. If 0.3‰ of the global ocean $\delta^{18}O$ signal were tied up in floating ice rather than in continental ice sheets, then the volume of water tied up in 18,000 B.P. continental ice sheets as compared to modern would be only 120 meters sea-level equivalent.

Other uncertainties concerning $\delta^{18}O$/glacio-eustatic sea-level calibration arise from estimation of local paleotemperatures and from problems of deep-sea diagenesis. Standard error of estimate for total faunal analysis paleotemperatures (Figure 16.3, for example) are on the order of 1°C. An error of 1°C in local temperature assumptions amounts to an error of about 25 meters in sea-level equivalent. Partial dissolution of foraminifer tests can alter their isotopic value by as much as 0.5‰. All things considered, we are probably safe to say that amplitude of late Pleistocene glacio-eustatic sea-level fluctuations is on the order of 120 meters ±50 meters.

Pleistocene Periodicity Consistent with Astronomical Theory of Glaciation. Most theories of Pleistocene glaciation are of little interest to the stratigrapher. Many are substantially *ad hoc* explanations intended only to "explain glaciation" —they make no statement concerning chronology or repeatability of glacial events. One theory, however, proposes an external control that must be continuous in time and predictable in its periodicity. This is the astronomical theory of climate change (Croll, 1867).

Stated most generally, the astronomical theory of climate change holds that variations in the tilt of the earth's axis, precession of the equinoxes, and variation in the eccentricity of the earth's orbit all added together to produce systematic variation in distribution of solar radiation between summer and winter and thereby to produce a major global climate change. If it can be demonstrated that perturbations of the earth's orbit have played a significant role in Pleistocene climatic change (glacial-interglacial cycles), then the stratigrapher is faced with fascinating prospects: These orbital changes have been a part of the earth's rotational history since time began! Their effects on climate may be recorded in the stratigraphic record as glacio-eustatic sea-level fluctuation or as faunal or floral alternations throughout time. In short, the perturbations of the earth's orbit are potentially a tuning fork for geologic time.

The tilt of the earth's axis (commonly referred to as the obliquity of the ecliptic) varies by about 1° with a periodicity of about 40,000 years. The higher the angle of tilt, the more pronounced the difference between summer and winter insolation in high latitudes. This particular effect is rather simple to understand. The two other effects—the precession of the equinoxes and variations in eccentricity of the earth's orbit—require a more elaborate explanation.

Figure 16.9 summarizes the combined effect of eccentricity and precession. The eccentricity of the earth's orbit varies with a periodicity of about 90,000 years. There is a virtually infinite array of scenarios which can be proposed by combining effects of eccentricity, precession, season, and location on the globe of regions chosen to be "climatically sensitive." Let's work our way through one such scenario, simply to see that reasonable arguments can be formulated.

At a time of high orbital eccentricity when the summer solstice occurs at or near the perihelion, the years will be times of exceedingly hot summers and exceedingly cold winters. One common version of the astronomical theory of climate

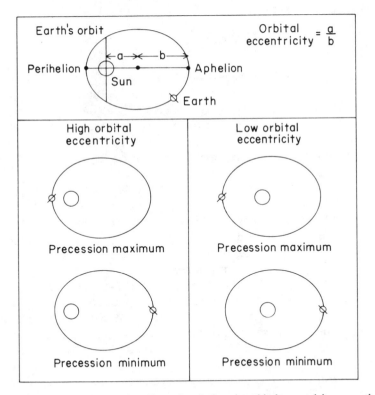

Figure 16.9 Diagram showing the effect of variations in orbital eccentricity upon the insolation intensity of precession maxima and minima. High orbital eccentricity combined with precession maximum result in maximum Northern Hemisphere summer insolation. High orbital eccentricity combined with precession minimum result in minimum Northern Hemisphere summer insolation and maximum Northern Hemisphere winter insolation. Low orbital eccentricity tends to modulate these effects. (After Mesolella *et al.*, 1969.)

change is that the earth has a climatic regime that generally favors the accumulation of Northern Hemisphere ice sheets. Only during times of unusually hot summers (high orbital eccentricity and precession maximum in Figure 16.9) will there be sufficient melting of continental ice sheets to cause a general retreat of the glaciers and resultant glacio-eustatic sea-level rise.

It has been conclusively demonstrated that the principal periodicities of the late Pleistocene $\delta^{18}O$ record are precisely the periodicities of the earth's eccentricity, tilt, and precession cycles (Hays *et al.,* 1976). Figure 16.10 compares the absolute chronology of calculated Northern Hemisphere summer insolation and orbital planktic $\delta^{18}O$ chronology from a core that includes the Brunhes-Matuyama boundary. Thus, the Pleistocene $\delta^{18}O$ signal is empirically linked to fluctuation in continental ice volume and to chronology of calculated values of variation in the earth's orbit. Thus armed, we are surely ready to take on glacio-eustatic sea-level—ice-volume fluctuations in older portions of the geologic record.

Sedimentologic-Stratigraphic Significance of Late Pleistocene Glacio-Eustatic Sea-Level Cycles

The late Pleistocene glacio-eustatic dynamic model prescribes both the amplitude and frequency of glacio-eustatic sea-level fluctuation. In order to gain some better understanding of the potential effects of glacio-eustatic events of this magnitude and frequency, let us return to two areas where we have previously given consideration to recent sedimentation and see if the late Pleistocene dynamic model would superimpose a large or a small effect on the region.

The 18,000 B.P. Texas-Louisiana Gulf Coast. Recall the general sedimentation pattern of the recent of the Texas-Louisiana Gulf Coast region (Figure 12.1) In Figure 16.11, a 150-meter sea-level lowering has been superimposed on the previous base map. The subaerial exposure surface created by a 150-meter glacio-eustatic sea-level lowering is on the order of 300 kilometers wide. Clearly, this is a feature of geologically significant magnitude.

Note further that the style of coastal clastic sedimentation would be quite different at glacio-eustatic lowstand as compared to glacio-eustatic highstand. Rivers emptying into the glacio-eustatic highstand coastal region must first fill flooded river valley estuaries before they begin to contribute sediment to construction of prograding barrier island—sandbar complexes. In the recent of the Gulf Coast, only the Mississippi, the Rio Grande, and the Brazos-Colorado rivers have completed this estuary-filling phase and have begun to contribute sediment to longshore transport and barrier-bar construction. Numerous other rivers are still building small deltas at heads of bays.

In sharp contrast, all rivers delivering sediment to glacio-eustatic lowstand coastal regions would begin immediately to dump their load as deltaic deposits protruding into open-marine conditions. There would be no bays, and every small river

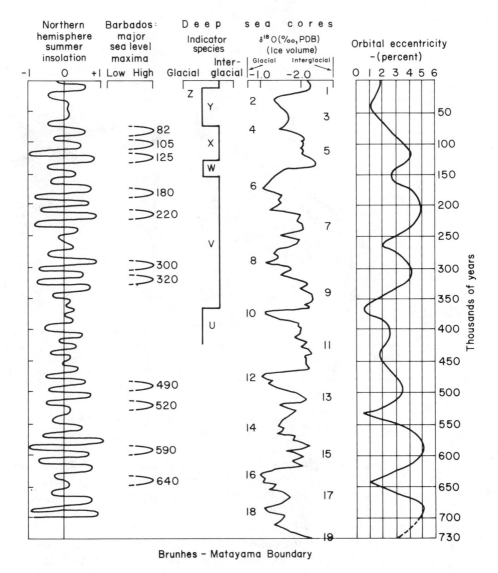

Figure 16.10 Comparison of the various types of oceanic Pleistocene climatic data with the Northern Hemisphere summer insolation curve and orbital eccentricity curve of the Milankovitch hypothesis. (After Mesolella *et al.*, 1969, with revision to reflect new data.)

would become an immediate point source of sediment supply to the open gulf. Note further, the geometry depicted in Figure 16.11 places delta sedimentation extremely close to the shelf-slope break. It is likely that newly deposited deltaic sediment would be gravitationally unstable and proceed to be transported by gravity processes into deep water. Thus, both geometry and sedimentation style would be profoundly af-

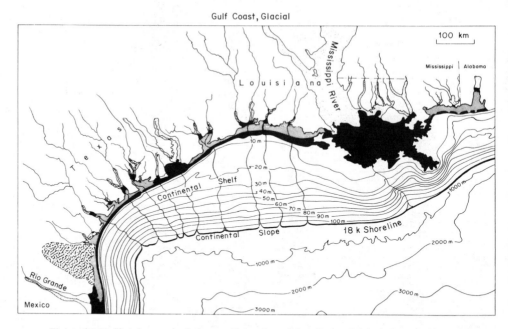

Figure 16.11 Sketch map depicting configuration of shoreline and deltaic features if a 150-meter glacio-eustatic sea-level lowering were imposed upon modern Gulf Coast bathymetry. (Previously presented in Figure 12.1.) Note that a 150-meter sea-level lowering produces subaerial exposure of the continental shelf several hundred kilometers wide and delivers river sediment discharge directly to the upper regions of the continental slope.

fected by a sea-level fluctuation on the order of those documented for late Pleistocene glacio-eustatic sea-level fluctuations.

Intermittent Highstand Coral-Reef Sedimentation of Belize. Examination of the Gulf Coast region provided convincing example that the aerial extent of an unconformity caused by a 150-meter sea-level lowering is geologically significant. Now let us utilize the Belize carbonate-sedimentation model to investigate the role of Pleistocene cyclicity in the modulation of sedimentation rate.

The maintenance of isolated carbonate patch reefs such as depicted in Figures 13.12, 13.17, and 13.18 is an extremely important phenomenon from the standpoint of petroleum exploration. Such structures are the basis for "pinnacle reef" exploration plays ranging in age from Miocene to Devonian. In our previous discussion of the patch reefs of Belize shelf lagoon, we noted that reef-complex facies tend to form on preexisting topographic highs whereas carbonate and terrigenous mud tends to accumulate in the deeper water adjacent to the patch reefs. However, both of these facies have relatively high sedimentation rates compared to the depth of water in which they occur. Recent lagoonal mud is accumulating at a rate of 20 centimeters per 1000 years. Thus, if sea level remains constant and existing sedimentation regimes remain in place, the whole Belize shelf would fill with sediment to sea level

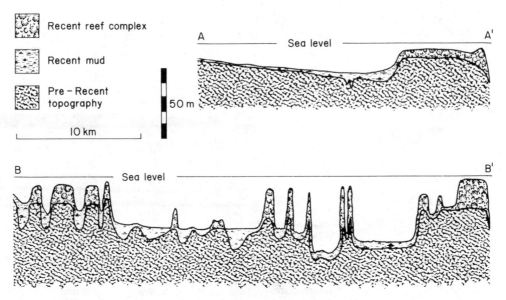

Recent reef complex

Recent mud

Pre – Recent topography

50 m

10 km

Figure 16.12 Cross sections indicating Pleistocene topography and recent sedimentation, southern Belize. Interaction between regional subsidence and late Pleistocene sea-level fluctuations generates the distinctive array of patch reefs in southern Belize.

in less than 200,000 years. However, because of late Pleistocene glacio-eustatic sea-level fluctuation, these very high sedimentation rates have occupied this shelf only intermittently during the last 200,000 years.

Figure 16.12 presents two cross sections through recent and late Pleistocene of Belize. The cross sections are based on a combination of high-frequency reflection seismic data, piston cores, and rotary drill cores. Consider Figures 16.12 and 16.6 simultaneously. Many of the topographic highs upon which recent reef-complex facies is being deposited are themselves the product of similar facies deposition during the last major glacio-eustatic highstand of the sea, isotope stage (5e), Rendez-vous Hill equivalent, 125,000 B.P. Throughout the intervening time from isotope stage 5e (125,000 B.P.) to the Holocene, glacio-eustatic sea level was sufficiently low that stage (5e) skeletal carbonates formed subaerial hills rising above the jungle that occupied what is now the floor of the lagoon.

Note also that interaction of glacio-eustatic sea-level fluctuations with regional driving subsidence has a profound effect on Belize reef geometry. Consider map (Figure 13.3) and cross sections (Figure 16.12) simultaneously. Isotope stage (5e) skeletal carbonates outcrop along the coast near the Belize-Mexico border. This same facies occurs at 30 meters below present sea level at shelf-margin in southern Belize, indicating a tendency to more rapid driving subsidence in the south. To the north, where driving subsidence is relatively low, here exemplified by 5 centimeters per 1000 years in the upper cross section of Figure 16.12, intermittent occupation of the bank margin with highstand high-sedimentation-rate reef-complex facies was sufficient to generate a broad bank with relatively uniform water depth throughout the Barrier Platform. To the south, where driving subsidence is relatively high, here taken as 10 centimeters per 1000 years in the lower cross section of Figure 16.12, the Barrier Platform breaks up into numerous isolated banks and pinnacle reefs.

Thus, glacio-eustatic exposure of isotope stage (5e) topographic highs of the Belize shelf has served two purposes that contribute to the perpetuation of an isolated patch-reef style of sedimentation. First, subaerial exposure serves to modulate the outrageously high sedimentation rates of reef-complex facies. Because these extremely high sedimentation rates are allowed to operate only approximately 5 percent of the time, the net result is a modulated sedimentation rate of only about 5 to 20 centimeters per 1000 years. Second, intermittent subaerial exposure allows for regional driving subsidence to continue with a zero sedimentation rate throughout the long periods of subaerial exposure that exist between the intermittent times of highstand reef-complex sedimentation. If *any* shallow-marine carbonate-sedimentation regime occupied the Belize shelf throughout the operation of a driving subsidence on the order of 5 or 10 centimeters per 1000 years, the carbonate sediments would simply keep pace with the new space created by subsidence. Sea-level fluctuations allow for new room to be created by driving subsidence at the time of a zero sedimentation rate. Each new highstand may thus accommodate meters to tens of meters of new high-sedimentation-rate skeletal carbonates in the space created by continual slow-driving subsidence throughout the previous glacio-eustatic lowstand.

The Pleistocene as the Key to the Tertiary. In summary, we have seen that the dynamics of the late Pleistocene glacio-eustatic sea-level fluctuations are indeed sufficient to profoundly affect a geological record of earth history. A sea-level fall of 150 meters creates an unconformity surface hundreds of kilometers wide in a region such as the modern Texas-Louisiana Gulf Coast. Southern Belize would surely be just one wide coastal swamp right out to the shelf edge if it were not for modulation of sedimentation rates by glacio-eustatic sea-level fluctuations. These are thoughts that we shall take with us as we approach the Tertiary.

Citations and Selected References

BENDER, MICHAEL L., RICHARD G. FAIRBANKS, F. W. TAYLOR, R. K. MATTHEWS, JOHN G. GODDARD, and WALLACE S. BROECKER. 1979. Uranium-series dating of the Pleistocene reef tracts of Barbados, West Indies. Bull. Geol. Soc. Amer. 90 (6): 577–594.
Defines terraces as morphostratigraphic units and names them; provides new helium-growth dates for the older terraces.

BERGER, W. H., and J. V. GARDNER. 1975. On the determination of Pleistocene temperatures from planktonic foraminifera. Foraminiferal Res. 5 (2): 102–113.
An interesting exposé of some of the pitfalls of total faunal analysis temperature regression equations.

BLOOM, A. L. 1978. Geomorphology. Prentice-Hall, Inc., Englewood Cliffs, N.J. 497 p.

BLOOM, A. L., W. S. BROECKER, M. A. CHAPPELL, R. K. MATTHEWS, and K. J. MESO-LELLA. 1974. Quaternary sea level fluctuations on a tectonic coast: new ^{230}Th/^{234}U dates from the Huon Peninsula, New Guinea. Quaternary Res. 4: 185–205.

BOWEN, D. Q. 1978. Quaternary geology, a stratigraphic framework for multidisciplinary work. Pergamon, London. 224 p.

BROECKER, W. S., D. L. THURBER, J. GODDARD, T. L. KU, R. K. MATTHEWS, and K. J. MESOLELLA. 1968. Milankovitch hypothesis supported by precise dating of coral reefs and deep-sea sediments. Science 159: 297–300.
A "major breakthrough" paper in its day; now, largely supplanted by Mesolella *et al.* (1969), Bender *et al.* (1979), and Hays *et al.* (1976).

BUDYKO, M. I. 1978. 5. The heat balance of the Earth, p. 85–113. *In* John Gribbin (ed.), Climatic change. Cambridge Univ. Press, London.
Easy reading concerning the basis for climatic change; highly recommended to those interested in climate models.

CHOI, D. R., and R. N. GINSBURG. 1982. Siliciclastic foundations of Quaternary reefs in the southernmost Belize Lagoon, British Honduras. Geol. Soc. Amer. Bull. 93: 116–126.

CHOI, D. R., and C. W. HOLMES. 1982. Foundations of Quaternary reefs in south-central Belize Lagoon, Central America. AAPG Bulletin (12): 2663–2671.

CLIMAP PROJECT MEMBERS. 1981. Seasonal reconstructions of the earth's surface at the last glacial maximum. Geol. Soc. Amer. Map and Chart Series MC-36.

CLINE, R. M., and J. D. HAYS (eds.). 1976. Investigation of late Quaternary paleo-ceanography and paleoclimatology. Geol. Soc. Amer. Mem. 145. 464 p.
A major volume reporting results of the CLIMAP Program.

CROLL, J. 1867. On the change in the obliquity of the ecliptic, its influence on the climate of the polar regions and on the level of the sea. Phil. Mag. 33: 426–445.
A very current topic has strong roots in the past.

DENTON, GEORGE H., and TERENCE J. HUGHES (eds.). 1981. The last great ice sheets. Wiley-Interscience, New York, 484 p.

EMILIANI, C., and J. GEISS. 1957. On glaciations and their causes. Sonderdruck aus der Geologischen Rundschau 46 (2): 576–601.
A thoughtful early paper.

FAIRBANKS, R. G., and R. K. MATTHEWS. 1978. The marine oxygen isotope record in Pleistocene coral, Barbados, West Indies. Quaternary Res. 10 (2): 181–196.

GARDNER, JAMES V., *et al.* 1980. Hydraulic piston coring of late Neogene and Quaternary sections in the Caribbean and equatorial Pacific: preliminary results of Deep Sea Drilling Project Leg 68. Bull. Geol. Soc. Amer. 91, (pt. 1, no. 7): 433–444.
A successful HPC leg complete with description of the coring device.

GATES, W. LAWRENCE. 1976a. Modeling the Ice-Age climate. Science 191: 1138–1143.
Global atmospheric circulation model for the 18,000 B.P. world inputting CLIMAP sea-surface temperature boundary conditions.

———. 1976b. The numerical simulation of Ice-Age climate with a global general circulation model. J. Atmospheric Sci. 33: 1844–1873.

HARMON, R. S., *et al.* 1981. Bermuda sea level during the last interglacial. Nature 289 (5797): 481–483.
A fascinating jigsaw puzzle of physical stratigraphy, amino acid stratigraphy, and radiometric dating.

HAYS, J. D., JOHN IMBRIE, and N. J. SHACKLETON. 1976. Variations in the earth's orbit. pacemaker of the ice ages. Science 194: 1121–1132.
Utilizes power spectrum analysis of deep-sea core isotopic and faunal records to convince most people that a causal relationship exists between perturbations of the earth's orbit and glacial-interglacial cyclicity.

HUGHES, T., G. H. DENTON, and M. G. GROSSWALD. 1977. Was there a late-Würm Arctic ice sheet? Nature 266: 596–602.
An easily accessible short version of Denton and Hughes (1981).

IMBRIE, JOHN, and NILVA G. KIPP. 1971. 5. A new micropaleontological method for quantitative paleoclimatology: application to a late Pleistocene Caribbean core, p. 71–181. *In* Karl K. Turekian (ed.), The late Cenozoic glacial ages. Yale Univ. Press, New Haven, Conn.

IMBRIE, JOHN, and J. Z. IMBRIE. 1980. Modeling the climatic response to orbital variations. Science 207 (4434): 943–952.

JOHNSON, R. G., and J. T. ANDREWS. 1979. Rapid ice-sheet growth and initiation of the last glaciation. Quaternary Res. 12: 119–134.

KIPP, NILVA G. 1976. New transfer function for estimating past sea-surface conditions from sea-bed distribution of planktonic foraminiferal assemblages in the North Atlantic, p. 3–41. *In* R. M. Cline and J. D. Hays (eds.), Investigation of late Quaternary paleoceanography and paleoclimatology. Geol. Soc. Amer. Mem. 145.

KUKLA, G. J. 1977. Pleistocene land-sea correlations, I: Europe. Earth-Science Rev. 13: 307–374.

MATTHEWS, R. K. 1973. Relative elevation of late Pleistocene high sea level stands: Barbados uplift rates and their implications. Quaternary Res. 3 (1): 147–153.

McINTYRE, A., and N. G. KIPP with A. W. H. BE, *et al.* 1976. Glacial North Atlantic 18,000 years ago: a CLIMAP reconstruction, p. 43–76. *In* R. M. Cline and J. D. Hays (eds.), Investigation of late Quaternary paleoceanography and paleoclimatology. Geol. Soc. Amer. Mem. 145.

MESOLELLA, K. J., R. K. MATTHEWS, W. S. BROECKER, and D. L. THURBER. 1969. The astronomical theory of climatic change: Barbados data. J. Geology 77: 250–274.

NEUMANN, A. CONRAD, and WILLARD S. MOORE. 1975. Sea level events and Pleistocene coral ages in the northern Bahamas. Quaternary Res. 5: 215–224.

NINKOVICH, DRAGOSLAV, and N. J. SHACKLETON. 1975. Distribution, stratigraphic position and age of ash layer "L," in the Panama Basin region. Earth Planet. Sci. Lett. 27: 20–34.

PETERSON, G. M., T. WEBB, III, J. E. KUTZBACH, T. VAN DER HAMMEN, T. A. WIJMSTRA, and F. A. STREET. 1979. The continental record of environmental conditions at 18,000 yr B.P.: an initial evaluation. Quaternary Res. 12 (1): 47–82.

POAG, C. WYLIE, and PAGE C. VALENTINE. 1976. Biostratigraphy and ecostratigraphy of the Pleistocene Basin Texas-Louisiana continental shelf. Trans. Gulf Coast Assoc. Geol. Soc. 26: 185–256.
Whereas more than twenty post-Olduvai glacial-interglacial cycles are recognized in deep-sea cores, this Gulf Coast study finds only seven phases of regression-transgression.

PORTER, STEPHEN C. 1979. Hawaiian glacial ages. Quaternary Res. 12 (2): 161–187.
A fascinating stratigraphy intertwining glacial and volcanic features.

PRELL, WARREN L. 1982. 20. Oxygen and carbon isotope stratigraphy for the Quaternary of Hole 502B: evidence for two modes of isotopic variability, p. 455–464. *In* W. L. Prell, J. V. Gardner *et al.*, Initial reports of the deep sea drilling project, vol. 68. U.S. Gov't. Printing Office, Washington, D.C.

———1983. Monsoonal climate of the Arabian Sea during the late Quaternary: a response to variations in low-latitude solar radiation (forthcoming). *In* A. Berger, *et al.* (eds.), Milankovitch and Climate. Understanding the response to orbital forcing. D. Reidel Publishers Co., Boston.

RUDDIMAN, W. F., and ANDREW MCINTYRE. 1981a. The mode and mechanism of the last deglaciation: oceanic evidence. Quaternary Res. 16 (2): 125–134.
To understand the North Atlantic is to understand late Pleistocene glacial cyclicity.

———1981b. Oceanic mechanisms for amplification of the 23,000-year ice-volume cycle. Science 212: 617–627.

SACHS, H. M. 1976. Evidence for the role of the oceans in climatic change: tests of Weyl's theory of the ice ages. J. Geophys. Res. 81: 3141–3150.
Companion reading to Weyl (1968).

SCHNEIDER, STEPHEN H., and STARLEY L. THOMPSON. 1979. Ice ages and orbital variations: some simple theory and modeling. Quaternary Res. 12 (2): 188–203.

SHACKLETON, N. J. 1967. Oxygen isotope analyses and Pleistocene temperatures re-assessed. Nature (London) 215: 15–17.
The classic paper which suggests the $\delta^{18}O$ signal in deep-sea core to be an ice-volume signal rather than a paleotemperature signal.

———1977. The oxygen isotope stratigraphic record of the late Pleistocene. Phil. Trans. Roy. Soc. London B. 280: 169–182.
Reports out a great deal of late Pleistocene $\delta^{18}O$ data.

SHACKLETON, N. J., and N. D. OPDYKE. 1973. Oxygen isotope and palaeomagnetic stratigraphy of equatorial Pacific core V28-238: oxygen isotope temperatures and ice volumes on a 10^5 and 10^6 year scale. Quaternary Res. 3: 39–55.
The classic study integrating $\delta^{18}O$ record, paleomagnetics, and the Barbados terrace sequence.

SHACKLETON, N. J., and R. K. MATTHEWS. 1977. Oxygen isotope stratigraphy of late Pleistocene coral terraces in Barbados. Nature 268 (5621): 618–620.

SUAREZ, MAX J., and ISAAC M. HELD. 1979. The sensitivity of an energy balance climate model to variations in the orbital parameters. J. Geophys. Res. 84 (C8): 4825–4836.

WATKINS, NORMAN D., JOHN KEANY, MICHAEL T. LEDBETTER, and TER-CHIEN HUANG. 1974. Antarctic glacial history from analyses of ice-rafted deposits in marine sediments: new model and initial tests. Science 186: 533–536.
Recognition of ice-rafted detritus around Antarctica in deep-sea core materials will be a major issue in the Tertiary. This study may be of interest as a late Pleistocene dynamical model.

WOILLARD, GENEVIEVE M. 1978. Grande pile peat bog: a continuous pollen record for the last 140,000 years. Quaternary Res. 9: 1–21.
See also discussion and reply in Quaternary Res. 12 (1): 152–155.

WEYL, PETER K. 1968. The role of the oceans in climatic change: a theory of the ice ages. Meteorological Monographs 8 (30): 37–62.
An early, thoughtful paper on "how the ocean works." See also Sachs (1976).

The Tertiary

17

The Tertiary comprises the time between the Pleistocene and the Cretaceous. In this chapter, we shall attempt to understand the Tertiary within the context of rather strict adherence to our knowledge concerning the dynamics of the late Quaternary earth. The late Quaternary glacio-eustatic sea level dynamic model developed in Chapter 16 is based largely upon detailed knowledge concerning the last 500,000 years of earth history. As we turn our attention to the Tertiary, we are attempting to understand the dynamics of a time interval two orders of magnitude greater than the time interval within which our dynamic model was constructed. It is qualitatively well known that the earth has been quite different from time to time within the Tertiary; surely, we do not expect 100 percent agreement between the late Quaternary dynamic model and the entire Tertiary. Nevertheless, it will be useful to describe variation within the Tertiary in terms of deviation from the late Quaternary dynamic model.

Continental Plates

In Chapter 8, we saw that a number of lines of evidence indicate that the position of continental plates has changed throughout geologic time. It is now time to begin systematically to put that information to use. Figure 17.1 presents Lambert equal-area and polar-stereographic projections depicting the present position of continental plates. The heavy outlines depict the major continental blocks which we shall try to keep track of as we go back in time. Recall that Figure 8.4 indicates the existence of numerous smaller continental blocks. Most of these blocks have been a feature of the earth's surface throughout Phanerozoic time. They are undoubtedly important to many regional geologic questions, but we simply do not have sufficient knowledge of the paleo-positions to keep track of them in the following discussions.

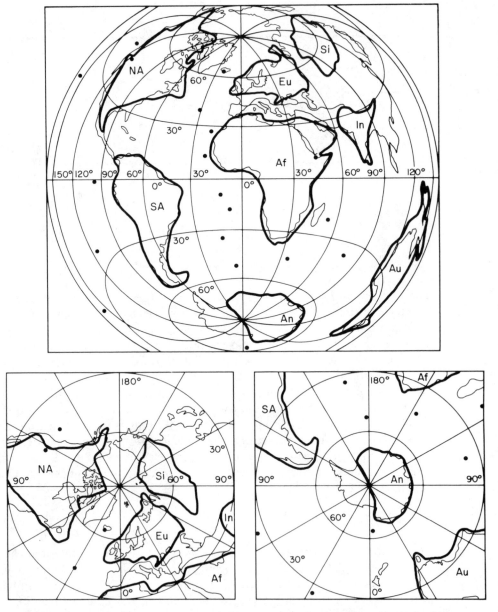

Figure 17.1 Maps indicating present position of major continental blocks which will be tracked back through time. Closed circles indicate hot spots which may be useful in determining the paleo-position of continents. Information contained in this map is analogous to that provided in Figure 8.4, but the format will allow us to more easily recognize opening and closing of ocean basins and rearrangement of continents around polar regions.

Figure 17.2 presents analogous maps for position of continental blocks in the Eocene, approximately 60 million years ago. In reconstructing the 60-million-year-ago world, several important lines of evidence are available to us. First and foremost, sea-floor magnetic stripes (Figure 8.1) are sufficiently well known for this time interval that we may basically fold the magnetic stripes back into mid-ocean ridges and obtain the paleo-relation of continents on opposite sides of the ridge. Secondly, apparent polar wandering data (Figures 8.5 and 8.6) is available for several continental blocks under consideration. Finally, it is to some extent possible to recognize the effect of hot spots on oceanic and continental lithosphere throughout this time interval. Presuming the hot spot to have a fixed location with regard to the pole of rotation of the earth, such data provide longitudinal fix on plate motions as well as independent confirmation of latitudinal information provided by apparent polar wandering data. This concept is especially well documented with regard to the Hawaiian Islands–Emperor Seamount chain of the north central Pacific (Dalrymple *et al.*, 1981). Similar arguments are less well documented for various hot spots in the North and South Atlantic (Crough *et al.*, 1980; Crough, 1981).

Paleo-Oceans. Comparison of Figures 17.1 and 17.2 indicates that major changes in continent ocean relationships occurred during the last 60 million years. Changes in continent ocean geometry will certainly have an effect on the sedimentary geology of the various continental blocks. Further, arrangement of continents may effect the average state of the globe with regard to temperature, continental ice volume, and thereby sea level. Let us, therefore, take the time to get these differences firmly in mind.

In order to better understand the comparison of maps 17.1 and 17.2, it will be useful at this point to glance ahead to similar maps concerning the Mesozoic (Figure 18.1) and late Paleozoic (Figure 19.1). Note in Figure 18.1 that South America, Africa, India, Antarctica, and Australia are all combined into a single Southern Hemisphere supercontinent, commonly referred to as Gondwanaland. Note in Figure 17.2 that only Australia and Antarctica remain connected and that India and Africa-Arabia have moved substantially northward.

With regards to activity in the Northern Hemisphere, it is useful to compare Figures 17.1, 17.2, and 18.1. Note that the size of the North Atlantic increases dramatically throughout this sequence of snapshots. Note that the Norwegian-Greenland Sea connection between the North Atlantic and Arctic Oceans does not exist at 60 million years. Note that a gap between North and South America appears to allow free exchange between the North Atlantic and Pacific Oceans in Figure 17.2. Similarly, there appears to have been an open seaway (even though probably cluttered with microcontinents!) between Europe-Siberia and Africa-Arabia-India. This is commonly referred to as the Tethys Seaway.

The continental configuration indicated in Figure 17.2 must surely produce a general oceanic circulation different from today's. Connection of Antarctica to Australia clearly rules out circumpolar circulation, which is a major feature of today's southern ocean. Further, the Norwegian-Greenland Sea, which is a major

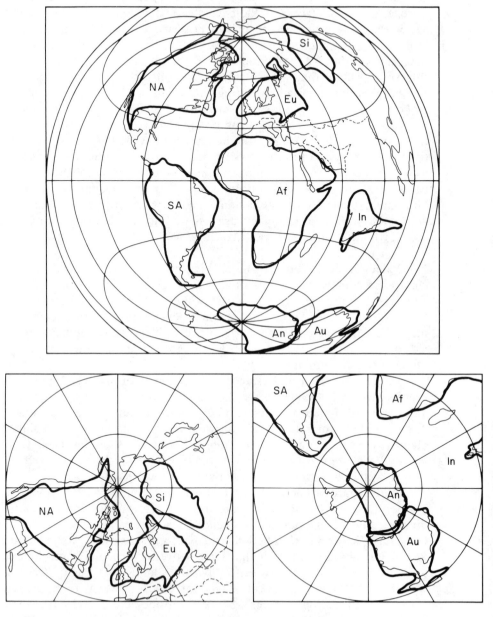

Figure 17.2 Maps indicating position of major continental blocks at 65 million years ago, compiled from sources cited in Chapter 8.

producer of deep water in today's ocean, does not exist at 60 million years ago. A throughgoing connection from the North Atlantic to the Pacific would surely decimate the North Atlantic salinity anomaly of today's ocean.

Thus, we have good reason to suspect that major changes have occurred within global ocean circulation over the last 60 million years in response to continual rearrangement of continents and oceans via plate tectonics. Relation of constantly unfolding plate geometry to other major events identifiable within the Tertiary stratigraphic record surely holds promise as a great unifying concept.

Polar Regions. With regard to the Quaternary, we have seen that accumulation of ice on continental polar regions is the world's most effective mechanism for modulating sea level. If we are to have ice on continents in the Tertiary, it seems almost axiomatic that we shall need to have continents within polar regions. Figure 17.2 indicates no great change in this regard from the present condition (Figure 17.1).

$\delta^{18}O$ and Continental Ice Volume

In the previous chapter, it was developed that $\delta^{18}O$ values for benthic and low-latitiude shallow-dwelling planktic foraminifers provide an excellent proxy indicator of continental ice volume within the late Quaternary. If we are to attempt to apply the late Quaternary glacio-eustatic dynamic model to the Tertiary, clearly it will be highly desirable for $\delta^{18}O$ to serve as one of the lead data sets.

The $\delta^{18}O$ Dilemma. Recall that the quantity that we measure, $\delta^{18}O_{calcite}$ for benthic or planktic foraminifers, is a function of both water temperature and the isotopic composition of the water in which the calcite grew. With regard to planktic foraminifers, local temperature can vary as a function of depth at which the foraminifer grew; likewise, $\delta^{18}O_{water}$ may vary because of local salinity effects or may reflect global $\delta^{18}O_{water}$ variation which is in turn related to global ice-volume fluctuation. Thus, random measurement of $\delta^{18}O$ on taxa of unknown affinity taken from samples of convenience may provide uninterpretable results. If we are to estimate temperature, we must constrain the $\delta^{18}O_{water}$ effect; if we are to estimate $\delta^{18}O_{water}$ (and thereby ice volume), we must somehow constrain temperature and local salinity effects.

With regard to development of the Pleistocene dynamic model, Shackleton (1967) developed the concept that $\delta^{18}O_{calcite}$ in deep-sea benthic foraminifers was primarily a measure of $\delta^{18}O_{water}$. He argued that modern bottom-water temperatures are sufficiently cold that 18,000 B.P. benthic foraminifer $\delta^{18}O$ values could not be explained by still colder bottom-water temperatures. Therefore, the benthic isotope signal is substantially a $\delta^{18}O_{water}$ signal. The similarity between benthic and planktic isotope records therefore indicates that the global ice-volume $\delta^{18}O_{water}$ effect was the primary signal.

In the Tertiary, preliminary data (Figure 17.3, for example) clearly indicate a tendency to warmer bottom water back through time. Thus, the arguments of

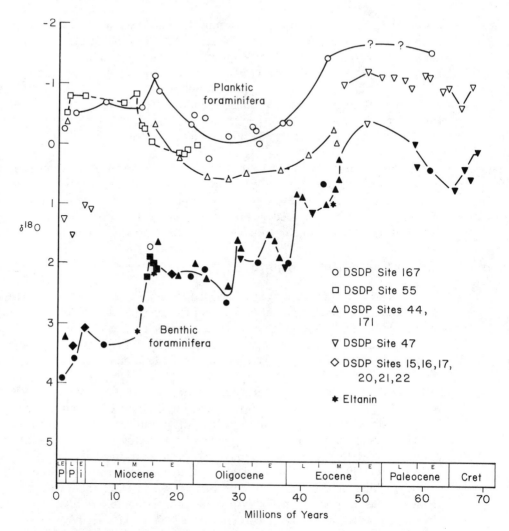

Figure 17.3 Tertiary $\delta^{18}O$ data for benthic and tropical planktic foraminifers. [After Samuel M. Savin, "The History of the Earth's Surface Temperature during the Past 100 Million Years," *Ann. Rev. Earth Planet. Sci.* 5, 319–355 (1977).] Reproduced with permission from the Annual Review of Earth and Planetary Sciences, Volume 5. Copyright 1977 by Annual Reviews Inc.

Shackleton (1967) cannot be applied to the Tertiary. We are left with but two choices. We may either constrain the global $\delta^{18}O_{water}$ effect by invoking geological arguments concerning continental ice volume or we may follow the lead of the late Quaternary dynamic model and constrain tropical sea-surface temperature. If we propose to constrain global ice volume, then the $\delta^{18}O_{calcite}$ record records variation in temperature; if we constrain tropical sea-surface temperature, then the tropical shallow-dwelling planktic $\delta^{18}O_{calcite}$ record records variation in global ice volume. These two alternatives lead to profoundly different interpretations concerning the general state of the planet throughout the Tertiary. For the time being, let us keep an open mind with regard to this choice and explore the alternative consequences.

The Oceanic Benthic and Tropical Shallow-Dwelling Planktic Foraminifer $\delta^{18}O$ Record. The upper portions of Figures 17.4 and 17.5 present a generalized representation of Tertiary $\delta^{18}O$ data for tropical sea-surface equilibrium calcite (shallow-dwelling planktic foraminifers) and for deep-sea benthic foraminiferal calcite. The curves are intended to represent average conditions. They surely smooth out real events on the scale of at least $\pm 0.3\%$, perhaps even larger. Figure 17.4 presents an interpretation scheme that seeks to constrain $\delta^{18}O_{water}$ by assuming the earth to be ice-free prior to the Middle Miocene. Figure 17.5 presents an interpretation scheme which proposes to constrain tropical sea-surface temperature to constant values.

Both interpretation schemes yield to similar results with regard to Tertiary bottom-water temperature. Either way, early Tertiary bottom water was considered warmer than modern. This result is consistent with our previous discussion of the effect of continental geometry on circumpolar circulation and the production of North Atlantic deep water.

With regard to interpretation of the tropical shallow-dwelling planktic foraminifer $\delta^{18}O_{calcite}$ record, the two interpretations are in stark disagreement. Insertion of an ice-free world assumption prior to the Middle Miocene forces the interpretation that tropical sea-surface temperatures were quite cool throughout the Lower Miocene and Oligocene. Assuming constant tropical sea-surface temperature, there appears to be a major shift in global ice volume around the Eocene-Oligocene boundary, but no "ice-free world" is indicated in the last 50 million years.

The Sedimentary Record of Continental Ice. If we want to have significant continental ice volume throughout the Tertiary, clearly Antarctica is a logical place to try to put it. Field checking the possibility of Eocene ice in Antarctica is not a simple task. The vast majority of our "field area" is presently beneath several kilometers of modern ice sheet. For this reason, let us regard global atmospheric circulation modeling experiments as a guide to what we might expect. With the modeling results in mind, we shall then examine the scattered and fragmentary geologic record of the Tertiary history of Antarctica.

Antarctica is beneath the South Pole today and there is an immense ice sheet on the continent. Reconstruction of plate motion indicates Antarctica has been beneath the South Pole throughout the Tertiary. Thus, an empirical, first-principle argument would assert that a Tertiary Antarctic ice sheet is to be anticipated until proven otherwise. Global atmospheric circulation modeling experiments lend credence to this assertion. The interior of the Antarctic continent remains quite cold regardless of specified changes in plate geometry and/or southern ocean sea-surface temperature (Barron *et al.,* 1981; Oerlemans, 1982; Barron and Washington, 1982). We are encouraged to anticipate the presence of ice somewhere on Antarctica

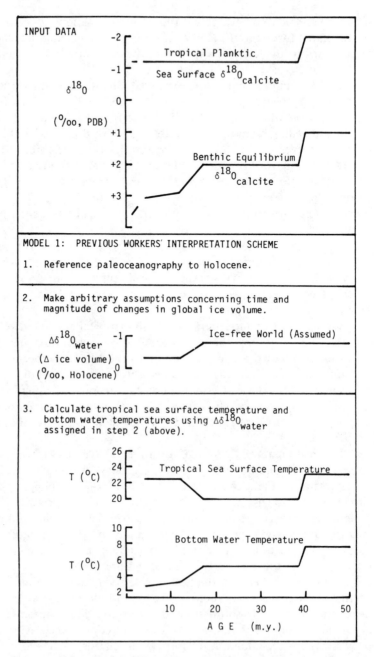

Figure 17.4 Diagram indicating generalized $\delta^{18}O$ record for the last 50 million years and an interpretation scheme based upon assumptions concerning global ice volume. Note especially that relatively cool tropical sea-surface temperatures indicated between 17 and 40 million years are an artifact of ice-volume assumptions. (From Matthews and Poore, 1980.)

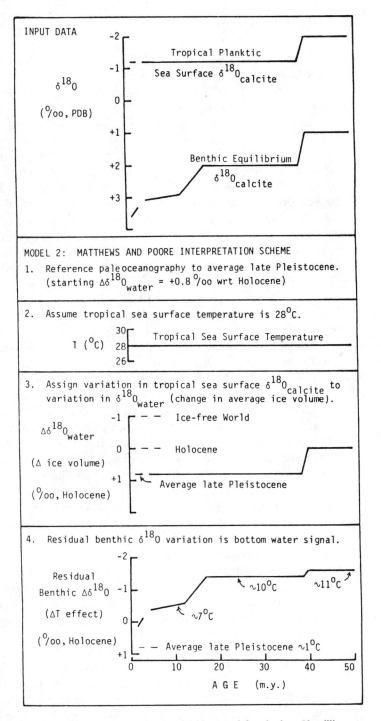

Figure 17.5 Diagram indicating generalized $\delta^{18}O$ record for the last 50 million years and an interpretation scheme based upon assumption of constant sea-surface temperature. In sharp contrast to the assumption presented in Figure 17.4, this interpretation scheme allows significant ice volume to exist even at 50 million years ago. (From Matthews and Poore, 1980.)

throughout the Tertiary. With this modeling result as a guide, let us examine some of the geological evidence.

Till sequences from the Dry Valleys of Antarctica have been studied by George Denton and his coworkers at the University of Maine (Prentice, 1982, for example). These sequences contain interbedded marine deposits that are biostratigraphically placed as Middle Miocene. At least one till lies below this biostratigraphic level and geologic arguments indicate even greater antiquity to initial glaciation of this region. They conclude that there was a pre–Middle Miocene Antarctic ice sheet larger than the modern or even the 18,000 B. P. ice sheet. They argue that West Antarctica stood as a plateau prior to the substantially Neogene episode of block faulting in the region. Such a configuration of the continent affords a firm, continental substrate for development of terrestrially based ice in Antarctica in contrast to the smaller, marine-based ice sheet that exists today. Ages on subglacially erupted volcanics in West Antarctica range back as far as 28 million years (LeMasurier and Rex, 1982). Owing to geochemical and geological complications, these dates probably represent maximum age for the emplacement of the volcanic material. However, in a geological sense, the dates in no way preclude existence of ice prior to the first subglacial volcanic eruption.

Figure 17.6 compares the modern configuration of Antarctic ice with the ice configuration consistent with the observations of LeMasurier and Rex (1982) and Prentice (1982). Thus, when we view the antiquity of the Antarctic ice sheet, we find that the data trail ends in the Neogene with a huge ice sheet sitting on the Antarctic continent. How long the ice sheet has been there and how it got there are substantially unknown. The Antarctic geology to answer these questions would now be buried beneath the East Antarctic ice sheet, if indeed the geological evidence was ever adequately recorded in a datable stratigraphy of that region. In this book, we shall presume ice in Antarctica back into the Paleogene and Cretaceous until some line of positive evidence (such as the δ^{18}O record) indicates the world to be ice-free.

Arguments against the existence of continental ice sheets throughout the Tertiary are rampant in the literature and therefore deserve comment. These arguments fall into two categories. First, there are arguments to the effect that the *absence* of positive evidence *for* glaciation somehow indicates there was no glaciation. Second, there are arguments that proceed from positve evidence for existence of seemingly warm conditions in places where one might expect Ancient continental ice.

Arguments based on the absence of positive evidence for early Tertiary glaciation can flow from the absence of glacial sediment types on continental areas or from the absence of ice-rafted detritus in deep-sea cores adjacent to potentially glaciated areas. It is easy to envision that the continental record of glaciation be removed by subsequent erosion. Continental glaciers tend to form in elevated regions; their stratigraphic record could easily be removed by erosion subsequent to deglaciation. Importantly, one of the most attractive spots to put early Tertiary glaciation must surely be East Antarctica. Here, unfortunately, the entire field area now lies beneath several kilometers of ice. We shall simply never know whether there is geological evidence for early Tertiary glaciation beneath all that ice in East Antarctica.

The second "negative evidence" argument concerning early Tertiary glaciation centers around the absence of ice-rafted detritus from certain high-latitude deep-sea cores. It is widely accepted that there is significant ice-rafted detitus in southern ocean cores from the Middle Miocene to Recent epochs. There are also reports of ice-rafted detritus in deep-sea cores tentatively ascribed to dates as old as Oligocene (Margolis and Kennett, 1971, for example). The existence of ice-rafted detritus is positive evidence for existence of glaciation on Antarctica. One can construct arguments to the effect that more abundant ice-rafted detritus in younger sediments has paleoenvironmental significance. However, the absence of ice-rafted detritus in older southern ocean sediments cannot be taken to indicate the absence of continental glaciation on Antarctica. The absence of ice-rafted detritus would simply indicate that continental ice sheets, if present, underwent phase change to water before reaching sea level. Antarctica is a huge continent; there is abundant space for large ice sheets that need not flow as ice to sea level.

The occurrence of faunas and floras of seemingly temperate affinity can be used to argue for the presence of relatively warm conditions at high latitude in the early Tertiary. In Antarctica, such an argument can be made on the basis of floral evidence from the Ross Sea region (Kemp and Barrett, 1975). With regard to the Tertiary of Antarctica, it is probably sufficient to note that Antarctica is a huge continent. Relatively warm coastal conditions around the Ross Sea in no way negate the possibility of extremely cold conditions to the interior of the continent. Given the polar position of Antarctica in early Tertiary (Figure 17.2), this is surely the best place to put early Tertiary ice.

Faunal and floral evidence from Ellesmere Island and elsewhere have been used to argue for temperate conditions in the early Tertiary and Cretaceous of the Arctic region. Let us hold that question in abeyance and return to it in our discussion of the Mesozoic. During the Mesozoic time, Antarctica is a far less attractive place to put ice. Thus, we shall be forced to direct our attention to the Arctic region.

Seismic Stratigraphy

On-lap and off-lap relationships recorded in the stratigraphic sequences at continental margin represent some combination of local basin subsidence and global sea-level history. In practice, the petroleum industry has made numerous studies of the seismic stratigraphy of basins around the world and has compiled the generalities that run through these studies under the name of "relative sea-level curve." This relative sea-level curve is taken to be a qualitative estimate of time and magnitude of major events in Tertiary eustatic history. Let us first examine the record from one study area and then move on to the compilation of a global relative sea-level curve.

The North Sea Tertiary. The general methodology of seismic stratigraphy has been presented earlier (Chapter 3). Let us now turn our attention to the classic seismic stratigraphy of the North Sea Tertiary.

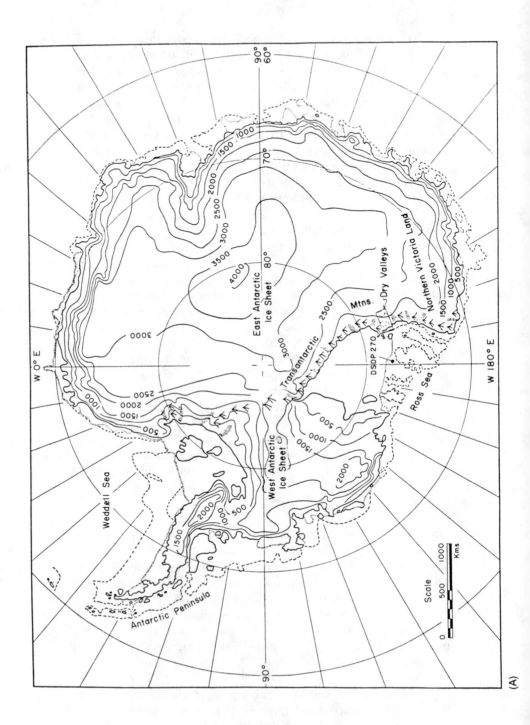

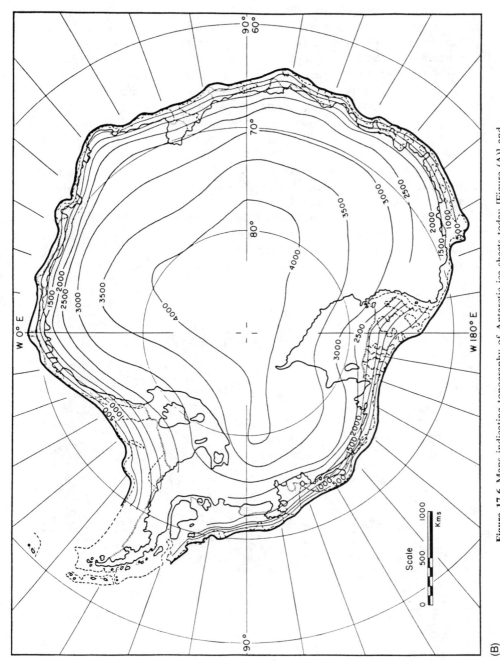

Figure 17.6 Maps indicating topography of Antarctic ice sheets today [Figure (A)] and during Middle Oligocene time [Figure (B)]. An ice sheet of the size depicted in Figure (B) is required to explain sub-ice emplacement of hyaloclastite deposits, given their present elevation, and is required to explain flow directions in till fabrics in the dry valleys region. (From M. L. Prentice and G. H. Denton; personal communication, 1983.)

413

Figure 17.7 presents data and interpretation for a seismic section taken more or less perpendicular to original strike. Land is to the right, basin to the left. Age assignments provided in the interpretation at left are presumably based on petroleum exploration wells that have penetrated the various strata, although identification of wells and paleontological control is not provided.

Outstanding features of this cross section are downlap and basinward thinning of unit (1) (early Oligocene), characteristic of progradation of highstand coastal clastics; classic on-lap and landward thinning of unit (2) (Middle Oligocene–early Miocene) indicative of basin fill from lowstand of the sea; and again, classic on-lap and landward thinning in unit (3) (late Miocene), indicative of basin fill from relative lowstand of the sea. Interpretation of seismic stratigraphic relationships within other time intervals requires additional data and/or the experienced eye of an expert.

Note well that the curve on the left is an interpretation of relative change of sea level. It is a schematic description of relation of sea level to subsiding basin. At this stage in the interpretation, we cannot say whether these relationships result from variation in eustatic sea level or variation in basin subsidence. To sort out global

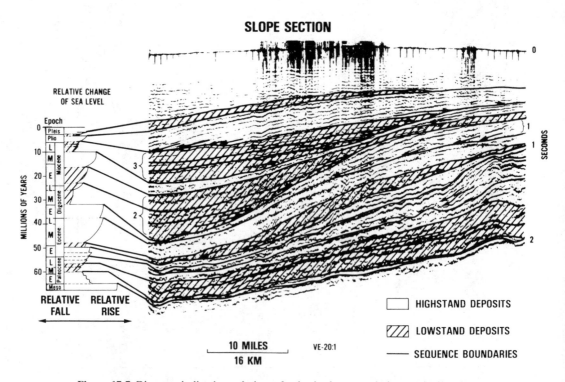

Figure 17.7 Diagram indicating relation of seismic data to relative sea-level curve abstracted from the seismic data; North Sea Tertiary seismic section taken perpendicular to paleobathymetry. (From Vail *et al.*, 1977.)

eustatic effects from local basin subsidence, we must compare similar studies from numerous regions.

Global Synthesis. Figure 17.8 presents the Tertiary relative sea-level curves from four continents and compares them to the generalized global compiled curve, taken to represent eustatic sea-level fluctuations. Note the fragmentary nature of the various local curves. Note large discrepancies concerning amplitude of major events. Such are the problems of sorting out the various local effects from the global eustatic effect. Most geologists probably agree the data indicate major falls in relative sea level somewhere around 30 to 40 million years ago and around 5 to 15 million years ago. These times of relatively low sea level correspond to sedimentation units (2) and (3) in Figure 17.8 and seem consistently represented in other study areas. We shall look forward to comparing these data to the $\delta^{18}O$ record in a moment.

Amplitude and Frequency of Tertiary Cyclicity

By analogy with the late Pleistocene dynamic model, we would be most pleased to demonstrate frequency and amplitude of a Tertiary glacio-eustatic sea-level cycle with deep-sea benthic and tropical planktic isotope data. Additionally, seismic stratigraphy should provide an independent check on isotopic estimates and also holds promise to elucidate cyclicity resulting from causes other than glacio-eustatic.

High-Frequency Variation in the Tertiary $\delta^{18}O$ Record. Whereas the late Pleistocene glacio-eustatic $\delta^{18}O$ record has been defined by sampling numerous cores at a less than 5000-year sample interval, the vast majority of isotopic data for the Tertiary has been gathered at something more like 500,000-year sample intervals (see Savin, 1977, summary article, for example). Clearly, a 500,000-year sample interval will not detect high-frequency cyclicity so well documented for the late Pleistocene (namely, approximately 20,000-, 40,000-, and 100,000-year cycles). Nevertheless, short strings of this data can be used to estimate standard deviation of the $\delta^{18}O$ signal within the time series. Such as exercise within the Tertiary $\delta^{18}O$ data set typically results in standard deviations on the order of $0.25°/oo$ throughout much of the post-Eocene Tertiary. By analogy with the late Pleistocene model, amplitude of glacio-eustatic sea-level fluctuation may be defined as ± 2 standard deviations; therefore, estimating about a $1°/oo$ amplitude to the glacio-eustatic $\delta^{18}O$ signal throughout much of the post-Eocene Tertiary. This amplitude is approximately two-thirds that documented for the late Pleistocene dynamic model.

Some Tertiary sequences have been sampled at approximately 5000-year sample intervals. Figure 17.9 presents a sequence of data from a Pliocene hydraulic piston core taken from the western Caribbean. Periodicities displayed are within range of the 20,000 to 100,000-year cycles well known from late Pleistocene dynamic model.

Thus, the Tertiary $\delta^{18}O$ record tends to suggest glacio-eustatic sea-level fluc-

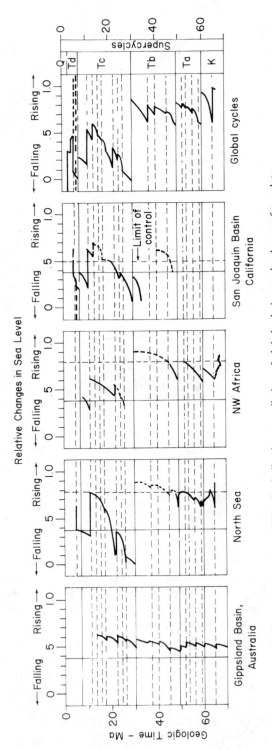

Figure 17.8 Diagram indicating compilation of global relative sea-level curve from data taken in four widely separated regions of the world. (From Vail *et al.*, 1977, with modification.)

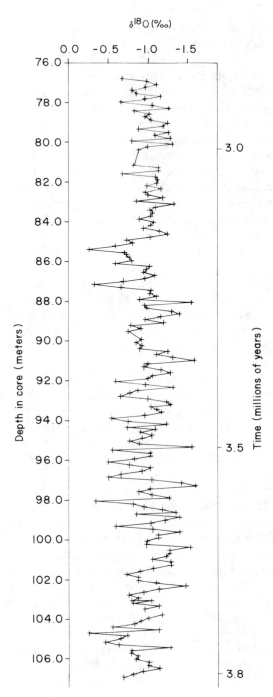

Figure 17.9 Pliocene tropical planktic $\delta^{18}O$ record from DSDP Hydraulic Piston Core 502A, western Caribbean. The data represent approximately 5000-year sample intervals throughout the time interval comprising approximately 2.9 to 3.8 million years before present. Late Pleistocene $\delta^{18}O$ signal in this region ranges from approximately $-1.5°/_{oo}$ (interglacial to $+0.5$ $°/_{oo}$ (glacial and average approximately -0.46 $°/_{oo}$. Thus, the Pliocene data depict $\delta^{18}O$ variation with an amplitude of approximately 70% that of late Pleistocene, and with an average condition approximately $0.5°/_{oo}$ lighter than late Pleistocene conditions. (From Prell, 1983, with modification.)

tuations of frequency comparable to that of late Pleistocene but of an amplitude considerably reduced from late Pleistocene.

Frequency of Regression within Seismic Stratigraphy Data. The global summary curve for the Tertiary indicates recognizable regressive events at intervals ranging from 2 to 7 million years. This represents a frequency of regressive events approximately two orders of magnitude less frequent than indicated by high-density sampling of the isotopic record. What is going on here? Should we expect the high-frequency cycles of the isotopic data to show up in the seismic stratigraphy data?

A simple calculation indicates that the seismic data should be incapable of resolving cyclicity on the scale suggested by the isotopic data. For reflectors to give clean, discrete echoes on the seismic record, they must be separated by at least one wave length. Assuming a sediment velocity of 3000 meters per second at a frequency of 30 cycles per second for the returning seismic signal, beds must be separated by 100 meters to provide discrete reflections. As we have seen elsewhere (Figure 16.12, for example), potential reflectors (subaerial exposure surfaces) in late Pleistocene highstand sequences tend to be separated by 30 meters or less. A sequence of such reflectors buried to a depth comparable to those associated with seismic-stratigraphy data acquisition would not produce discrete seismic events for each discrete subaerial exposure surface. Thus, we should not expect seismic stratigraphy to resolve details of glacio-eustatic sea-level fluctuations that may be apparent within the isotopic data.

It is probable that the minor regressions within the seismic-stratigraphy data separate time intervals during which high-frequency glacio-eustatic highstands of the sea recurred to approximately the same level. High-frequency recurrence of highstands superimposed upon general subsidence of the basin would produce classical relative costal on-lap. A major shift in the level of glacio-eustatic highstands would produce the discontinuity surface against which the next package of relative coastal on-lap would be apparent.

Synthesis of Tertiary Eustatic Sea-Level History. It is attractive to combine these various sets of data according to a few simple rules and thereby produce a curve that may approximate Tertiary eustatic sea-level history. Figure 17.10 presents a Tertiary eustatic sea-level curve based on three rules of construct. First, the calculations of Pitman (1978) are taken to represent the eustatic component ascribable to changes in the rate of sea-floor spreading. (This calculation flows largely from observations concerning the late Mesozoic and is discussed in greater detail in Chapter 18.) Secondly, the tropical planktic $\delta^{18}O$ record is taken to represent the glacio-eustatic component. This data set is given preference over seismic stratigraphy with regard to amplitude of the glacio-eustatic signal. Finally, seismic stratigraphy is used to locate major events not sufficiently defined in the $\delta^{18}O$ record. Amplitude of events is scaled in accordance with the $\delta^{18}O$ calibration provided by events that the two data sets have in common.

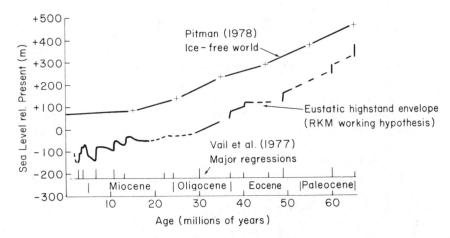

Figure 17.10 Tertiary eustatic sea-level curve compiled from (1) calculated mid-ocean ridge volume, (2) the tropical and temperate shallow planktic $\delta^{18}O$ record, and (3) seismic-stratigraphy indications of transgressions and regressions at continental margin. See text for discussion.

A general tendency to eustatic regression throughout the Tertiary is inherited from a major change in spreading rate in the late Mesozoic. This tendency is probably also enhanced by continual buildup of still larger continental glaciers throughout the Tertiary. High-frequency glacio-eustatic sea-level fluctuations are not depicted in Figure 17.10. The curve is intended to represent a running average of level attained by highstands.

Overview of the Tertiary

In the preceding pages, we have seen that the position of continental plates has changed dramatically since 60 million years ago (Figures 17.1 , 17.2), and we have seen that both the $\delta^{18}O$ record and seismic stratigraphy suggest that major eustatic events have occurred. If we would approach the Tertiary with a simple big-picture model of the last 60 million years of earth history, surely we would like to see plate motion and eustatic history placed into a single "event correlation" framework. To accomplish this, let us first examine the probable impact of plate motion on global ocean circulation and then consider the climatic eustatic consequences of such changes.

Plate Tectonics and Global Ocean Circulation. In considering the change in continental plate geometry from 60 million years ago to the present, three regions are of particular importance with regard to changes in global ocean circulation. These are the opening of Australia from Antarctica, the opening of the Norwegian-Greenland Sea between Greenland and Europe, and the opening of the Drake Passage between Antarctica and South America. At least two of these events involve the rifting of continents by developmentof new axes of sea-floor spreading. Thus, if we are seeking to identify "event correlation" based upon these major rifting events, it is perhaps well to review briefly the rates at which these events typically occur.

With regard to lateral motion of continents away from one another, recall that this process is measured in centimeters per year. Ten centimeters per year would be a convenient, reasonable number for the opening of such gaps between continents. At this rate, 1 million years would be required to form a channel 100 kilometers wide through the middle of a former continental block. With regard to depth of this channel, recall that such rifting sites would follow the general thermal subsidence curve (Figures 8.7) withrates not exceeding 10 centimeters per 1000 years. If the separation between the continents occurred as brittle fracture, new ocean crust forming at 2800 meters might provide the initial depth of channel. More likely, hundreds of kilometers of ocean between the continental blocks will be accomplished by ductile deformation, resulting in relatively shallow seas. Alternatively, hot spots located within the rifting zone create volcanic ridges that might initially stand above sea level and subside at rates on the order of 10 centimeters per 1000 years. Thus, millions of years may be required for a relatively shallow seaway to become a relatively deep seaway. Thus, if we are to think of event correlation with the event being the breakup of a continent, we shall be speaking in terms of "events" that have their begining and ending separated by millions to tens of millions of years.

Of the three events under consideration here, the separation of Australia from Antarctica is by far the simplest. During the Eocene, there was a major reorganization of sea-floor spreading in the eastern Indian Ocean (Veevers, 1977). The major axis of spreading shifted from northwest of Australia to the separation of Australia from Antarctica. From Eocene to Recent times, a more or less constant spreading rate and concurrent generation of oceanic lithosphere has occurred along the Southeast Indian Ridge. Separation of Antarctica from Australia, now approximately 2000 kilometers, has occurred at the rate of 30 to 50 kilometers per million years. Judging from the record of sea-floor spreading in the Indian Ocean, there was no tendency for Australia to separate from Antarctica prior to 53 million years ago. By 32 million years ago, there was a gap of some 1000 kilometers, and water depths on the flanks of the Southeast Indian Ridge were approaching a respectable oceanic water depth of 4000 meters. Thus, we are talking about an "event" that is probably within the time frame of 40 million years ±10 million years.

There is no *a priori* way to judge "how big is big" with regard to the paleoclimatic significance of this plate tectonics "event." The opening of this gap is surely a giant step toward creation of modern circumpolar circulation and the resultant thermal isolation of Antarctica, but the exact time when this opening attained global significance and how fast it attained global significance are questions to be asked within the context of the stratigraphic record.

The opening of the Norwegian-Greenland Sea is likewise an early Tertiary phenomenon, but it is further complicated by the existence of a hot spot (Iceland) at ridge crest. Magnetic anomalies in oceanic crust of the Norwegian-Greenland Sea suggest spreading began around 55 to 60 million years ago (Vogt, 1972; Talwani and Eldholm, 1977). However, the existence of volcanic ridges associated with the Iceland hot spot may have prevented significant interaction between the basins of the Arctic and North Atlantic Oceans until much later. As with the opening of Australia from Antarctica, we see a big plate-tectonic "event" here, but there is no *a priori* way to judge when or how it began to effect North Atlantic circulation. These are questions to be asked within the context of the stratigraphic record.

The status of the Drake Passage (between South America and Antarctica) throughout the Tertiary is a subject of considerable uncertainty. Delineation of sea-floor magnetic anomalies indicate coherent spreading within the Drake Passage from anomaly 8 to anomaly 5 (from 29 to 5 million years ago) (Barker and Burrell, 1977). The history of the region prior to 29 million years ago cannot be determined from sea-floor magnetics; there may or may not have been older oceanic lithosphere between Antarctica and South America. For older time intervals, the question of ocean circulation in this region is further complicated by debate concerning the relation of South America to the Antarctic Peninsula (Harrison *et al.*, 1979; Dalziel, 1982). Finally, there is the question of whether the Scotia Arc (Dalziel and Elliot, 1971) to the east of the Drake Passage may have served as the principal constriction in the region. Alas! Once again, we see a plate tectonics "event" that is surely significant, but questions of "when and how" must be addressed to the stratigraphic record.

Climate-Eustatic Response. Both the $\delta^{18}O$ record and seismic stratigraphy can be generalized to suggest major regressive events centered more or less around the Oligocene-Eocene boundary and the Middle to late Miocene. Regression at or around the Oligocene-Eocene boundary is surely in the right time interval for the opening of Australia from Antarctica and the opening of Greenland from Europe. The opening of Australia from Antarctica is surely a major step toward thermal isolation of Antarctica; the opening of Greenland from Europe is surely a major step toward production of North Atlantic deep water. Both of these conditions would tend to give rise to eustatic sea-level lowering. The mid-Miocene event is marked by increased production of ice-rafted detritus around Antarctica (Hayes and Frakes, 1975, for example) and additional cooling of bottom water (for example, the data of Woodruff *et al.*, 1981). Whether these events relate to plate tectonics within the area of the Drake Passage remains largely a matter of speculation.

As a simple first approximation, it is reasonable to place the majority of Tertiary continental ice accumulation in East Antarctica. Antarctica is a huge continent and the paleo–South Pole is apparently situated within the continent throughout the Tertiary (Smith and Briden, 1977, among others). Additionally, the best geological evidence against an East Antarctic ice sheet throughout the Tertiary is presently situated beneath 3 kilometers of ice and is therefore not likely to be uncovered in the near future by even the most enthusiastic opponent of this hypothesis. Using $\delta^{18}O$ data as calibration, the average state of the East Antarctic ice sheet probably grew by approximately 50 meters sea-level equivalent around the Oligocene-Eocene boundary and by approximately 30 meters sea-level equivalent during the mid-Miocene. Further, cooling of the whole deep ocean from the Eocene to late Miocene would produce another 20 meters sea-level lowering from thermal contraction of the water column. The data of Woodruff *et al.* (1981) suggests this effect could be largely accommodated within the Middle Miocene, but the data of Miller and Curry (1982) suggest the effect may have begun as early as the Oligocene-Eocene boundary.

Thus, when we view the Tertiary from the viewpoint of "event correlation," we see that a strong case can be made for plate tectonics and the changing intercontinental geometry as the major cause of eustatic events. Both the $\delta^{18}O$ record and seismic stratigraphy can be viewed as recording two major events, but the events may be spread over quite a long time and/or both data sets are quite "noisy." Whereas a strong case can be made that both of these major regressive events are related to continental ice volume, the high-frequency fluctuations in ice volume, so well known from the late Pleistocene, are almost lost among the noise of the Tertiary data sets. With regard to the $\delta^{18}O$ record, this is due in large part to inadequate sampling. With regard to seismic stratigraphy, it may be physically impossible for low-frequency seismic reflections to resolve the phenomena that we suspect to be represented.

We see the last 65 million years to be roughly divided into three major times of high sea level with intervening times of low sea level. Surely, the Paleocene involves higher stands of eustatic sea level than does Oligocene. Perhaps, the entire Eocene is something of a time of transition. Surely, the highstands of Lower Miocene are followed by generally lower stands of the sea in the Upper Miocene. The Plio-Pleistocene appears to once again involve relatively high highstands, but in this young portion of the Cenozoic, high-frequency intervening lowstands are quite well documented.

It is tempting to think of these generalites concerning the Tertiary as a 20- or 30-million-year "cycle." However, if we are thinking about major eustatic events resulting from plate tectonic events, we probably have no justification to propose an underlying plate tectonics cyclicity. While we could propose that drifting continents can only drift so far without banging into another continent, such collisions would probably have to occur at critical latitudes in order to generate a global eustatic response. Further, there is no well-founded theory for prediction of periodicity to the breakup of continents, such as we are describing for Australia-Antarctica and for Greenland-Europe. Thus, rather than thinking of periodicity on the scale of tens of millions of years, we should probably think of these events as random and infrequent. This is the principal lesson of the Tertiary that we should take with us in our dynamic model as we approach the Mesozoic and Paleozoic.

Sedimentologic/Stratigraphic Consequences of the Tertiary Model

The next step in our deductive approach to the stratigraphy of the Tertiary is to test the general overview against specific examples of detailed geologic studies. Unfortunately, the literature does not conveniently lend itself to this task because the vast majority of literature concerning detailed studies of Tertiary strata were not written with any global overview in mind. Most geologists who write about the details of local stratigraphic situations are operating in the inductive mode, developing their own syntheses from the data of their immediate study areas.

Nevertheless, several studies do lend themselves to reinterpretation in light of the overview presented above. For example, by analogy with modern sedimentation rates in Belize, it is unlikely that the Miocene patch reefs reported by Vincelette and Soeparjadi (1976) could have remained as isolated reef structures if sedimentation rates were not attenuated by intermittent subaerial exposure resulting from glacio-eustatic sea-level fluctuations.

With regard to Tertiary deltaic sedimentation, there are outstanding examples of cyclic sedimentation within both the Niger and the Mississippi delta regions (for example, Weber, 1971; and Curtis, 1970). However, it is traditional to "explain" this cyclicity in terms of shifting delta depocenters and/or rapid subsidence associated with faulting. The overview presented here suggests that glacio-eustatic sea-level fluctuations should be added to the smorgasbord of mechanisms available to explain cyclicity in Tertiary deltaic sequences. The situation may be too complex to yield definitive results, however. On a more general level, the extreme regressive nature of the Frio Formation, subsurface of the Texas Gulf Coastal Plain (Galloway et al., 1982, for example), is surely consistent with major glacio-eustatic lowstands of the sea during early (to perhaps middle) Oligocene.

Citations and Selected References

ALVAREZ, L. W., W. ALVAREZ, F. ASARO, and H. V. MICHEL. 1980. Extraterrestrial cause for the Cretaceous-Tertiary extinction: Science 208 (4448): 1095–1108.

BARKER, P. F., and J. BURRELL. 1977. The opening of the Drake Passage. Marine Geol. 25: 15–34.

BARRON, E. J., S. L. THOMPSON, and S. H. SCHNEIDER. 1981. An ice-free Cretaceous: results from climate model simulations. Science 212: 501–508.

BARRON, E. J., and W. M. WASHINGTON. 1982. Atmospheric circulation during warm geologic periods: Is the equator-to-pole surface-temperature gradient the controlling factor? Geology 10: 633–636.

BERGGREN, W. A., and C. D. HOLLISTER. 1977. Plate tectonics and paleocirculation—commotion in the ocean. Tectonophysics, 38: 11–48.
Popular "overview."

BRASS, G. W., W. W. HAY, W. T. HOLSER, W. H. PETERSON, E. SALTZMAN, J. L. SLOAN, II, and J. R. SOUTHAM. 1980. Ocean circulation, plate tectonics and climate. Manuscript.

BUCHARDT, BJØRN. 1978. Oxygen isotope palaeotemperatures from the Tertiary period in the North Sea area. Nature 275: 121–123.
Extremely light $\delta^{18}O$ values may mean warm temperature or reduced salinity.

CROUGH, S. T. 1981. Mesozoic hotspot epeirogeny in eastern North America. Geology 9: 2–6.
Traces interaction of continents with hot spots.

CROUGH, S. THOMAS, W. J. MORGAN, and R. B. HARGRAVES. 1980. Kimberlites: their relation to mantle hotspots. Earth Planet. Sci. Lett., 50: 260–274.
More interaction of continents with hot spots.

CURTIS, D. M. 1970. Miocene deltaic sedimentation, Louisiana Gulf Coast, p. 293–308. *In* J. P. Morgan (ed.), Deltaic sedimentation Modern and Ancient. Soc. Econ. Paleontologists and Mineralogists Spec. Publ. 15.

DALRYMPLE, G. B., D. A. CLAGUE, M. O. GARCIA, and S. W. BRIGHT. 1981. Petrology and K-Ar ages of dredged samples from Laysan Islands and Northampton Bank volcanoes, summary. Bull. Geol. Soc. Amer. 92 (pt. 1, no. 6): 315–318.
Dating of the world's best example of a hot-spot trace the Hawaiian Islands–Emperor Seamount chain.

DALZIEL, I. W. D. 1982. West Antarctica: Problem child of Gondwanaland. Tectonics, 1: 3–20.

DALZIEL, I. W. D., and D. H. ELLIOT. 1971. Evolution of the Scotia arc. Nature 233: 246–252.

DAVIES, T. A., and T. R. WORSLEY. 1981. Paleoenvironmental implications of oceanic carbonate sedimentation rates, p. 169–179. *In* John E. Warme, Robert G. Douglas, and Edward L. Winterer, (eds.), The deep sea drilling project: a decade of progress. Soc. Econ. Paleont. Mineral. Spec. Pub. 32.
Carbonate flux to deep ocean considered to be highly dependent of sea level to continents.

DONN, W. L., and D. NINKOVICH. 1980. Rate of Cenozoic explosive volcanism in the North Atlantic Ocean inferred from deep sea cores. J. Geophys. Res. 85 (B10): 5455–5460.

DOUGLAS, R. G., and S. M. SAVIN. 1971. Isotopic analyses of planktonic foraminifera from the Cenozoic of the Northwestern Pacific, Leg 6, p. 1123–1127. *In* A. G. Fisher *et al.*, Initial reports of the deep sea drilling project. U. S. Gov't. Printing Office, Washington, D. C.

DOUGLAS, ROBERT, and F. WOODRUFF. 1980. Deep sea benthic foraminifera. p. 1233–1327. *In* C. Emiliani (ed.), The sea, vol. 7. Wiley-Interscience, New York.

EJEDAWE, J. E. 1981. Patterns of incidence of oil reserves in Niger Delta Basin. Bull. Amer. Assoc. Petrol. Geol. 65 (9): 1574–1585.

ESTES, RICHARD, and J. HOWARD HUTCHISON. 1980. Eocene lower vertebrates from Elles-

mere Island, Canadian arctic archipelago. Palaeogeog. Palaeoclimat. Palaeoecol. 30: 325–347.
Faunal evidence cited in favor of warm conditions in the Eocene of the high Arctic region.

FRAKES, L. A. 1979. Climates throughout geologic time. Elsevier, New York. 294 p.

HAQ, BILAL U., ISABELLA PREMOLI-SILVA, and G. P. LOHMANN. 1977. Calcareous plankton paleobiogeographic evidence for major climatic fluctuations in the early Cenozoic Atlantic Ocean. J. Geophys. Res. 82 (27): 3861–3876.

HARRISON, C. G. A., E. J. BARRON, and W. W. HAY. 1979. Mesozoic evolution of the Antarctic peninsula and the southern Andes. Geology 7 (8): 374–378.

HAYES, D. E., L. A. FRAKES *et al.* 1975. Initial reports of the deep sea drilling project, Volume 28. U. S. Government Printing Office, Washington, D. C.

HSÜ; KENNETH J., LUCIEN MONTADERT, DANIEL BERNOULLI MARIA BIANCA CITA, ALBERT ERICKSON, ROBERT E. GARRISON, ROBERT B. KIDD, FREDERIC MELIERES, CARLA MÜLLER, and RAMIL WRIGHT. 1977. History of the Mediterranean salinity crisis. Nature 267: 399–403.
Convenient reference to the ever-popular "Messinian salinity crisis."

IMBRIE, JOHN, and NILVA G. KIPP. 1971. 5. A new micropaleontological method for quantitative paleoclimatology: application to a late Pleistocene Caribbean core. p. 71–181. *In* Karl K. Turekian (ed.), The late Cenozoic glacial ages. Yale Univ. Press, New Haven, Conn.

KEIGWIN, L. D., JR. 1980. Palaeoceanographic change in the Pacific at the Eocene-Oligocene boundary. Nature 287: 722–725.
Thorough citation of previous studies. Good starting point.

KEMP, ELIZABETH M. 1978. Tertiary climatic evolution and vegetation history in the southeast Indian Ocean region. Palaeogeog. Palaeoclimat. Palaeoecol. 24: 169–208.

KEMP, ELIZABETH M., and PETER J. BARRETT. 1975. Antarctic galciation and early Tertiary vegetation. Nature 258: 507–508.

KENNETT, J. P. 1978. The development of planktonic biogeography in the southern ocean during the Cenozoic. Marine Micropaleont. 3: p. 301–345.
A popular account of Southern Hemisphere Cenozoic history. Also available at J. Geophys. Res. 82: 3843–3860.

KVASOV, D. D., and M. YA. VERBITSKY. 1981. Causes of Antarctic glaciation in the Cenozoic. Quaternary Res. 15 (1): 1–17.
Interesting graphics concerning one possible scenario for Antarctic glaciation.

LEMASURIER, W. E., and D. C. REX. 1982. Volcanic record of Cenozoic glacial history in Marie Byrd Land and western Ellsworth Land, II: revised chronology and evaluation of tectonic factors, p. 725–734. *In* C. Craddock (ed.), Antarctic geoscience. Univ. of Wisconsin Press, Madison.
K-Ar dating of volcanics erupted beneath ice indicates a large Antarctic ice cap was in existence at 28 million years ago.

MARGOLIS, S. V., and J. P. KENNETT. 1971. Cenozoic paleoglacial history of Antarctica recorded in subantarctic deep-sea cores. Amer. J. Sci. 271: 1–36.

MATTHEWS, R. K., and R. Z. POORE. 1980. Tertiary $\delta^{18}O$ record and glacio-eustatic sea-level fluctuations. Geology 8 (10): 501–504.

MERCER, J. H. 1983. Cenozoic glaciation in the Southern Hemisphere: Ann. Rev. Earth Planet. Sci. 11: 99–132.

MILLER K. G., and W. B. CURRY. 1982. Eocene to Oligocene benthic foraminiferal isotopic record in the Bay of Biscay. Nature 296: 347–352.

MOORE, T. C. JR., and G. ROSS HEATH. 1977. Survival of deep-sea sedimentary sections. Earth Planet. Sci. Lett. 37: 71–80.
Is the phenomenon missing section, or simply the imperfection of the biostratigraphic record?

MOORE, T. C., JR., H. VAN ANDEL, C. SANCETTA, and N. PISIAS. 1978. Cenozoic hiatuses in pelagic sediments. Micropaleontology 24 (2): 113–138.
As above.

MOORE, T. C., JR., N. G. PISIAS, and L. D. KEIGWIN, JR. 1981. Ocean basin and depth variability of oxygen isotopes in Cenozoic benthic foraminifera. Marine Micropaleont. 6: 465–481.

OERLEMANS, J. 1980. Continental ice sheets and the planetary radiation budget. Quat. Res. 14: 349–359.

OERLEMANS, J. 1982. A model of the Antarctic ice sheet. Nature 297: 550–553.

OLSSON, RICHARD K., KENNETH G. MILLER, and TIMOTHY E. UNGRADY. 1980. Late Oligocene transgression of middle Atlantic Coastal Plain. Geology 8 (11): 549–554.
Coastal plain geology which should be correlative with events in "seismic stratigraphy." See also discussion and reply in Geology 9 (7): 290–292.

PITMAN, W. C., III. 1978. Relationship between eustacy and stratigraphic sequences of passive margins. Bull. Geol. Soc. Amer. 89 (9): 1389–1403.

PRELL, W. L. 1983. A reevaluation of the initiation of northern hemisphere glaciation at 3.2 MY: new isotopic evidence. Geology 11 (forthcoming).

PRENTICE, M. L. 1982. Surficial geology and stratigraphy in central Wright Valley, Antarctica: Implications for Antarctic Tertiary glacial history. M. Sc. Thesis, Univ. of Maine, Orono, 248 p.
Basic data; look for subsequent publications by Prentice, G. H. Denton, and others.

QUILTY, PATRICK G. 1977. Cenozoic sedimentation cycles in western Australia. Geology 5: 336–340.
Continental margin unconformities which surely contain eustatic information.

SAVIN, SAMUEL M. 1977. The history of the earth's surface temperature during the past 100 million years. Ann. Rev. Earth Planet. Sci. 5: 319–355.

SAVIN, SAMUEL M., ROBERT G. DOUGLAS, and FRANCIS G. STEHLI. 1975. Tertiary marine paleotemperatures. Bull. Geol. Soc. Amer. 86: 1499–1510.

SHACKLETON, N. J. 1967. Oxygen isotope analyses and Pleistocene temperature reassessed. Nature 215: 15–17.
A "breakthrough" paper which proposes for the first time that the $\delta^{18}O$ signal in late Pleistocene deep-sea cores is primarily an ice-volume $\delta^{18}O_{water}$ signal rather than a temperature signal.

SHACKLETON, N. J., and J. P. KENNETT. 1974. 17. Paleotemperature history of the Cenozoic and the initiation of Antarctic glaciation: oxygen and carbon isotope analyses in DSDP Sites 277, 279, and 281. p. 743–755. *In* J. P. Kennett, R. E. Houtz, *et al.,* Initial reports of the deep sea drilling project, Vol. 24.
Can one really sort out paleotemperature effects from ice-volume effects using cores from such high latitude?

SHACKLETON, N. J., and N. D. OPDYKE. 1977. Oxygen isotope and paleomagnetic evidence for early Northern Hemisphere glaciation. Nature 270: 216–219.
Can one really talk about ice-volume events on the basis of only a benthic isotope curve?

SMITH, A. G., and J. C. BRIDEN. 1977. Mesozoic and Cenozoic paleocontinental maps. Cambridge Univ. Press, Camb. 63 p.

SOUTHAM, JOHN R., and WILLIAM W. HAY. 1977. Time scales and dynamic models of deep-sea sedimentation. J. Geophys. Res. 82 (27): 3825–3842.
Big-picture approach.

TALWANI, M., and O. ELDHOLM. 1977. Evolution of the Norwegian-Greenland Sea. Bull. Geol. Soc. Amer. 88: 969–999.

THIERSTEIN, HANS R., and WOLFGANG H. BERGER. 1978. Injection events in ocean history. Nature 276: 461–466.

VAIL, P. R., R. M. MITCHUM, JR., and S. THOMPSON, III. 1977. Seismic stratigraphy and global changes of sea level, part 4: global cycles of relative changes of sea level, p. 83–97. *In* C. E. Payton (ed.), Seismic stratigraphy—applications to hydrocarbon exploration. Amer. Assoc. Petrol. Geol. Mem. 26.

VEEVERS, J. J. 1977. 6. Models of the evolution of the eastern Indian Ocean, p. 151–164. *In* J. R. Heirtzler, H. M. Bolli, T. A. Davies, J. B. Saunders, J. G. Sclater (eds.), Indian Ocean geology and biostratigraphy (studies following deep sea drilling legs 22-29). Amer. Geophys. Union, Washington, D. C.

VINCELETTE, R. R., and R. A. SOEPARJADI. 1976. Oil-bearing reefs in the Salwati Basin of Irian Jaya, Indonesia: Amer. Assoc. Petrol. Geologists Bulletin. 60 (9): 1448–1462.

VOGT, P. R. 1972. The Faeroe–Iceland–Greenland aseismic ridge and the Western Boundary Undercurrent. Nature 238: 79–81.

WATTS, A. B. 1982. Tectonic subsidence, flexure, and global changes of sea level. Nature 297: 469–474.
Proposes that seismic stratigraphy relative sea level curves contain a strong component of local passive margin subsidence as opposed to bona fide eustatic transgression. Construction of an allegedly erroneous eustatic sea-level curve is proposed to result from a sampling problem. Interesting reading!

WEBER, K. J. 1971. Sedimentological aspects of oil fields in the Niger Delta. Geologie en Mignbouw 50 (3): 476–599.

WEISSEL, JEFFREY K., DENNIS E. HAYES, and ELLEN M. HERRON. 1977. Plate tectonics synthesis: the displacements between Australia, New Zealand, and Antarctica since the late Cretaceous. Marine Geol. 25: 231–277.

WOLFE, JACK A. 1980. Tertiary climates and floristic relationships at high latitudes in the Northern Hemisphere. Palaeogeog. Palaeoclimat. Palaeoecol. 30: 313–323.

WOODRUFF, FAY, SAMUEL M. SAVIN, and ROBERT G. DOUGLAS. 1981. Miocene stable isotope record: a detailed deep Pacific Ocean study and its paleoclimatic implications. Science 212: 665–668.

The Mesozoic

18

The Mesozoic era is the middle unit of the tripartite division of the last 570 million years of earth history. The names Paleozoic, Mesozoic, and Cenozoic were erected during the 1840's and generally reflect striking differences in fossil content among the various eras in Europe. The fossil records of the Mesozoic contrast with the "Ancient life" of the Paleozoic and the "Recent life" of the Cenozoic. In the context of our overview of the physical history of the globe, we shall acknowledge that these changes in characteristics of life-forms may in part reflect changes in the physical conditions of the globe, but this discussion will not be closely bound to the fossil record.

The Mesozoic spans approximately 155 million years from basal Triassic (approximately 220 million years ago) to the top of the Cretaceous (approximately 65 million years ago). Thus, we are seeking to generalize for a time interval of earth history that is on the order of two-and-a-half times as long as the Tertiary. As we go back in time, we shall find that our view of the physical situation becomes less and less precise. It will be good to carry with us a sense of the complexity that can develop within a time interval such as the Tertiary. As our view of the past becomes more and more clouded by the veil of antiquity, it will be well to vividly recall the Tertiary and realize that something on the order of those dramatic gyrations may have gone on throughout geologic time. Whether or not we shall perceive them properly is what makes the Mesozoic and Paleozoic so much fun.

Continental Plates

As we continue our reconstruction of plate positions back through time, we encounter serious degradation of the information available to us. In reconstructing the 60-million-year-ago world (Figure 17.2), one of the most useful exercises was simply to fold younger oceanic crust back into the mid-ocean ridges and thereby require that connected continental plates be relocated. As we move now to reconstruct the world of 140 million years ago, note well in Figure 8.4 that we really have a very small amount of 140-million-year-old oceanic crust to work with.

Similarly, our tracking of the interaction between hot spots and plate is seriously degraded by 140 million years ago. The world's finest example of a hot-spot trace on a lithospheric plate, the Hawaiian Islands–Emperor Seamount chain, provides us no record prior to 70 million years ago. Whatever record there was has long since been eaten by the Aleutian Trench. For these reasons, we are forced to rely more and more on apparent polar wandering data taken from the continents themselves.

Paleo-Oceans. Figure 18.1 presents a reconstruction of paleogeography at around 140 million years ago (late Jurassic). Once again, it is useful to glance forward and backward in time by comparing Figure 18.1 with Figures 17.2 and 19.1. Clearly, one of the big paleo-oceanographic studies of the Mesozoic is the opening of the Atlantic. Sea-floor magnetics suggest a significant North Atlantic existed by the Middle Jurassic. Opening of the South Atlantic appears to have commenced in the south around the mid-Cretaceous. By the late Cenozoic, the separation of equatorial South America from West Africa was complete. Thus, substantially the entire perimeter of the North and South Atlantic begins a long, relatively uncomplicated history of postrifting thermal subsidence of passive continental margins sometime during Mesozoic time.

The separation of Africa and India from Antarctica-Australia is substantially a late Mesozoic event. Note especially the tendency of Africa to rotate counterclockwise toward Europe-Siberia (compare Figures 18.1 and 17.2). Thus, closing of the seaway between Africa and Europe will be a continuing theme throughout the Mesozoic.

Throughgoing equatorial ocean circulation is another hallmark of the Mesozoic. Whereas the Mesozoic begins with a virtually pole-to-pole supercontinent (Smith and Briden, 1977), the opening of North America from Africa, Europe from Africa, and North America from South America creates a throughgoing tropical ocean among the fragments of the former supercontinent. Throughgoing tropical circulation of the Tethys Sea remains an important feature of the world ocean until closure of Arabia toward Siberia creates a congested situation in the mid-Tertiary. We shall find this tropical, throughgoing seaway to be an especially attractive place for bank-margin carbonate deposition throughout the Mesozoic. Combination of this geographic position with thermal subsidence of passive margins, notably the northeast portion of the Arabian Shield, resulted in the formation of some of the world's most important petroleum reserves.

Polar Regions. If we are to have a global ice volume to account for Mesozoic glacio-eustatic sea-level fluctuations, we had best find ourselves a continent appropriately situated within a polar region. With regards to the Tertiary, we found

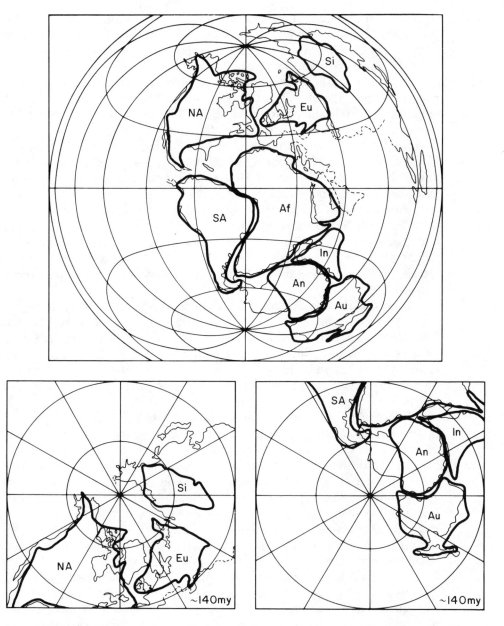

Figure 18.1 Plate tectonics configuration of the major continents at 140 million years ago (late Jurassic). Note the Gondwanaland supercontinent in the Southern Hemisphere. Note the opening of the North Atlantic basin and the tendency for a throughgoing equatorial circulation. Both poles have large continents at high latitude. (Compiled from data cited in Chapter 8.)

Antarctica to maintain a conveniently polar position and thus to be a likely spot for continental ice accumulation. Taking things one step at a time, we were satisfied with an Antarctic fossil site for ice accumulation and simply deferred consideration of possible sites for Northern Hemisphere ice accumulation.

Mesozoic Southern Hemisphere continent geometry presents us with a somewhat different problem. Whereas large areas of Antarctica and eastern Australia are at fairly high latitude during the Mesozoic (Figure 18.1), the Australian portion of this region is presently situated under temperate conditions. Thus, whereas we could point to Antarctica and say that the evidence of early Tertiary continental glaciation was not conveniently buried beneath 3 kilometers of ice, we must acknowledge that any evidence of Mesozoic ice in eastern Australia should be readily accessible for field documentation.

Unfortunately, the Lower Mesozoic of eastern Australia appears to be rather uncooperative. Triassic through earliest Cretaceous of Tasmania, New South Wales, and Queensland is dominated by nonmarine clastics that do not appear to be particularly diagnostic (Ludbrook, 1978). Thus, thoughtful consideration of Northern Hemisphere prospective sites for ice accumulation is called for with regard to the Mesozoic.

Where might one put Northern Hemisphere continental ice during the Mesozoic? A quick comparison of Figure 18.1 with Figure 17.2 suggests eastern North America, Greenland, and Europe to occupy a consistently lower-latitude position. In contrast, eastern Siberia maintains a consistent high-latitude position. Surrounded by ocean on three sides, it is tempting to draw an analogy between eastern Siberia in Mesozoic time and the growth of the Laurentide ice sheet over eastern Canada during the late Pleistocene.

Does the geological literature concerning eastern Siberia provide any indication of glaciation within the time frame? The Russian literature concerning the Mesozoic of eastern Siberia is ambiguous on the question of Mesozoic glaciation of the region. No one comes right out and reports extensive tillite. However, Epshteyn (1977) reports widespread occurrence of pebbles in shales throughout Mesozoic and Cenozoic sediments of the region and calls upon ice rafting of the pebbles to deliver these coarse sediments to the otherwise fine-grained shale in which they are found. Contrariwise, there is an extensive literature concerning Cretaceous "warm floras" in the region (Vachrameev, 1978, for example). Unfortunately, the ice-rafted dropstone literature does not reference the warm floras literature, and vice versa.

With regards to Mesozoic high-latitude Europe and North America, there is extensive biogeographic literature concerning land bridges between the continents and general climatic conditions of the region. Cretaceous to Eocene floras and faunas of Spitsbergen and Ellesmere Island (Dawson, 1980, for example) are cited as particularly compelling evidence for relatively warm climate. Whereas arguments based on habitat of individual taxa are subject to question (Ostrom, 1969), arguments based on diversity of assemblages appear well founded.

It may be possible to reconcile warm flora and fauna with nearby accumulation of continental ice sheet by proposing an increased role for ocean-surface currents in the transport of heat to polar regions. Modeling studies indicate that extremely cold temperatures to the interior of continents can be compatible with relatively warm coastal temperatures (Barron *et al.*, 1981; Barron and Washington, 1982).

$\delta^{18}O$ and Continental Ice Volume

Recall that we developed a research strategy for looking at $\delta^{18}O$ and continental ice volume simultaneously in the late Pleistocene. We convinced ourselves that the $\delta^{18}O$ record was a good indicator of ice dynamics and glacio-eustatic sea-level fluctuations in the late Pleistocene. When we turned our attention to the Tertiary, we found the direct record of continental ice volume to be extremely difficult to read, but the $\delta^{18}O$ record in deep-sea cores apparently intact. As we approach the late Paleozoic (Chapter 19), we shall find a transposition of these data sets; the $\delta^{18}O$ record will be virtually nonexistent, but there will be an abundant literature concerning stratigraphic evidence for glaciation of Gondwanaland.

Unfortunately, the Mesozoic data on this subject are transitional between the Tertiary and late Paleozoic. That is, $\delta^{18}O$ data of the quality to which we are accustomed from the Tertiary are quite scarce, while at the same time there is scant stratigraphic documentation of Mesozoic continental ice sheet.

Relevant isotopic studies have been undertaken on a relatively few western Pacific DSDP sites. The data in Figure 18.2 indicate continuation of the tendency that we observed in the Tertiary (Figure 17.3). The $\delta^{18}O$ values for both benthic and shallow-dwelling tropical planktic foraminifers tend to become lighter with increasing age. Following the interpretation scheme of Matthews and Poore (1980), shallow-dwelling planktic $\delta^{18}O$ values of $-3\,°/\infty$ are taken as indicative of a world with no ice sheets on continents. There are a scant four data points indicating this condition at around 100 million years before present (Albian-Aptian). Since that time, the scant data available are consistent with progressively larger amounts of continental ice volume. An especially noteworthy shift in both planktic and benthic values occurs between 80 and 65 million years before present (being approximately the transition from Mesozoic to Cenozoic era).

Abundant Mesozoic $\delta^{18}O$ data exists for European molluscs. However, these data are not suited for the interpretation scheme employed here. To begin with, Mesozoic Europe is at a sufficiently high latitude that temperature cannot be reasonably constrained by global atmospheric circulation considerations. Second, seaways connecting mid-latitude central Europe to the Arctic region could lead to large variations in $\delta^{18}O_{water}$ as a function of relatively small variations in salinity. This is because low-salinity water from a high-latitude Arctic source might have extremely negative $\delta^{18}O_{water}$. This problem was clearly recognized by Lowenstam (1964), but has subsequently been ignored by various writers who have used the data to infer warm conditions for Europe during the Mesozoic. The data are here regarded as second-order information. In order to interpret the data, one must offer external constraint on global ice volume and either local temperature or local salinity variation.

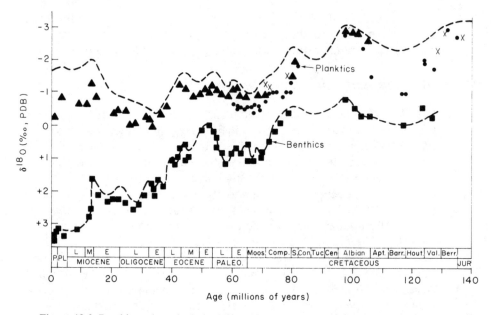

Figure 18.2 Benthic and tropical planktic $\delta^{18}O$ record for the last 140 million years. Note that the theoretical value of $-3^{\circ}/\circ\circ$ for tropical sea surface in an ice-free world is attained by several data points around 100 million years ago. Throughout the rest of the curve, there was either water tied up in ice to make $\delta^{18}O_{water}$ of the world ocean heavier or tropical sea-surface temperatures were cooler. See text for discussion. [From R. Douglas and F. Woodruff, "Deep Sea Benthic Foraminifer," in C. Emiliani (ed.), *The Sea* 8, Fig. 34.] Copyright 1981 John Wiley & Sons, Inc.

Seismic Stratigraphy

The Mesozoic contains some of the earth's most prolific petroleum reserves. Thus, the seismic stratigraphy of this time frame has been worked in great detail. Moreover, the opening of the North Atlantic throughout Mesozoic time is reasonably understood. Seismic crosssections off West Africa provide an especially vivid picture of on-lap—off-lap relationships on a subsiding passive margin.

The Mesozoic of Offshore West Africa. Figure 18.3 represents a spectacular seismic cross section from offshore northwest Africa. Note various Jurassic on-lap–down-lap relationships, culminating in the development of pronounced shelf-slope escarpment by J3.2 time. Cretaceous deposition begins with K1.1 in a decidedly regressive position to the former J3.2 bank margin. Approximately 3 kilometers of Lower Cretaceous sediments showing interval down-lap and top-lap relationships culminate with a shelf-slope boundary approximately 20 kilometers seaward of its Jurassic position. Throughout much of the region, major unconformity separates Lower Cretaceous from earliest Tertiary deposits.

Figure 18.4 presents a seismic cross section from offshore West Africa, somewhat to the south of Figure 18.3. Comparison of the two cross sections emphasizes

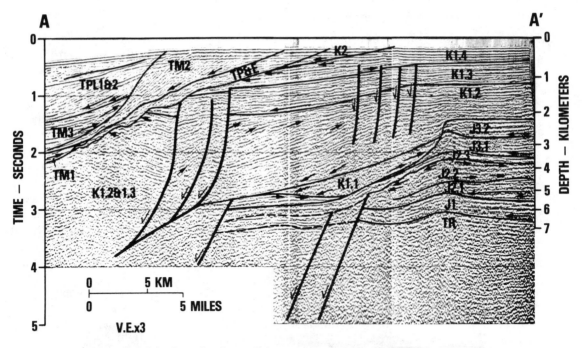

Figure 18.3 Seismic section from offshore northwest Africa indicating sequence nomenclature defined in terms of seismic reflections. Arrows indicate on-lap–off-lap relationships. Compare this section to Figure 18.4. (From Mitchum *et al.*, 1977.)

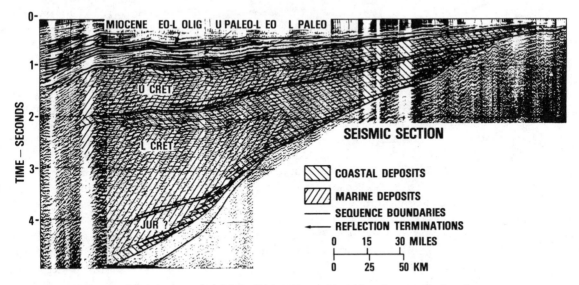

Figure 18.4 Seismic section from offshore West Africa. Note that a general tendency to on-lap is recorded from Jurassic through Upper Cretaceous sediments. Compare this with Figure 18.3. One of the major problem in seismic stratigraphy is to sort out local tectonic effects which vary from one region to another from global eustatic effects which must be common to all regions. An attempt to abstract the eustatic component from these and other records is presented in Figure 18.5. (From Vail *et al.*, 1977a)

the complexity of attempting to compile a global eustatic curve from seismic data on even relatively simple subsiding passive margins. In Figure 18.4, note the consistent observation of an on-lap relationship from probably Jurassic sediments at a depth of 4 seconds in the lower left-hand portion of the diagram to Upper Cretaceous sediments near the surface at the upper right-hand portion of the diagram. Superimposed upon the seismic cross section are generalized facies encountered by petroleum exploration drilling. Note especially that uppermost Lower Cretaceous regression of coastal deposits toward the basin (west) occurs concurrently with continuing on-lap onto the pre-Mesozoic unconformity (east). The onset of Upper Cretaceous sedimentation is indicated by on-lap of marine strata onto Lower Cretaceous coastal deposits near the center of the diagram. Note that Upper Cretaceous marine deposits overlie Lower Cretaceous coastal deposits for a distance of approximately 200 kilometers. Once again, we see Upper Cretaceous regression of shoreline facies concurrent with continued on-lap of Upper Cretaceous nonmarine deposits onto the pre-Mesozoic unconformity.

Combining the data in Figures 18.3 and 18.4 into a consistent picture of regional relative sea-level change, we can extract several important generalities to carry forward toward an emerging picture of global eustacy. To begin with, relationships in Figure 18.3 strongly suggest a major regressive event between the Jurassic and Lower Cretaceous. However, on-lap relationships in Figure 18.4 suggest that the general tendency throughout Jurassic–late Cretaceous times is toward high relative sea levels within this region. Finally, both records indicate major regressive events separating Lower from Upper Cretaceous and Upper Cretaceous from Tertiary. The question remains how much of these various events to ascribe to local subsidence history and how much to ascribe to global eustacy.

Global Synthesis. The global eustatic curve for the Mesozoic is presented in Figure 18.5. Note that the highest highstands of the sea throughout the last 220 million years are indicated to have occurred in the Middle to Late Cretaceous. Throughout the earlier Mesozoic, there is a tendency toward higher and higher highstands of the sea; throughout the Tertiary, there is a tendency to lower and lower highstands of the sea. Thus, the Middle to Late Cretaceous appears to be both unusual and pivotal when compared to history 100 million years to either side of it. Indeed, we shall see that there is apparently a fairly simple explanation for this anomalous condition.

Relation of Sea-Level Highstands to Rate of Sea-Floor Spreading. In Chapter 8, we saw that the elevation of sea floor underlain by newly formed oceanic lithosphere stands about 2.8 kilometers below sea level, whereas sea floor underlain by older oceanic lithosphere more typically resides at greater than 5 kilometers below sea level. The explanation for this phenomenon is that newly formed lithosphere is

---

Figure 18.5 (*opposite*) Relative sea-level curve abstracted from global synthesis of seismic stratigraphy. (From Vail *et al.*, 1977b, with modifications.)

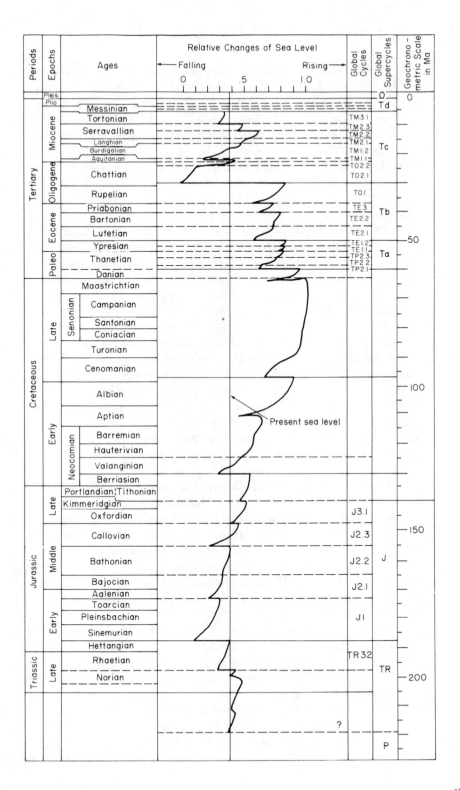

437

hotter and therefore thermally expanded relative to older lithosphere. During its cooling history after formation of approximately 80 million years, oceanic lithosphere undergoes exponential thermal contraction and subsides by approximately 3 kilometers. Clearly, if the rate of sea-floor spreading were to change, the relative proportion of ocean basin underlain by relatively young, thermally expanded oceanic lithosphere would likewise change. If the sea floor were comprised of a larger proportion of relatively young lithosphere and if the larger proportion of relatively young lithosphere were compensated for by subduction of relatively old lithosphere, then the volume of the ocean basin would become smaller and sea level would rise onto the continents.

Pitman (1978) has estimated the volume of new oceanic lithosphere produced during 10-million-year intervals back to 85 million years ago. This estimate is obtained by correlating sea-floor magnetic anomalies around the various major spreading centers of the world and simply multiplying distance between specific magnetic anomalies times the estimated length of the spreading ridge for that time frame. The global sum of these calculations for each time frame is then the estimate of volume of new oceanic lithosphere produced per time frame. The results of this calculation indicate more rapid production of oceanic lithosphere in Cretaceous time. Resulting change in the volume of mid-ocean ridges should have caused sea level to rise onto the continents on the order of 350 meters or greater, depending on our frame of reference. The calculated values of this effect from 85 million years ago to present at 10-million-year intervals is presented as the solid dots in the upper portion of Figure 18.6. Note the generally good agreement between the Pitman-calculated points and the general tendency toward lower and lower highstands of the sea on the seismic-stratigraphy generalized curve.

Comparison to $\delta^{18}O$ Ice-Volume Estimates. The lower portion of Figure 18.6 presents generalized isotopic curves for oceanic benthic foraminifers and shallow-dwelling tropical planktic foraminifers. Recall that the interpretation scheme of Matthews and Poore (1980) suggests the tropical planktic curve to be an indicator of global continental ice volume and the benthic curve, corrected for ice-volume effect, carries the additional message of variation in bottom-water temperature. The two graphs in Figure 18.6 are plotted to approximately the same scale, using the Fairbanks and Matthews (1978) calibration of $\delta^{18}O$ variation to late Pleistocene sea-level fluctuations.

The upper portion of Figure 18.6 superimposes the Pitman (1978) calculations and the generalized seismic-stratigraphy relative sea-level curve in accordance with Vail *et al.* (1977b, p. 91). The Pitman (1978) calculations of paleo–sea level are based on an ice-free world. Thus, planktic isotopic values less than $-3°/oo$ in the lower diagram may be superimposed on the upper diagram and taken to indicate glacio-eustatic sea-level lowering. Comparing the upper and lower portions of Figure 18.6, we clearly see some similarities and some differences among these independent records of eustatic history. The isotopic data are consistent with an ice-free world at approximately 100 million years ago. The isotopic data for the Lower Tertiary gen-

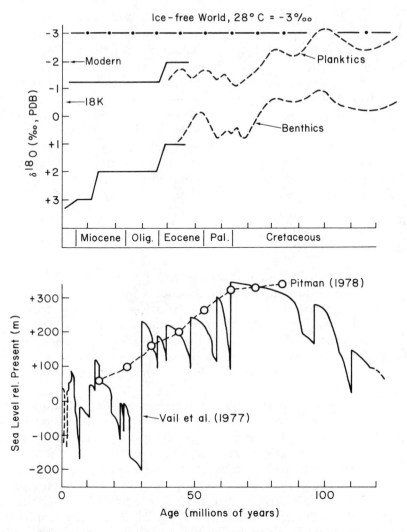

Figure 18.6 Comparison of the seismic-stratigraphy sea-level curve with the $\delta^{18}O$ record of global ice volume. The Y-axes of the two graphs are intercalibrated by way of Fairbanks and Matthews, (1978). Calibration from late Pleistocene Barbados terrace $\delta^{18}O$ elevation data. Note similarities and differences between the two curves. The most striking difference between the two curves is the tendency of seismic-stratigraphy data to estimate much larger sea-level fluctuations than the $\delta^{18}O$ data. These tendencies are corrected for in Figure 18.7. (From Matthews, 1983.)

erally reflect lower glacio-eustatic sea levels than for the Cretaceous. This generality is also captured in the seismic-stratigraphy curve within this general time interval. Indeed, the difference between average late Cretaceous and average Lower Tertiary sea-level stands is approximately the same scale as average isotopic differences.

Note, however, that major discrepancies exist with regard to timing and magnitude of the major mid-Tertiary regression. Whereas the seismic-stratigraphy curve calibrates this major regression at approximately 400 meters sea-level fall and places it in Middle Oligocene (30 million B.P.), the isotope data calibrate the event at around 80 to 100 meters sea-level fall, and the deep-sea biostratigraphy places the major isotopic shift near the Oligocene-Eocene boundary (approximately 38 million B.P.). Thus, both records recognize a major mid-Tertiary regression, but the two records are in serious disagreement as to the magnitude and timing of that event. The principle of least astonishment clearly suggests that we regard this as a clear case of "event correlation." If there is a relatively distinct, single glacio-eustatic regression within the mid-Tertiary, surely that is the major event within the seismic-stratigraphy record. Discrepancies concerning biostratigraphic placement of such an event simply point up once again the frailties of *in vacuo* biostratigraphy.

Amplitude and Frequency of Mesozoic Cyclicity

In the Pleistocene and Tertiary, we relied primarily on the tropical planktic $\delta^{18}O$ record to provide amplitude and frequency information concerning glacio-eustatic sea-level history. In the Tertiary, we supplemented this with observations from seismic stratigraphy. In the Mesozoic, abundant, relative isotopic data is yet to be gathered. Some of the higher-latitude studies have sufficient data points per time interval to cite an amplitude for the signal, but, as noted above, interpretation of that signal is not constrained to glacio-eustatic sea-level fluctuation.

Frequency of Regression within Seismic-Stratigraphy Data. Detailed studies of seismic stratigraphy have been used to pick apart the various periods of the Mesozoic. An especially well-presented study is that of Vail and Todd (1982). They recognize twelve cycles of relative coastal on-lap within the Jurassic of the North Sea. Unconformities separating the twelve cycles are considered to be the result of eustatic sea-level falls, nine of which are estimated to have occurred rapidly, presumed to be suggestive of glacio-eustatic control. Thus, seismic stratigraphy suggests major transgressive-regressive packages to be on the order of 5 or 6 million years per unit. This is somewhat longer than we observed for the Tertiary, perhaps suggesting more time between regressive events, or perhaps suggesting loss of resolution of detail as we go back in time.

It is worthy of note that several authors have pointed to the Milankovitch hypothesis with regard to periodicities observed within Mesozoic sediments. Whereas the work of Van Houten (1964) has good internal chronology by way of annual varves within Triassic lake sediment, the data bear no clear-cut relation to glacio-eustatic sea-level fluctuations. They record climatic variation not necessarily associated with continental ice-volume variation. On the other hand, the work of Fischer (1964) speaks clearly of relative sea-level changes, but in the strictest sense neither magnitude nor temporal significance of these events can be demonstrated. Thus, these data are not strictly admissible to our analysis. On the other hand, the farther back we go in time, the stronger will become the temptation to take the data that is available to us and not quibble. Admissibility of evidence is a question that individual scientists must answer for themselves.

Synthesis of Mesozoic Eustatic Sea-Level History. Figure 18.7 presents an hypothesis concerning eustatic sea-level history for the past 100 million years. This figure is constructed in accordance with the rules outlined in Chapter 17 for the construction of Figure 17.10. This hypothesis can be extended by taking into account late Paleozoic glaciation and resultant sea-level lowering. However, the magnitude of late Paleozoic ice volume is not well known and no estimates exist for ridge volume for this older time interval. Nevertheless, the generalities of the last 100 million years can serve to guide our thoughts concerning older time frames.

Overview of the Mesozoic

When we looked at the generalities of the Tertiary, we found that a lot of what we saw could be explained as paleo-oceanographic reconfiguration and resultant adjustment of cryosphere size following major plate tectonics events. As we continue our journey back in time through the Mesozoic, we shall not be able to pick out such clear-cut event-response mechanisms with regard to plate configurations and the cryosphere. However, we do obtain our first look of an astounding new dynamical phenomenon: the response of the global ocean-atmosphere system to increased spreading rates.

Plates and Global Ocean Circulation. The Mesozoic sees the inception of modern plate configuration. The era begins (Figure 19.1) with a pole-to-pole continent

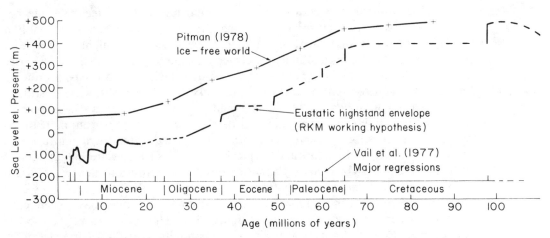

Figure 18.7 Eustatic sea-level curve for the last 100 million years. The curve is constructed from sea-floor spreading data, tropical planktic $\delta^{18}O$ data, and seismic stratigraphy by rules outlined in Chapter 17 text. (From Matthews, 1983.)

occupying one side of the earth and a huge ocean with occasional poorly known continental blocks occupying the other side of the earth. The era ends with a throughgoing equatorial circulation and with prominent mid-ocean gyres occupying the newly formed North and South Atlantic Oceans (Figure 17.2). Throughout the Mesozoic, a united Africa-Antarctica-Australia remnant of Gondwanaland remains at relatively high latitude, but does not appear to occupy a truly polar position. In contrast, North America, Europe, and Siberia generally trend from a more temperate position in early Mesozoic to a circumpolar configuration by late Mesozoic. From our experience with the Tertiary, we surely recognize opportunity for these rearrangements of continents to produce recognizable shifts within global ocean-atmospheric circulation. However, it is at present difficult to write event-effect scenarios for the Mesozoic with anything like the clarity obtainable for Tertiary.

Just when our Tertiary geo-game (plate configuration event-response) is becoming rather dull, along comes a brand-new geo-game with an even larger time interval. The truly big story of Mesozoic plate tectonics is the concept of long-term variation in the rate of production of oceanic lithosphere. A late Cretaceous world with sea level 300 to 500 meters above present would present quite a different picture of land-sea distribution than we see today. This in turn would have numerous positive feedbacks concerning global climate and eustacy.

Climate-Eustatic Response. Tropical to mid-latitude areas covered by water absorb more solar radiation than continental areas similarly situated. Thus, if a highstand of the sea puts shallow water where there was previously land, that portion of the earth will absorb more solar radiation. This increased energy is very likely to be distributed widely over the planet before ultimately escaping back to space. Thus, highstands of the sea should tend to produce a warmer global climate (Barron *et al.*, 1980, for example).

This phenomenon carries with it several important positive feedback mechanism. If there is a tendency for global climate to be warmer, there is probably a tendency to melt any high-latitude continental ice sheets that might exist at that time. Such melting of continental ice sheets would produce still further sea-level rise and thus would cover still larger areas of the continents with shallow seas, producing a tendency to capture even more energy and make the planet even warmer still.

Another potential positive feedback would be production of "warm saline bottom water" by low-latitude marginal seas (Brass *et al.*, 1982, for example). In today's ocean circulation, we are accustomed to seeing water made more dense primarily by cooling at high latitude. Typical temperate surface water might equally well be made dense by a relatively small amount of evaporation, hence the origin of the concept of "warm saline bottom water." Indeed, the late Cretaceous $\delta^{18}O$ record from benthic foraminifers suggests bottom water may have been considerably warmer in the late Cretaceous (Figure 18.2). This may present yet another positive feedback concerning warm climate. Thermal expansion of the ocean by warming of the water would add another 10 or 20 meters to sea level.

Thus, if there has ever been an ice-free world, it would be appropriate that it should occur at a time when sea levels are anomalously high for some other reason. The late Cretaceous appears to offer such an opportunity and the available tropical shallow planktic foraminifer isotope data appear to confirm an ice-free condition (Figures 18.2 and 18.6).

Finally, it is worthy of note that a salinity-stratified ocean might be much harder to overturn than a thermally stratified ocean. If Cretaceous oceans did indeed tend to be salinity-stratified rather than thermally stratified, we might expect large-scale stagnation of whole ocean basins. Indeed, such observations are the subject of an extensive literature (Arthur and Schlanger, 1979, for example).

We shall see in Chapter 19 that late Paleozoic glaciation of Gondwanaland is well documented. It is difficult to assess the transition from glaciated late Paleozoic to probable ice-free mid-Cretaceous. We have seen in the seismic-stratigraphy data a general tendency to higher relative sea level throughout the Mesozoic. However, data concerning pre-Cretaceous rates of production of oceanic lithosphere are substantially lacking. Thus, we cannot evaluate the effect of changing rate of sea-floor spreading independent from the general transition toward an ice-free world.

The positioning of Africa-Antarctica-Australia at relatively high latitude throughout the Mesozoic indeed raises a problem concerning the likelihood of a truly ice-free world. Model calculations (Barron and Washington, 1982) suggest that the interior of Antarctica would remain substantially below freezing even if surface water around the continent were somehow miraculously maintained at a warm level. Indeed, warm seas adjacent to a cold continent should be conducive to development of a continental ice sheet (Ruddiman 1979, for example).

Sedimentologic/Stratigraphic Consequences of the Mesozoic Model

The overriding generality of the Mesozoic is the transition from relatively low sea levels associated with late Paleozoic glacio-eustasy to the relatively high sea levels associated with mid-Cretaceous rapid sea-floor spreading rates and probable ice-free world. Thus, the tendency of earth history is generally transgressive from approximately 200 million years ago to approximately 100 million years ago. This flooding of the continents encouraged development of bank-margin carbonates. Two such geologic settings are particularly noteworthy.

Mesozoic carbonate rocks of Saudi Arabia contain approximately one half of the world's petroleum reserves. These carbonates accumulated under passive margin subsidence conditions from Jurassic through mid-Cretaceous times (Powers, 1962; Haynes and McQuillan, 1974; Gealy, 1977; Berberian and King, 1981; Ayers *et al.*, 1982; and others). It is interesting to consider the paleogeography of these vast oil deposits and think about other places in the world that might have similar geologic setting.

As indicated in Figure 18.1, the petroleum-producing regions of Saudi Arabia have a paleogeographic setting that is equatorial and on the western side of an ocean basin. The other major site that has a similar middle Mesozoic paleogeography is the eastern margin of the North American continent. As noted in Chapter 10, the eastern margin of the North American continent is a site of passive margin subsidence throughout this time interval. Indeed, several of the major oil fields of Mexico produce from Mesozoic bank-margin carbonates, and these facies are considered a prime target throughout the offshore area of the eastern United States and Canada (Mattick *et al.,* 1978). It is especially intriguing to note this analogy with regard to the geology of the Blake Plateau (Figure 10.9). The thick accumulations of Jurassic and lower Cretaceous bank margin carbonates must surely be considered prime targets for discovery of giant gas fields. However, one must drill in water depths of 1 to 3 kilometers to get at these targets!

Citations and Selected References

ANGEVINE, C. L., and D. L. TURCOTTE. 1981. Thermal subsidence and compaction in sedimentary basins: application to Baltimore Canyon trough. Bull. Amer. Assoc. Petrol. Geol. 65 (2): 219–225.

ARTHUR, MICHAEL A., and SEYMOUR O. SCHLANGER. 1979. Cretaceous "oceanic anoxic events" as causal factors in development of reef-reservoired giant oil fields. Bull. Amer. Assoc. Petrol. Geol. 63 (6): 870–885.

AYRES, M. C., M. BILAL, R. W. JONES, L. W. SLENTZ, M. TARTIR, and A. O. WILSON. 1982. Hydrocarbon habitat in main producing areas, Saudi Arabia: AAPG Bulletin 66 (1): 1–19.

BARRON, E. J., J. L. SLOAN II, and C. G. A. HARRISON. 1980. Potential significance of land-sea distribution and surface albedo variations as a climatic forcing factor; 180 m.y. to the present. Palaeogeog. Palaeoclimat. Palaeoecol. 30: 17–40.

BARRON, E. J., STARLEY L. THOMPSON, and STEPHEN H. SCHNEIDER. 1981. An ice-free Cretaceous? Results from climate model simulations. Science 212 (4494): 501–508.

BARRON, E. J., C. G. A. HARRISON, J. L. SLOAN, II, and W. W. HAY. 1981. Paleogeography, 180 million years ago to the present. Eclogae Geol. Helv. 74 (2): 443–470. Reconstructions done specifically as input to global circulation models. Especially noteworthy for attempting to reconstruct the extent of continental shallow seas.

BARRON, E. J., and W. M. WASHINGTON. 1982. Atmospheric circulation during warm geologic periods: is the equator-to-pole surface-temperature gradient the controlling factor?. Geology 10 (12): 609–688.

BEBOUT, D. G., and R. G. LOUCKS (eds.). 1977. Cretaceous carbonates of Texas and Mexico: applications to subsurface exploration. Univ. of Texas at Austin, Bureau of Economic Geology, Report of Investigations 89. 332 p.

BERBERIAN, M., and G. C. P. KING. 1981. Towards a paleogeography and tectonic evolution of Iran. Canadian Jour. of Earth Sci. 18: 210–265.

BRASS, G. W., J. R. SOUTHAM, and W. H. PETERSON. 1982. Warm saline bottom water in the Ancient ocean. (Submitted to Nature.)

CLARK, DAVID L., and JENNIFER A. KITCHELL. 1981. Terminal Cretaceous extinctions and the Arctic spillover model. Science, 212 (4494): 577.

CROWELL, J. C., and L. A. FRAKES. 1971. Late Paleozoic glaciation, part 4: Australia. Bull. Geol. Soc. Amer. 82: 2515–2540.

DAWSON, M. R. 1980. Cenozoic history in and around the northern Atlantic and Arctic Oceans [special issue]. Palaeogeog., Palaeoclimat. Palaeoecol. 30: 217–362.
Collection of eight papers. Especially good starting point concerning paleontological evidence for relatively warm high-latitude climate in late Mesozoic and early Tertiary.

DEBOER, JELLE, and FREDERIC G. SNIDER. 1979. Magnetic and chemical variations of Mesozoic diabase dikes from eastern North America: evidence for a hotspot in the Carolinas? Bull. Geol. Soc. Amer. 90 (pt. 1, no. 2): 205–215.

DONN, WILLIAM L., and DAVID M. SHAW. 1977. Model of climate evolution based on continental drift and polar wandering. Bull. Geol. Soc. Amer. 88: 390–396.

DOUGLAS, ROBERT G., and SAMUEL M. SAVIN. 1971. 37. Isotopic analyses of planktonic foraminifera from the Cenozoic of the northwest Pacific, Leg 6, p. 1123–1127. *In* A. G. Fischer *et al.*, Initial reports of the deep sea drilling project, vol. 6. U.S. Gov't. Printing Office, Washington, D.C.

——. 1973. 20. Oxygen and carbon isotope analyses of Cretaceous and Tertiary foraminifera from the central North Pacific, p. 591–605. *In* E. L. Winterer, J. I. Ewing, *et al.*, Initial reports of the deep sea drilling project, vol. 17. U.S. Gov't. Printing Office, Washington, D.C.

——. 1975. Oxygen and carbon isotope analyses of Tertiary and Cretaceous microfossils from Shatsky Rise and other sites in the North Pacific Ocean, p. 509–520. *In* R. L. Larson, R. Moberly, *et al.*, Initial reports of the deep sea drilling project, vol. 32. U.S. Gov't. Printing Office, Washington, D.C.

——. 1978. Oxygen isotopic evidence for the depth stratification of Tertiary and Cretaceous planktic foraminifera. Marine Micropaleont. 3: 175–196.

DOUGLAS, ROBERT, and FAY WOODRUFF. 1981. Deep-sea benthic foraminifera, p. 1233–1327. *In* C. Emiliani (ed.), The sea, Vol. 7. Willey-Interscience, New York.

EPSHTEYN, O. G. 1977. Mesozoic-Cenozoic climates of northern Asia and glacial-marine deposits. Internat. Geol. Rev. 20 (1): 49–58.

EXON, N. F., and J. B. WILLCOX. 1978. Geology and petroleum potential of Exmouth Plateau area off western Australia. Bull. Amer. Assoc. Petrol. Geol. 62 (1): 40–72.

FAIRBANKS, RICHARD G., and R. K. MATTHEWS. 1978. The marine oxygen isotope record in Pleistocene coral, Barbabos, West Indies. Quat. Res. 10: 181–196.

FISCHER A. G. 1964. The Loafer cyclothems of the Alpine Triassic, p. 107–150. *In* D. F. Merriam (ed.), Symposium on cyclic sedimentation, State Geol. Survey of Kansas, Bull. 169.
Recognizes approximately 200 cycles in subtidal to supratidal carbonates. Five to 8 cycles often compose a grand cycle.

FRAKES, L. A. 1979. Climates throughout geologic time. Elsevier, New York. 301 p.

GEALY, W. K. 1977. Ophiolite obduction and geologic evolution of the Oman Mountains and adjacent areas. Bull. Geol. Soc. Amer: 1183-1197.

HALLMAN, A. 1978. Eustatic cycles in the Jurassic. Palaeogeog. Palaeoclimat. Palaeoecol. 23: 1-32.

HANCOCK, J. M., and E. G. KAUFFMAN. 1979. The great transgressions of the late Cretaceous. J. Geol. Soc. Lond. 136: 175-186.

HAYNES, S. J., and H. McQUILLAN. 1974. Evolution of the Zagros suture zone, southern Iran. Bull. Geol. Soc. Amer. 85: 739-744.

HEDBERG, H. D., J. D. MOODY, and R. M. HEDBERG. 1979. Petroleum prospects of deep offshore. Bull. Amer. Assoc. Petrol. Geol. 63 (3): 286-300.

HICKEY, LEO J. 1981. Land plant evidence compatible with gradual, not catastrophic, change at the end of the Cretaceous. Nature 292 (5823): 529-531.

IBRAHIM, M. W., and R. J. MURRIS. 1981. Middle East: stratigraphic evolution and oil habitat: discussion and reply. Bull. Amer. Assoc. Petrol. Geol. 65 (3): 540-543.

KENT, D. V. 1977. An estimate of the duration of the faunal change at the Cretaceous-Tertiary boundary. Geology 5 (12): 769-771.

LaBRECQUE, JOHN L., and PETER BARKER. 1981. The age of the Weddell Basin. Nature 290 (5806): 489-492.

LANTZY, RONALD J., MICHAEL F. DACEY, and FRED T. MACKENZIE. 1977. Catastrophe theory: application to the Permian mass extinction. Geology 5 (12): 724-728.

LePICHON, XAVIER, and JEAN-CLAUDE SIBUET. 1981. Passive margins: a model of formation. J. Geophys. Res. 86 (B5): 3708-3720.

LIVACCARI, RICHARD F., KEVIN BURKE, and A. M. C. SENGÖR, 1981. Was the Laramide orogeny related to subduction of an oceanic plateau? Nature 289: 276-278.

LOWENSTAM, HEINZ A. 1964. Palaeotemperatures of the Permian and Cretaceous periods, p. 227-248. In A. E. M. Nairn (ed.), Problems in Palaeoclimatology. Wiley-Interscience, New York.

LUDBROOK, N. H. 1978. Australia, p. 209-244. In M. Moullade, and A. E. M. Nairn (eds.), The phanerozoic geology of the world, II: The Mesozoic A. Elsevier, New York.

McHONE, J. GREGORY, and THOMAS CROUGH. 1981. Mesozoic hotspot epeirogeny in eastern North America: comment and reply. Geology 9 (8): 341-343.

MARGOLIS, S. V., P. M. KROOPNICK, and D. E. GOODNEY. 1977. Cenozoic and late Mesozoic paleoceanographic and paleoglacial history recorded in circum-Antarctic deep-sea sediments. Marine Geol. 25: 131-147.

MATTHEWS, J. L., B. C. HEEZEN, R. CATALANO, A. COOGAN, M. THARP, J. NATLAND, and M. RAWSON. 1974. Cretaceous drowning of reefs on mid-Pacific and Japanese guyots. Science, 184: 462-464.

MATTHEWS, R. K. 1983. The oxygen isotope record of ice volume history: 100 million years of glacio-eustatic sea-level fluctuation. *In* John S. Schlee (ed.), Interregional uncomformities. Amer. Assoc. Petrol. Geol. Mem. (forthcoming).

MATTHEWS, R. K., and R. Z. POORE. 1980. Tertiary $\delta^{18}O$ record and glacio-eustatic sea-level fluctuations. Geology 8: 501–504.

MATTICK, R. E., O. W. GIRARD, JR., P. A. SCHOLLE, and J. A. GROW. 1978. Petroleum potential of U. S. Atlantic slope, rise, and abyssal plain: AAPG Bulletin 62 (4): 592–608.

MITCHUM, R. M., JR., P. R. VAIL, and S. THOMPSON III. 1977. Seismic stratigraphy and global changes of sea level, part 2: the depositional sequence as a basin uplift for stratigraphic analysis. p. 53–62. *In* C. E. Payton (ed.), Seismic stratigraphy—applications to hydrocarbon exploration. Amer. Assoc. Petrol. Geol. Mem. 26.

NEILL, WILLIAM M. 1976. Mesozoic epeirogeny at the South Atlantic margin and the Tristan hot spot. Geology 4: 495–498.

OSTROM, JOHN H. 1969. Terrestrial vertebrates as indicators of Mesozoic climates. Proceedings of the North American Paleontological Convention, part D: 347–376.

PAYTON, C. E. (ed.). 1977. Seismic stratigraphy—applications to hydrocarbon exploration. Amer. Assoc. Petrol. Geol. Mem. 26. 516 p.

PITMAN, WALTER C., III. 1978. Relationship between eustacy and stratigraphic sequences of passive margins. Bull. Geol. Soc. Amer. 89 (9): 1389–1403.

POWERS, R. W. 1962. Arabian upper Jurassic carbonate reservoir rocks, p. 122–192. *In* W. E. Ham (ed.), Classification of carbonate rocks. Amer. Assoc. Petrol. Geol. Mem. 1.

RAUP, D. M. 1979. Size of the Permo-Triassic bottleneck and its evolutionary implications. Science 206 (4415): 217–218.

RUDDIMAN, W. F. 1979. Warmth of the subpolar North Atlantic Ocean during Northern Hemisphere ice-sheet growth. Science 204: 173–175.
Well-reasoned argument and empirical data suggest warm ocean feeds moisture to a cold continent to grow an ice sheet.

RUSSELL, D. A. 1979. The enigma of the extinction of the dinosaurs. Ann. Rev. Earth Planet. Sci. 7: 163–182.

RYAN, WILLIAM B. F., and MARIA B. CITA. 1977. Ignorance concerning episodes of ocean-wide stagnation. Marine Geol. 23: 197–215.

SAVIN, SAMUEL M. 1977. The history of the earth's surface temperature during the past 100 million years. Ann. Rev. Earth Planet. Sci. 5: 319–355.

SCHLAGER, WOLFGANG. 1981. The paradox of drowned reefs and carbonate platforms. Bull. Geol. Soc. Amer. 92 (4): 197–211.

SCHLANGER, SEYMOUR O. 1981. Shallow-water limestones in ocean basins as tectonic and paleoceanic indicators, p. 209–226. *In* John E. Warme, Robert G. Douglas, and Edward L. Winterer (eds.), The deep sea drilling project: a decade of progress. Soc. Econ. Paleont. Mineral. Spec. Pub. 32.

SCHLANGER, SEYMOUR O., HUGH C. JENKYNS, and ISABELLA PREMOLI-SILVA. 1981. Volcanism and vertical tectonics in the Pacific Basin related to global Cretaceous transgressions. Earth Planet. Sci. Lett. 52: 435–449.

SLEEP, N. H. 1976. Platform subsidence mechanisms and "eustatic" sea-level changes. Tectonophysics 36: 45–56.

SLOSS, L. L. 1979. Global sea level change: a view from the craton, p. 461–467. *In* J. S. Watkins (ed.), Geological and geophysical investigations of continental margins. Amer. Assoc. Petrol. Geol.

SMIT, J., and G. KLAVER. 1981. Sanidine spherules at the Cretaceous-Tertiary boundary indicate a large impact event. Nature 292 (5818): 47–49.

SMITH, A. G., and J. C. BRIDEN. 1977. Mesozoic and Cenozoic Paleocontinental Maps. Cambridge Univ. Press. Cambridge. 63 p.

THIEDE, JORN. 1981. Reworked neritic fossils in upper Mesozoic and Cenozoic Central Pacific deep-sea sediments monitor sea-level changes. Science 211 (4489): 1422–1424.

THIERSTEIN, HANS R. 1981. Late Cretaceous nannoplankton and the change at the Cretaceous-Tertiary boundary, p. 355–394. *In* John E. Warme, Robert G. Douglas, and Edward L. Winterer (eds.), The deep sea drilling project: a decade of progress. Soc. Econ. Paleont. Mineral. Spec. Pub. 32.

THIERSTEIN, HANS R., and WOLFGANG H. BERGER. 1978. Injection events in ocean history. Nature 276: 461–466.

VACHRAMEEV, V. A. 1978. The climates of the Northern Hemisphere in the Cretaceous in the light of paleobotanical data. (Trans. from Russian.) Paleont. Zhur., no. 2, p. 3–17.

VAIL, P. R., R. M. MITCHUM, JR., and S. THOMPSON III. 1977a. Seismic stratigraphy and global changes of sea level, part 3: relative changes of sea level from coastal onlap, p. 63–81. *In* C. E. Payton, (ed.), Seismic stratigraphy—applications to hydrocarbon exploration: Amer. Assoc. Petrol. Geol. Mem. 26.

———. 1977b. Seismic stratigraphy and global changes of sea level, part 4: global cycles of relative changes of sea level, p. 83–97. *In* C. E. Payton (ed.), Seismic stratigraphy—applications to hydrocarbon exploration. Amer. Assoc. Petrol. Geol. Mem. 26.

VAIL, P. R., and R. G. TODD. 1983. Cause of northern North Sea Jurassic unconformities. *In* J. S. Schlee (ed.), Interregional unconformities and hydrocarbon accumulations. Amer. Assoc. Petrol. Geol. Mem. (forthcoming).

VAN HOUTEN, F. B. 1964. Cyclic lacustrine sedimentation, Upper Triassic Lockatong formation, central New Jersey and adjacent Pennsylvania, p. 497–532. *In* D. F Merriam (ed.), Symposium on cyclic sedimentation, State Geol. Survey of Kansas, Bull. 169. Recognizes three scales of cyclicity, presumably relating to climatic fluctuations. Short cycles, 14 to 20 feet in thickness, are related to the 21,000-year precession cycle. Intermediate cycles of 70 to 90 feet, and long cycles of 325 to 350 feet are also recognized.

———. 1980. Latest Jurassic–Early Cretaceous regressive facies, northeast Africa craton. Bull. Amer. Assoc. Petrol. Geol. 64 (6): 857–867.

WATTS, A. B., and W. B. F. RYAN. 1976. Flexure of the lithosphere and continental margin basins. Tectonophysics 36: 25–44.

The Paleozoic

19

The Paleozoic comprises approximately 350 million years. Its lower boundary distinguishes between diverse and abundant marine invertebrates (above) and the general paucity of higher life-forms in Precambrian strata (below). The upper boundary is traditionally taken to mark a major crisis in the history of life on earth with relatively few taxa escaping across the Permian-Triassic boundary to begin life anew in the Mesozoic.

On our trip back through time, at every new time interval we have picked up new insight as to "how the earth works." In the Pleistocene, we become thoroughly grounded in glacio-eustatic sea-level fluctuation. In the Tertiary, we found that plate tectonics could radically alter continent-ocean geometry within a scant 60 million years and that plate configuration could likewise alter glacio-eustatic sea level. Upon reaching the late Mesozoic, we were able to get our first look at the profound effect of a changing spreading rate on a eustatic sea level. As we approach the Paleozoic, we shall encounter no new surprises, just variations on themes we have already seen. Perhaps this is because our model of global dynamics is becoming rather complete; or perhaps it is because our record of earth history is becoming so blurred that we would not recognize something profoundly new if it were placed in front of us.

Continental Plates

Gone are the sea-floor magnetic stripes; largely unrecognized are any footprints of hot spots. Our Paleozoic plate reconstructions rely entirely upon apparent pole position data taken from the continents. Corroborative evidence can be found in large pieces of field geology consistent with plate-collision tectonics.

Paleo-Oceans. Figures 19.1 and 19.2 continue our trip back in time. The major change from our look at the Mesozoic world is that the late Paleozoic world (Figure

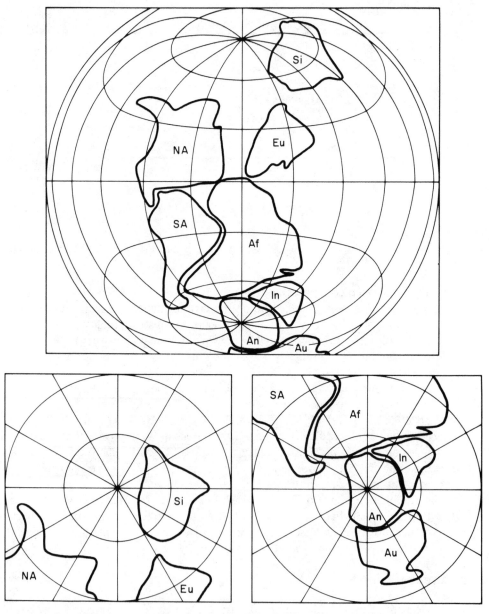

Figure 19.1 Maps indicating position of major continental blocks approximately 300 million years ago (Carboniferous). Note closure of North America onto South America-Africa. Note polar position of a major portion of the Gondwanaland supercontinent. (Constructed from data cited in Chapter 8.)

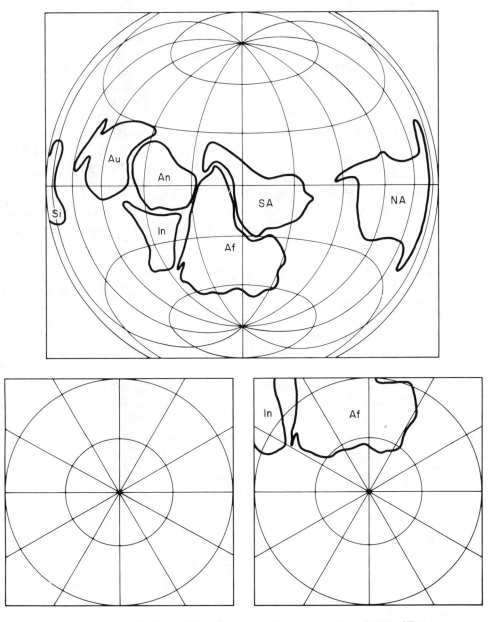

Figure 19.2 Maps indicating position of major continents approximately 500 million years ago (Cambrian). Note the absence of continents from polar regions. To arrive at Figure 19.1 from 19.2, allow Africa (with the rest of Gondwanaland attached, of course!) to pass southward along the 30°E meridian, across the South Pole, and emerge passing northward along the 150°W meridian toward collision with North America. In this manner, it appears that the supercontinent of Gondwanaland occupied a polar position throughout most of the Paleozoic. (Complied from data cited in Chapter 8.)

19.1) consists of a substantially pole-to-pole supercontinent with no throughgoing equatorial ocean circulation. Paleo-oceanographic understanding of the significance of such a huge continent awaits quantitative modeling.

The transition from early Paleozoic plate configuration to late Paleozoic plate configuration (Figures 19.1 and 19.2) requires some explanation. To begin with, there is the lack of a European plate in Figure 19.2. Strictly speaking, these various figures of plate configuration are constructed from apparent pole position data and the Cambrian pole position data for Europe are simply inadequate to this purpose.

So much for Europe. Surely, we are still curious that Africa and South America are "upside down" in Figure 19.2. The transition from Cambrian continental plate configuration to Carboniferous continental plate configuration is actually rather simple. Throughout the Paleozoic, the Gondwanaland supercontinent takes a trip across the South Pole and comes up again with northwest Africa in equatorial position by the Carboniferous (Figure 19.1). The drift of Gondwanaland is summarized in Figure 19.3.

The other major aspect of the transition from Cambrian to Carboniferous continental configuration is the closing of the ocean basin between North America and South America–Africa. By Devonian times, Gondwanaland and North America were in relative positions not unlike that shown in Figure 18.1, but with everything shifted somewhat southward. There was at this time substantial throughgoing equatorial ocean circulation. Subsequent closing of this seaway delivers us to the plate configuration depicted in Figure 19.1.

Polar Regions. The major transgression of Cambrian seas onto the continents provides us the first opportunity to obtain paleomagnetic data within a closely de-

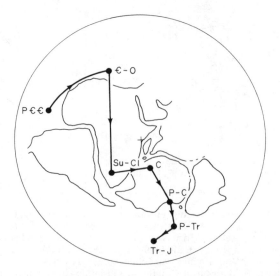

Figure 19.3 Cartoon indicating the passage of the South Pole across Africa and Antarctica from Cambro-Ordovician through Permo-Triassic time. (From Frakes, 1979.)

fined stratigraphic time interval. The picture we get (Figure 19.2) is that of a marked tendency to circumequatorial distribution of major continental plates. Importantly, there is no data to indicate even minor continental plates in polar position during the Cambrian. With regard to the Northern Hemisphere, this generality remains reasonably intact throughout the Paleozoic. With regards to the Southern Hemisphere, however, a marvelous story unfolds throughout the Paleozoic.

As indicated in Figure 19.3, the supercontinent of Gondwanaland appears to drift across the South Pole throughout virtually the entire Paleozoic. Thus, whereas Tertiary glacio-eustatic history is profoundly influenced by the presence of a continent the size of Antarctica beneath the South Pole, the Paleozoic presents us with the specter of a continent six times the size of Antarctica in a relatively polar position for 200 to 300 million years.

The Continental Record of Ice

It is always darkest before the dawn! We began our journey back in time with the $\delta^{18}O$ record as our primary sensor of global ice volume. As we journeyed back through the Mesozoic, that record gave out on us at an "ice-free world" in the late Cretaceous, and we found no geological literature making much of a case for early Mesozoic ice on the continents. Now, as we hit the late Paleozoic, we come back into a time of great insight concerning ice volume on continents.

Late Paleozoic Continental Glaciation in Australia, South Africa, South America, Antarctica, and India. Clear evidence for the former existence of continental ice sheets in regions presently situated at relatively low latitude has fascinated geologists for many years. Commonly cited evidence of late Paleozoic glaciation includes apparent glacial striations on bedrock, tillite sedimentary fabric, and occurrence of large pebbles (presumably ice-rafted dropstones) in otherwise fine-grained marine strata. Taken individually, arguments can be raised contrary to glacial interpretation of each of these phenomenon. However, taken together and in a well-defined paleogeography, the combination is rather convincing of the previous existence of continental ice sheets. Extensive descriptions of such phenomenon in Australia (Crowell and Frakes, 1971, for example), South Africa (Hamilton and Krinsley, 1967; Crowell and Frakes, 1972, for example), South America (Frakes and Crowell, 1969, for example), India (Banerjee, 1966, for example), and Antarctica (Frakes *et al.*, 1971, for example) are attributed to existence of late Paleozoic continental ice sheets of sizable proportion. Figure 19.4 summarizes the evidence for late Paleozoic ice sheets on a reconstructed Gondwanaland.

The age of late Paleozoic Gondwanaland glaciation is generally assigned to Middle Carboniferous through Permian time. However, there are major problems with the biostratigraphic context of these deposits. Much of the dating rests upon terrestrial or marginal-marine floras and faunas. These fossils are subject to the classic arguments concerning environmental versus temporal significance. The reader is referred to extensive descriptive reveiw of Frakes (1979) for further details.

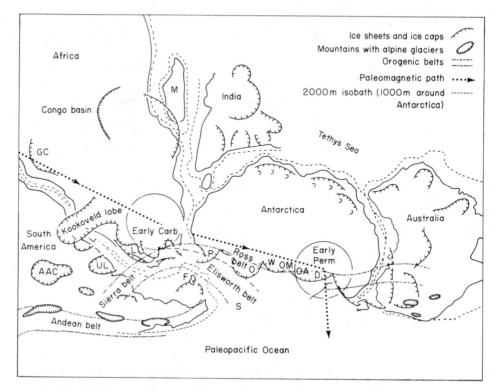

Figure 19.4 Major late Paleozoic glaciated regions of Gondwanaland. (From Crowell and Frakes, 1975.)

Early Paleozoic Continental Glaciation in North Africa. Evidence of lower Paleozoic continental glaciation in North Africa is especially attractive because of the extremely flat-lying paleotopography and the availability of marine invertebrate fossils below and above sediments representing the glaciated time interval. An extensive Cambro-Ordovician blanket sandstone, taken to represent fluvial to marine deposition on a huge epicontinental shelf, provides evidence of the extremely flat paleotopography and provides a lower bound for the age of glaciation in the region (Burollet and Byramjee, 1969). Above the glacial sediment are laterally extensive graptolite-bearing shales of Silurian age. Thus, the intervening glacial unit (Cambro-Ordovician "unit IV" of Beuf *et al.*, 1969; Tamadjert formation of Frakes, 1979) is defined by marine biostratigraphy to be Ordovician-Silurian in age and occupy an extremely flat paleogeography.

The glacial unit is characterized by tillite fabric, intraformational glacial striations and polished surfaces, microconglomeratic shales, and U-shaped cut and filled structures commonly associated with glacial excavation. As indicated in Figure 19.5, Beuf *et al.* (1969) maps the aerial extent of Ordovician-Silurian North African glaciation to exceed 2000 kilometers in width. Tucker and Reid (1973) report dropstones in rhythmically bedded shales of the Waterfall formation, late Ordovician

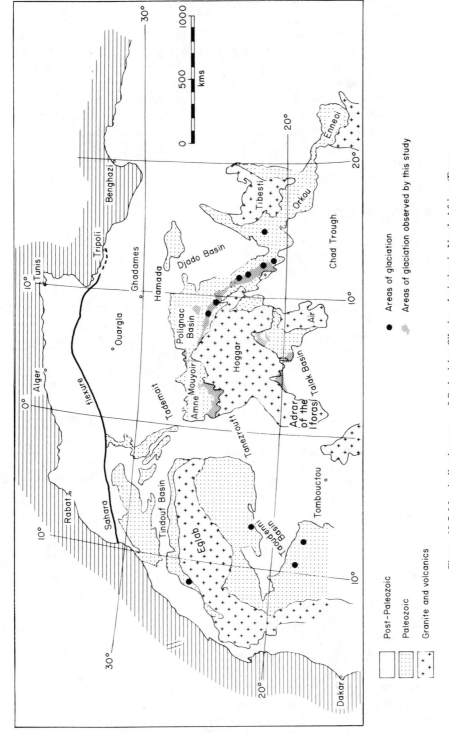

Figure 19.5 Map indicating areas of Ordovician-Silurian glaciation in North Africa. (From Beuf *et al.,* 1969.)

Areas of glaciation

Areas of glaciation observed by this study

Post-Paleozoic

Paleozoic

Granite and volcanics

of Sierra Leone. To the east, scattered evidence exists for extension of the Ordovician North African ice sheet into Saudi Arabia (Frakes, 1979). If we accept these bounds to an Ordovician North African ice sheet, we are talking about a feature comparable to the present Antarctic ice cap.

Thus, we see two very clear glimpses of the geologic record of Paleozoic continental ice on the African continent: (1) the Ordovician-Silurian of North Africa, and (2) the late Paleozoic of South Africa. What can be said of the intervening 100 or so million years and 5000 or so kilometers that separate these observations? Unfortunately, the intervening region can be generally characterized as Precambrian overlain by post-Dwyka (South African glacial time-rock) sediments (White, 1972).

Seismic Stratigraphy

Global supercycles for the Paleozoic are depicted in the lower portion of Figure 19.6. Vail *et al.* (1977b) acknowledge the existence of numerous higher-frequency cycles within the late Paleozoic but consider knowledge of the cycles to be too fragmentary for inclusion in a purportedly global curve at this time.

The general tendency depicted in Figure 19.6 is that of a major Cambrian transgression culminating in the greatest extent of shallow seas onto the continents in late Cambrian–early Ordovician times. Note that the general tendency to Cambro-Ordovician on-lap is spread over approximately 70 million years. Following the Cambro-Ordovician transgression, the generalized curve depicts a series of major regressive events followed by on-lap. Note the tendency for successive on-laps to top out somewhat short of the previous on-lap sequence and note the tendency for the classical period boundaries to occur at or near major regressive events.

Frequency and Amplitude of Cyclicity

Paleozoic supercycles depicted in Figure 19.6 are on the order of 50 to 80 million years in duration. Cyclic sedimentation with much higher frequency is well known throughout the Paleozoic in the form of repetitive lithologic sequences. Among the most striking examples are the cyclothems of the Pennsylvanian of the central United States, made popular by the voluminous works of R. C. Moore, Harold R. Wanless, Sr., and their legions of students (see Moore, 1964; Wanless, 1964; for example). Cyclicity in bank-margin carbonates of the Permian basin of West Texas is well known and of fundamental importance to petroleum production from the region (Silver and Todd, 1969). The Mississippian carbonates of the Rocky Mountains are depicted as cyclic (Rose, 1976) and the "kickback" event correlation scheme of Irwin (1965) is a powerful unifying concept.

The overriding problem with all of these schemes is the lack of sufficiently precise time control to allow quantification of cyclicity. When modern sedimentation models are put to many outstanding examples of Paleozoic cyclicity, it can be calculated that the deposits could have accumulated within a few tens of thousands of years. Yet, when one evaluates the time constraints on these sequences, it can

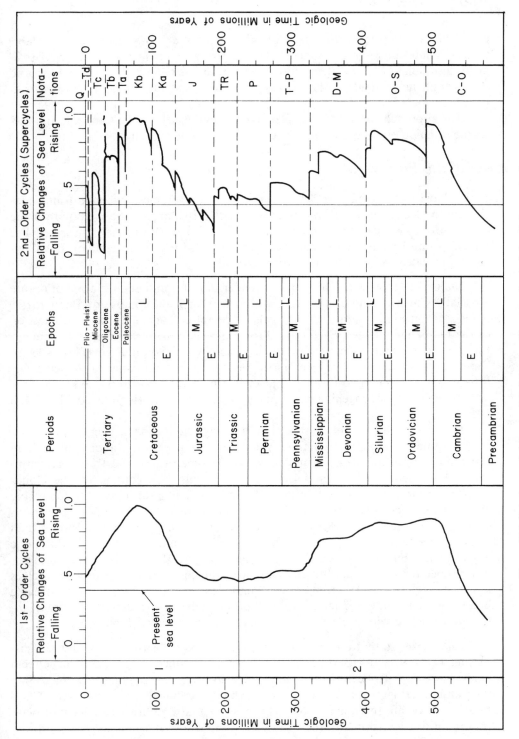

Figure 19.6 Relative sea-level curve for the entire Phanerozoic, as deduced from seismic stratigraphy. (From Vail *et al.*, 1977a,b.)

equally be demonstrated that tens of millions of years may have been available to accumulate these same sediments! Do the strata represent a short burst of sediment accumulation or do they represent intermittent sedimentation widely spaced throughout the available time? Such are the vagaries of life for the Paleozoic stratigrapher.

The approach suggested here is to allow experience with the Cenozoic and Mesozoic to guide our expectations concerning Paleozoic cyclicity.

Overview of the Paleozoic

As we have stepped back to larger and larger time frames, we have found new elements added to our model of the earth at each step along the way. Upon reaching the Paleozoic, we are at last looking at a time interval with no new surprises. The Paleozoic can be generally understood in the light of major elements of global dynamics revealed to us in previous time intervals.

Major Plate Events. Three major tendencies of continental plate motion are especially worthy of note. To begin with, our first good glimpse of Phanerozoic plate configuration (Figure 19.2) reveals a general circumequatorial configuration to the major landmasses. Secondly, throughout the Paleozoic there is a tendency for the Gondwanaland continent to drift across the South Pole (Figure 19.3). Thirdly, the ocean basin between North America and South America–Africa closes throughout the upper two-thirds of Paleozoic time, giving rise to the classic geology of the Appalachian orogenic belt of eastern North America (Figures 10.14 through 10.17).

In terms of global ocean circulation, we begin the Paleozoic (Figure 19.2) with a single superocean and no continents in polar regions. Understanding of such an ocean must await quantitative modeling. By Devonian times, North America, Europe, and Siberia are substantially Northern Hemisphere continents, the Gondwanaland continent occupies Southern Hemisphere, and a substantial throughgoing equatorial circulation exists. This configuration is not unlike that of large portions of Mesozoic time (Figure 18.1). To the extent that we are beginning to understand Mesozoic global ocean circulation, then we perhaps understand the mid-Paleozoic global ocean. Finally, by late Paleozoic times (Figure 19.1) we are faced with another supercontinent situation, this time with the continents arranging themselves longitudinally and substantially from South Pole to North Pole with no oceanic break among them. Once again, we shall probably be wise to defer to future quantitative modeling with regards to global circulation within such an ocean.

Climatic-Eustatic Response. From our accumulated experience with the Pleistocene, Tertiary, and Mesozoic, certain generalities concerning the Paleozoic seem readily apparent. To begin with, if there were no continents in polar position during the Cambrian (Figure 19.2), then we are probably looking at an ice-free latest Cambrian–Ordovician. In Figure 19.6, we likewise see that Cambro-Ordovician displays maximum transgression strikingly parallel to the late Cretaceous ice-free world.

Three independent lines of evidence tell us that the ice-free condition of Cambro-Ordovician times is not long-lived. Plate reconstructions depict a long history of the Gondwanaland continent passing across the South Pole (Figure 19.3). Ordovician-Silurian tillite of North Africa (Figure 19.5) tells us the Gondwanaland continent became glaciated as it began to pass across the South Pole. Finally, seismic stratigraphy (Figure 19.6) tells us that numerous regressions occurred and that subsequent transgressions did not reach as far onto the continents as the Cambro-Ordovician ice-free world transgression.

Ordovician-Silurian glaciation of North Africa is especially important because it demonstrates that the earth-sun relationship was conducive to ice accumulation by early Paleozoic time. If a continent exists beneath a pole, there should be ice on it at least intermittently. North African early Paleozoic glaciation appears to demonstrate that this generality from the late Pleistocene has been operative throughout the Phanerozoic. Indeed, numerous reports of Precambrian glacial deposits seem to demonstrate the generality well back into earth history (Frakes, 1979).

It is clear that the Gondwanaland continent was beneath the South Pole from the Ordovician right on through to the late Paleozoic. Did a polar ice cap continuously occupy this huge polar continent throughout that time? As with the question of Lower Tertiary glaciation of Antarctica, we once again face the question of admissibility of negative evidence in geologic thought. North African glaciation is placed at the Ordovician-Silurian boundary. The evidence of glaciation is clearly capped by Silurian graptolite-bearing shales. The next look that we get at clear evidence of glaciation is at the other end of the continent: the "late Paleozoic" glaciation of South Africa, Australia, and South America. It seems virtually inescapable that ice occupied some portion of the Gondwanaland continent throughout the time of its passage across the South Pole. Thus, glacio-eustatic sea- level fluctuation may be called upon to explain major regressions and transgressions (Figure 19.6) throughout this time interval.

Can we extend this argument for continuous Paleozoic continental ice volume into the frequency domain? Surely, we can make this hypothesis! Maybe we'll be wrong—but nothing ventured, nothing gained!

Experience with the Pleistocene and Tertiary $\delta^{18}O$ record and the Antarctic continent may be transferable to the Paleozoic situation with regards to a polar Gondwanaland. High tilt melts polar ice; that gave us lots of 40,000-year power. High eccentricity of the earth's orbit further accentuates the effect of tilt; that gave us 100,000- to 400,000-year power. Perhaps we should expect these periodicities to Paleozoic glacio-eustatic sea-level fluctuation until the concept is refuted by solid data.

Sedimentologic/Stratigraphic Consequences of the Paleozoic Model

As noted above, several generalities concerning the Paleozoic lead us to anticipate glacio-eustatic sea-level fluctuation with periodicities comparable to those observed in the Pleistocene. Does the stratigraphic record lend credence to this assertion? With regard to the Devonian, the observation of Punctuated Aggradational Cycles (PAC's) does lend credence to numerous high-frequency eustatic events (Anderson and Goodwin, 1980). Anderson and Goodwin propose that the classic Devonian strata of New York State contain numerous discontinuity surfaces which can be correlated throughout the region. Typical cycles are from 1 to 5 meters in thickness. For many years, these relatively small subdivisions have been overlooked because they are a second-order phenomenon subtly displayed within overriding tendencies to extended periods of transgression or regression.

With regard to the late Paleozoic, the strikingly cyclic nature of the sedimentary record on the North American continent has been recognized for a long time. The kickback correlation scheme devised by Irwin (1965) for the Mississippian of the Williston Basin almost certainly rests upon rapid glacio-eustatic transgression. The Pennsylvanian and Permian cyclothems of Kansas (Moore, 1964, for example) are of such vast lateral extent as to demand a eustatic causal mechanism. Finally, the concept of reciprocal sedimentation (highstand deposits alternating with lowstand deposits) has served well to make sense out of otherwise complicated stratigraphic relationships throughout the late Paleozoic of New Mexico and west Texas (Wilson, 1967; Silver and Todd, 1969, for example). Matthews (1974) provides a convenient overview of these and other studies of late Paleozoic cyclicity.

Citations and Selected References

ANDERSON, E. J., and P. W. GOODWIN. 1980. Application of the PAC hypothesis to limestones of the Helderberg Group, eastern section. Soc. Econ. Paleont. Mineral. Guidebook, Tulsa, Oklahoma. 32 p.

BAMBACH, R. K., C. R. SCOTESE, and A. M. ZIEGLER. 1980. Before Pangea: the geographies of the Paleozoic world. Amer. Scientist 68: 26–38.
Good graphics concerning Paleozoic continent configurations through time.

BANERJEE, I. 1966. Turbidites in a glacial sequence: a study from the Talchir formation, Raniganj coalfield, India. J. Geology 74: 593–606.

BEERBOWER, J. R. 1961. Origin of cyclothems of the Dunkard Group (Upper Pennsylvania-Lower Permian) in Pennsylvania, West Virginia, and Ohio. Bull. Geol. Soc. Amer. 72: 1029–1050.

BEUF, S., B. BIJU-DUVAL, J. STEVAUX, and G. KULBICKI. 1969. Extent of "Silurian" glaciation in the Sahara: its influences and consequences upon sedimentation, p. 103–116. *In* W. H. Kanes (ed.), Geology, archaeology and prehistory of the southwestern Fezzan, Libya. Petrol. Exploration Soc. of Libya, Eleventh Annual Field Conference.

BRENNER, ROBERT. 1980. Construction of process-response models for ancient epicontinental seaway depositional systems using partial analogs. Bull. Amer. Assoc. Petrol. Geol. 64 (8): 1223–1244.

BUROLLET, P. F., and R. BYRAMJEE. 1969. Sedimentological remarks on lower Paleozoic sandstones of south Libya, p. 91–102. *In* W. H. Kanes (ed.), Geology, archaeology and prehistory of the southwestern Fezzan, Libya. Petrol. Exploration Soc. of Libya, Eleventh Annual Field Conference.

CHUBER, S., and W. C. PUSEY. 1969. Cyclic San Andres facies and their relationship to diagenesis, porosity, and permeability in the Reeves Oil Field, Yoakum County, Texas, p. 136–151. *In* J. G. Elam and S. Chuber (eds.), Cyclic sedimentation in the Permian Basin. West Texas Geol. Soc. Pub. 69–56.

CROSTELLA, A., C. BOWIN, and C. JOHNSTON. 1981. Arc-continent collision in Banda Sea region: discussion and reply. Bull. Amer. Assoc. Petrol. Geol. 65 (5): 866–867.

CROWELL, J. C., and L. A. FRAKES. 1970. Phanerozoic glaciation and the causes of ice ages. Amer. J. Sci. 268: 193–224.

CROWELL, J. C., and L. A. FRAKES. 1971. Late Paleozoic glaciation: Part IV, Australia. Bull. Geol. Soc. Amer. 82: 2515–2540.

———. 1972. Late Paleozoic glaciation: Part V, Karroo Basin, South Africa. Bull. Geol. Soc. Amer. 83: 2887–2912.

———. 1975. The Late Paleozoic glaciation, p. 313–331. *In* K. S. W. Campbell (ed.), Gondwana Geology. Australian National University, Canberra, A.C.I.

DAPPLES, E. C., and M. E. HOPKINS (eds.). 1969. Environments of coal deposition. Geol. Soc. Amer. Spec. Paper 114. 204 p.
Six papers concerning accumulation of organic sediments in Recent and Ancient sedimentary environments.

DENNISON, J. M. 1976. Appalachian Queenston Delta related to eustatic sea-level drop accompanying late Ordovician glaciation centred in Africa, p. 108–120. *In* M. G. Bassett, (ed.), The Ordovician system: proceedings of a Palaeontological Association Symposium, Birmingham, Sept. 1974. Univ. of Wales Press and Nat. Museum of Wales, Cardiff.

DENNISON, J. M., and J. W. HEAD. 1975. Sealevel variations interpreted from the Appalachian Basin Silurian and Devonian. Amer. J. Sci. 275: 1089–1120.

DERAAF, J. F. M., H. G. READING, and R. G. WALKER. 1965. Cyclic sedimentation in the Lower Westphalian of No. Devon, England. Sedimentology 4: 1–52.
Clastic cycles of turbidite through deltaic depositional environment.

DUNHAM, R. J. 1969. Vadose pisolite in the Capitan Reef (Permian), New Mexico and Texas, p. 182–191. *In* G. M. Friedman (ed.), Depositional environments in carbonate rocks, Soc. Econ. Paleontologists and Mineralogists Spec. Pub. 14.

———. 1970. Stratigraphic reefs vs. ecologic reefs. Bull. Amer. Assoc. Petrol. Geol. 54: 1931–1932.

FRAKES, L. A. 1979. Climates throughout geologic time. Elsevier, New York. 301 p.

FRAKES, A., and J. C. CROWELL. 1969. Late Paleozoic glaciation I: South America. Bull. Geol. Soc. Amer. 80: 1007–1042.

FRAKES, L. A., J. L. MATTHEWS, and J. C. CROWELL. 1971. Late Paleozoic glaciation, III: Antarctica. Bull. Geol. Soc. Amer. 82: 1581–1604.
Darwin tillite of Beardmore Basin.

GALLOWAY, W. E., D. K. HOBDAY, and K. MAGARA. 1982. Frio Formation of Texas Gulf coastal plain: depositional systems, structural framework, and hydrocarbon distribution. AAPG Bulletin 66 (6): 649–688.

GILL, DAN. 1979. Differential entrapment of oil and gas in Niagaran pinnacle-reef belt of northern Michigan. Bull. Amer. Assoc. Petrol. Geol. 63, 608–620.

HAMILTON, W., and D. KRINSLEY. 1967. Upper Paleozoic glacial deposits of South Africa and southern Australia. Bull. Geol. Soc. Amer. 78: 783–800.
Field, petrographic, and electron microscope documentation of glacial and glacio-marine sediments.

HANDFORD, C. R. 1981. Sedimentology and genetic stratigraphy of Dean and Spraberry formations (Permian), Midland Basin, Texas. Bull. Amer. Assoc. Petrol. Geol. 65 (9): 1602–1616.

HANDFORD, C. R., and S. P. DUTTON. 1980. Pennsylvania–early Permian depositional systems and shelf-margin evolution, Palo Duro Basin, Texas. Bull. Amer. Assoc. Petrol. Geol. 64 (1): 88–106.

HANKEN, NILS-MARTIN. 1979. Sandstone pseudomorphs of aragonite fossils in an Ordovician vadose zone. Sedimentology 26 (1): 135–142.

HECKEL, PHILIP H., and JOHN F. BAESEMANN. 1975. Environmental interpretation of conodont distribution in upper Pennsylvanian (Missourian) megacyclothems in eastern Kansas. Bull. Amer. Assoc. Petrol. Geol. 59 (3): 486–509.

IRWIN, M. L. 1965. General theory of epeiric clear water sedimentation. Bull. Amer. Assoc. Petrol. Geol. 49: 445–459.
Kickback correlation on the basis of a theoretical model.

KREBS, WOLFGANG. 1979. Devonian basinal facies: the Devonian system. Spec. Papers in Palaeont. 23: 125–139.

LANTZY, RONALD J., MICHAEL F. DACEY, and FRED T. MACKENZIE. 1977. Catastrophe theory: application to the Permian mass extinction. Geology 5 (12): 724–728.

MARKELLO, J. R., and J. F. READ. 1981. Carbonate ramp-to-deeper shale shelf transitions of an Upper Cambrian intrashelf basin, Nolichucky formation, Southwest Virginia Appalachians. Sedimentology 28 (4): 573–598.

MATTHEWS, R. K. 1974. Dynamic stratigraphy. Prentice-Hall, Inc., Englewood Cliffs, N.J. 370 p.

MERRIAM, D. F. (ed.). 1964. Symposium on cyclic sedimentation. State Geol. Survey of Kansas, Bull. 169. 636 p.
Numerous papers by a spectrum of authors representing many divergent views.

MITCHUM, R. M., JR., P. R. VAIL, and S. THOMPSON III. 1977. Seismic stratigraphy and global changes of sea level, part 2: the depositional sequence as a basin uplift for stratigraphic analysis, p. 53–62. *In* C. E. Payton (ed.), Seismic stratigraphy—applications to hydrocarbon exploration. Amer. Assoc. Petrol. Geol. Mem. 26.

MOMPER, J. A. 1966. Stratigraphic principles applied to the study of the Permian and Penn-

sylvanian systems in the Denver Basin. Twentieth Annual Conference, Wyoming Geol. Assoc., p. 87–90r.
Deduces by independent lines of evidence that there is a 100,000- and a 500,000-year-cycle in these sequences.

MOORE, R. C. 1964. Paleoecological aspects of Kansas, Pennsylvanian, and Permain cyclothems, p. 287–380. *In* D. F. Merriam (ed.), Symposium on cyclic sedimentation. State Geol. Survey of Kansas, Bull. 169.

MOREL, P., and E. IRVING. 1981. Paleomagnetism and the evolution of Pangea. J. Geophys. Res. 86 (B3): 1858–1872.

MOUNTJOY, ERIC W., and ROBERT RIDING. 1981. Foreslope stromato-poroid-renalcid bioherm with evidence of early cementation, Devonian Ancient Wall reef complex, Rocky Mountains. Sedimentology 28 (3): 299–320.

NEWELL, N. D., J. K. RIGBY, A. G. FISCHER, A. J. WHITEMAN, J. E. HICKOX, and J. S. BRADLEY. 1953. The Permian reef complex of the Guadelupe Mountains region, Texas and New Mexico. W. H. Freeman and Co., San Francisco. 236 p.

PAYTON, C. E. (ed.). 1977. Seismic stratigraphy—applications to hydrocarbon exploration. Amer. Assoc. Petrol. Geol. Mem. 26. 516 p.

PLAYFORD, PHILLIP, E. 1980. Devonian "Great Barrier Reef" of Canning Basin, Western Australia. Bull. Amer. Assoc. Petrol. Geol. 64 (6): 814–840.

ROSE, P. R. 1976. Mississippian carbonate shelf margins, western United States. J. Res. U.S. Geol. Surv. 4: 449–466.

SAUNDERS, W. BRUCE, W. H. C. RAMSBOTTOM, and W. L. MANGER. 1979. Mesothemic cyclicity in the mid-Carboniferous of the Ozark shelf region? Geology 7 (6): 293–296.

SCHOLL, D. W. 1969. Modern coastal mangrove swamp stratigraphy and the ideal cyclothem, p. 37–62. *In* E. C. Dapples and M. E. Hopkins (eds.), Environments of coal deposition. Geol. Soc. Amer. Spec. Pub. 14.
A thoughtful comparison of Recent South Florida swamp stratigraphy and the ideal cyclothem of Pennsylvanian strata of the mid-continent.

SCHOPF, T. J. M. 1980. Paleoceanography. Harvard Univ. Press. Cambridge, Mass. 341 p. Reviewed by James P. Kennett in Nature 293 (5828): 170.

SCOTESE, C. R., R. K. BAMBACH, C. BARTON, R. VAN DER VOO, and A. M. ZIEGLER. 1979. Paleozoic base maps. J. Geology 87 (3) 217–277.

SCOTT, G. R. 1979. Paleomagnetic studies of the Early carboniferous St. Joe Limestone, Arkansas. J. Geophys. Res. 84 (B11): 6277–6285.

SILVER, B. A., and R. G. TODD. 1969. Permian cyclic strata, Northern Midland and Delaware basins, West Texas and Southeastern New Mexico. Bull. Amer. Assoc. Petrol. Geol. 53: 2223–2251.

SLEEP, NORMAN H., JEFFREY A. NUNN, and LEI CHOU. 1980. Platform basins. Ann. Rev. Earth Planet. Sci. 8: 17–34.

SUTHERLAND, PATRICK K., and THOMAS W. HENRY. 1977. Carbonate platform facies and new stratigraphic nomenclature of the Morrowan Series (Lower and Middle Pennsylvanian), northeastern Oklahoma. Bull. Geol. Soc. Amer. 88: 425–440.

Tucker, M. E., and P. C. Reid. 1973. The sedimentology and context of late Ordovician glacial marine sediments from Sierra Leone, West Africa. Palaeogeog. Palaeoclimat. Palaeoecol. 13: 289–307.

Vail, P. R., R. M. Mitchum, Jr., and S. Thompson III. 1977a. Seismic stratigraphy and global changes of sea level, part 3: relative changes of sea level from coastal onlap, p. 63–81. *In* C. E. Payton (ed.), Seismic stratigraphy—applications to hydrocarbon exploration. Amer. Assoc. Petrol. Geol. Mem. 26.

——. 1977b. Seismic stratigraphy and global changes of sea level, part 4: global cycles of relative changes of sea level, p. 83–97. *In* C. E. Payton (ed.), Seismic stratigraphy—applications to hydrocarbon exploration. Amer. Assoc. Petrol. Geol. Mem. 26.

Wanless, H. R. 1964. Local and regional factors in Pennsylvanian cyclic sedimentation. State Geol. Surv. of Kansas, Bull. 169: 593–607.

Wanless, H. R., and J. R. Cannon. 1966. Late Paleozoic glaciation. Earth-Science Rev. 1: 247–286.

White, W. A. 1972. Deep erosion by continental ice sheets. Bull. Geol. Soc. Amer. 83 (4): 1037–1056.

Wickham, John, Dietrich Roeder, and Garrett Briggs. 1976. Plate tectonics models for the Ouachita foldbelt. Geology 4: 173–176.

Wilson, J. L. 1967. Cyclic and reciprocal sedimentation in Virgilian strata of southern New Mexico. Bull. Geol. Soc. Amer. 78: 805–818.

Glossary

This glossary is intended only as a convenient supplement to the text. Explanations and definitions are brief and are limited to the meanings that are relevant to the context in which the words or phrases are used in this book. See the index for words which are defined in the text. For more complete definitions of these and other geological terms, consult the *Glossary of Geology*, published by the American Geological Institute.

Abyssal plain Extremely flat regions at the bottom of major ocean basins. Water depth commonly greater than 4000 meters.

Active continental margin A continental margin at or near plate boundaries, as opposed to a passive margin.

Alluvial fan A cone-shaped deposit of sediment that occurs where intermittent mountain streams flow into broad, flat valleys.

Aquifer Permeable rocks that will produce fresh water from wells drilled into them.

Arc-trench system Common topographic expression of a subduction zone. Volcanic island arc develops on the up plate and a deep-sea trench is evident where the down plate begins to plunge under the subduction complex.

Argillaceous Containing clay minerals; as in, an argillaceous silt stone.

Asthenosphere That portion of the earth's mantle immediately interior to the lithosphere. The asthenosphere is much less rigid than the lithosphere. Transition from lithosphere to asthenosphere occurs generally around a depth of 100 kilometers.

Aulocogen An aborted pull-apart basin.

Authigenic Formed in place as opposed to transported, as in, authigenic calcite cement.

Azimuth   Horizontal direction measured clockwise from north; east equals 90 degrees, south equals 180 degrees, and so on.

Back-arc basin   Basin reflecting formation of new oceanic lithosphere by a secondary convecting cell immediately behind the arc-trench system.

Bar   (1) A unit of pressure more or less equal to atmospheric pressure near sea level. (2) A positive topographic feature formed by sand or gravel deposits in rivers, near beaches, etc.

Bay   A general term for an indentation in a shoreline. Smaller than a gulf, larger than a cove.

Benthic (benthonic)   Denoting sediment water interface; as in benthic organisms, which live in or on the sediment.

Bentonite   A clay layer presumably formed by alteration of a volcanic ash.

Bioclastic sedimentary particles   Broken fragments of skeletal remains, such as molluscs, corals, and coralline algae. Common components in limestones.

Biogenic   Formed by organisms; for example, mollusc shells are a common biogenic component of shales.

Bird's-eye fabric   A common pattern in supratidal carbonates in which former gas bubbles become preserved as open or calcite-filled cavities. Cavities are typically 2 to 5 millimeters in diameter and may constitute 50% of the rock.

Bottom stability   The degree to which the sediment-water interface is immobile. Many marine orgainisms have life styles adapted to stability or lack of stability of the substrate.

Bryozoa   A phylum of attached and incrusting marine invertebrates. Locally important contributors to bioclastic limestones.

Burrow-mottled fabric   The burrow of an organism typically leaves features discernible in the sediment. When a number of burrows are superimposed, the identity of the individual burrow may be lost but the characteristic mottled appearance is easily recognizable on outcrop, in slab, or in thin section.

Caliche   Calcium carbonate precipitated by evaporative processes in or near a soil zone.

Chicken-wire anhydrite   A common recrystallization fabric in which impurities are concentrated along surfaces bounding large aggregates of white anhydrite. In slab, the zones of impurities produce a resemblance of wire mesh.

Comminution   Particle-size reduction by physical processes.

Contemporaneous   Existing or occurring at the same time.

Continental drift   An old phrase for plate tectonics.

Coquina   A clean sediment composed largely of shells of marine invertebrates, usually molluscs.

Craton   Relatively old continental lithosphere.

Creep   Slow, more or less continuous, downslope movement.

Cryosphere   The solid ice portion of the world's water budget; as in, hydrosphere-atmosphere-cryosphere interaction produces glacio-eustatic sea-level fluctuations.

Cumulative probability scale A scale on a graph so devised that a normal distribution plots as a straight line.

Curie point The temperature at which a cooling mineral acquires permanent magnetic properties that record the surrounding magnetic-field orientation and strength at the time of cooling. Above the Curie point, the magnetic properties of a mineral change as the surrounding field changes. Below the Curie point, magnetic properties of the mineral do not change.

Cut-and-fill structures The sedimentary record of local alternation between erosion and sedimentation. Applied particularly to sandbars with complicated internal structure.

Cyclothem A series of beds deposited during a single sedimentary cycle of the type that prevailed during the Pennsylvanian period.

Datum A reference point, like, or plain from which other quantities are measured. A stratigraphic section may be hung from an unconformity datum and other measurements represented as distance above or below the unconformity.

Deep-sea fan A submarine equivalent of an alluvial fan.

Deflation surface A layer of coarse sediment derived from the sediment below after wind has transported the fine fraction.

"Del notation" The common scheme for reporting isotope data relative to a standard. Measurement is parts per thousand ($^\circ/_{oo}$) deviation from any common standard. $\delta^{18}O$ ($^\circ/_{oo}$ PDB) is read "del eighteen O, per mil, relative to the PDB standard."

Density current Any current that flows downslope because it is more dense than the fluid occupying the lower elevation.

Diachronous Differing in geologic age. Thus, a diachronous sedimentary unit is one that crosses time lines, be they real or hypothetical.

Diagenesis Those physical and chemical changes that occur in a sediment after initial deposition but before metamorphism.

Diamicton Fancy name for poorly sorted sediments, commonly taken to be of glacial origin.

Downlap Seismic stratigraphy term indicating inclined strata terminate downdip against a surface which dips less severely than the downlapping strata.

Driving subsidence rate That portion of subsidence rate that is attributable to subsidence mechanism as opposed to isostatic subsidence under the weight of new sedimentary load. Total subsidence rate is commonly 2.5 to 3.0 times as fast as the driving subsidence rate because of isostatic compensation for newly accumulated sediment.

Dropstones The coarsest example of ice-rafted detritus. Pebbles or cobbles in otherwise fine-grained sediment, presumbably dropped into the muddy bottom by the melting of rock-laden icebergs.

Ebb tide The outgoing portion of the tidal cycle.

Ecology The study of the mutual relationships between organisms and their environments.

Effective wave base   The depth at which movement of sediment by wave action ceases to be a significant geologic process. Generally on the order of 30 to 100 meters, it varies with wave intensity and grain size of the sediment.

Embayment   Indentation of the shoreline (or former shoreline) into the continent.

Emergence   The upward movement of a point within the sediment relative to sea level. Sediment may be lifted (tectonic emergence) or sea level may be lowered (eustatic emergence) to produce the same relative motion.

Environmental datum   Any lithology that must have formed under specific conditions and that in turn provide an indication of depositional conditions of associated lithologies.

Epeirogeny   Regional uplift or subsidence of relatively undeformed continental lithosphere, as distinct from orogeny, which involves pronounced syntectonic deformation.

Equilibrium profile   Here taken to mean that surface of a sedimentary environment that would be attained if tectonic and eustatic activity totally ceased. If an equilibrium profile were attained, all sediment entering a region would eventually leave the region and no net sediment accumulation would be recorded.

Estuary   A drowned river valley now under the influence of coastal tidal action.

Eustasy (eustacy)   Pertaining to worldwide sea level; as in, eustatic sea-level rise affects all coastlines of the world, whereas tectonic subsidence affects only that particular subsiding coastline.

Evaporites   Sediments precipitated from aqueous solution as a result of the evaporation of the water. Anhydrite ($CaSO_4$) and halite ($NaCl$) are common evaporite minerals.

Facies   The sedimentological record of a depositional environment.

Factor analysis   A data analysis technique that identifies natural groupings of parameters within a data matrix, thereby reducing the number of terms needed to describe variation among the samples.

Fauna   The animals of any given environment or geologic age.

Fissile bedding   Bedding plains consisting of myriad thin sheets of clay minerals; here taken as distinct from burrow-mottled clays which will appear massively bedded.

Flexural rigidity   A measure of the stiffness of an elastic plate under the influence of a bending couple.

Flexure zone   In Gulf Coast geology, that line downdip from which a formation thickens rapidly.

Flood tide   The incoming portion of the tidal cycle.

Flora   The plants of any given environment or geologic age.

Flume   A laboratory apparatus constituting a long, straight channel in which sediment transport studies are carried out.

Fluvial   Produced by rivers.

Foraminifer   Single-celled organism which secretes a calcium carbonate shell (test). There are both planktic and benthic foraminifers.

Foreland basin Basin formed between an emerging mountain range and the adjacent craton, relatively late in the mountain-building process.

Forereef The environment immediately seaward of the reef flat environment, as opposed to backreef.

Gene pool Genes are the fundamental units governing hereditary characters. The gene pool is the sum of all genes within an interbreeding population.

Glacial striations Small-scale grooves cut into rock by the rasping action of rock-laden glaciers. It is especially nice to find glacial striations on dropstones or on cobbles in diamicton, thus providing independent evidence of glacial origin.

Glauconite A green clay mineral commonly found in marine sandstones. It usually appears to be authiogenic and so is potentially useful for radiometric dating, paleosalinity estimations, etc.

Graben A tectonic block down-faulted with respect to rocks on either side.

Gravity tectonics Tectonic deformation in which the driving force is downslope transport under the influence of gravity.

Hemipelagic Deep-sea sediments that contain abundant clay derived from the continents.

Hiatus Missing time-rock section represented by a surface of nondeposition; as distinct from an erosional unconformity.

Horizon A surface or distinctive layer within layered sedimentary rocks.

Hotspots Regions of persistent localized melting beneath the base of the lithosphere. They are commonly recognized by their surficial expression, volcanic activity superimposed upon the overriding lithosphere.

Hyaloclastites Literally "glass clastics," the term refers to substantial mountains of fractured glass which are presumed to represent rapid quenching of magma in volcanic eruption beneath glacial ice.

Hypersaline Having a salinity substantially greater than that of seawater.

Ice-rafted detritus Coarse-grained sediment dropped into otherwise fine-grained sediment by the melting of icebergs.

In situ In the situation in which it was originally formed or deposited, as opposed to being transported from one situation to its final resting place. The term is usually italicized.

Intertidal The benthic zone near sea level that lies between normal high and low tides.

Intraclast A large, composite sedimentary particle formed contemporaneously in nearby sedimentary environments, as in mud chips on a supratidal flat.

Isopach map A contour map depicting the thickness of a sedimentary unit.

Lagoon The relatively quiet-water environment behind a barrier beach or reef.

Leaching Selective removal of soluble minerals by throughgoing water.

Lithosphere The relatively rigid outer shell of the earth comprising the crust and upper mantle. It averages approximately 100 kilometers in thickness.

Loess A wind-deposited sediment consisting mostly of silt, the silt commonly derived from finely ground rock washed out of continental glaciers.

Longshore current   A persistent nearshore current moving essentially parallel to the coast. Usually generated by waves breaking obliquely to the shoreline.

Mantle   The layer of earth between crust and core.

Massif   A French word for mountain; here used to indicate igneous and metamorphic terrain which forms a topographic high smaller than a craton but bigger than a dome.

Maxwell solid   A material that behaves elastically under the stress of "short" duration and plastically under the stress of "long" duration. "Silly putty" is a Maxwell solid.

Melobesia   Delicate red algae that live on the surface of marine grasses. A common source of high-magnesium calcite mud in the Recent epoch.

Nepheloid layer   A very dilute suspension of clay minerals commonly observed near the sediment-water interface in the deep sea.

Net sand map   A contour map indicating the total thickness of sand units within a stratigraphic interval.

Niche   That subdivision of an environment occupied by a single species. Some attributes of the environment may be shared among several niches, but each niche is a unique combination of attributes.

Novaculite   A very distinct bedded white chert.

Obduction   The adding of material onto the front edge of the up block at a subduction zone; as opposed to subduction, where the material in question goes down the subduction zone.

Offlap   Seismic stratigraphy term indicating that successively younger reflectors wedge out successively basinward, indicating relative emergence.

Oncolith   Calcium carbonate nodule formed of fine-grained sediment bound together by soft-bodied algae.

Onlap   Seismic stratigraphy term indicating inclined strata that terminate updip against a surface which more severely inclined than the onlapping strata; indicative of submergence.

Oolites   Spherical, carbonate sand particles composed of concentric laminae of microscopic aragonite needles.

Orogeny   Mountain building involving significant structural deformation; as opposed to epeirogeny.

Oxygen minimum   A region within the oceanic thermocline that is characterized by relatively low-dissolved oxygen values resulting from oxidation of particulate organic matter below the zone of photosynthesis.

Paleogeography   A geography that existed at some specific time in the past. Commonly reconstructed as the physiographic context within which sedimentation of a specific unit is believed to have occurred.

Passive continental margin   A continental margin that is not the boundary between two lithospheric plates; as opposed to an active margin.

Patch reef   A reef structure of limited aerial extent, commonly occurring on continental shelves; as opposed to barrier reefs, commonly occurring at shelf margin.

Pellet Silt or sand-sized aggregation of carbonate mud. Generally fecal in origin, it is a common grain type in nonskeletal carbonate sediments.

Penecontemporaneous At nearly the same time.

Peneplain surface A geomorphic feature representing the erosion of a former landscape to very low relief equilibrium profile.

Permeability A measure of the ease with which a fluid can be passed through the pore space of a rock.

Phanerozoic The last 570 million years of geologic time.

Phreatic lens Hydrologic environment where pore space of the rock is saturated with fresh water; as opposed to the vadose environment which is above the phreatic lens. The phreatic lens floats on more dense fluid, commonly seawater or brine of continental origin.

Physiography The spatial arrangement of the upper surface of the lithosphere and the conditions existing upon it.

Pisolith An outrageously large "oolite"; some are surely oncoliths, but others are of uncertain origin.

Planktic (planktonic) Anything that floats freely in seawater, as in planktic foraminifers, which live in the upper few hundred meters of the oceanic water column.

Point bar The sandbar deposited on the inside of a meander loop.

Polar wandering An archaic name for the study of remnant magnetism in continental materials. As the science developed, it is the continents that wander (plate tectonics) and not the magnetic poles.

Porosity The portion of the total volume of a rock that is not occupied by solid mineral matter.

Progradation The seaward advance of the shoreline resulting from sediment deposition.

Provenance The terrane or parent rock from which a sediment was derived.

Red beds Red sedimentary rocks; usually sand stones and shales of fluvial origin.

Reef A wave-resistant structure constructed by sedentary carbonate-secreting organisms. Note that this definition has straightforward application in the Recent epoch but must be inferred from related sediments in Ancient sequences.

Reef apron The shallow-water zone of reef-derived sand lagoonward of the reef flat.

Reef flat The shallow-water zone of cobble-to boulder-sized rubble immediately behind a living-coral barrier reef.

Regression A nonspecific term meaning that the shoreline moved toward the center of the basin. It may be caused by the prograding of sediments or by tectonic or eustatic emergence.

Relaxation time A measure of the rate at which stress is released by plastic deformation of a material.

Relict sediment Surface sediment representative of an environment that previously occupied the area under discussion. Not deposited under present conditions.

Rhodolith   Calcium carbonate nodules formed of stoney red algae; as distinct from oncoliths.

Rise   Sloping ocean-floor bathymetry which rises above the abyssal plain. Passive continental margins have distinct continental rises. Also synonymous with ridge when used to describe topography related to sea-floor spreading, as in the East Pacific Rise and the Mid-Atlantic Ridge; both are spreading centers.

Sabkha   Supratidal, evaporative, deflation surfaces on the Arabian peninsula. Site of extensive evaporite sedimentation.

Sessile benthos   Bottom-dwelling aquatic organisms that live permanently attached to the bottom.

Shelf   Ocean margin bathymetry between the shoreline and the continental slope. The break between continental shelf and continental slope is usually defined by a pronounced increase in slope, commonly occurring at about 200 meters below modern sea level.

Slope   Ocean bathymetry between the continental shelf and the continental rise.

Slope basin   A small closed bathymetric low resulting from tectonic deformation of the continental slope. A common feature of the subduction complex. Also, a common feature of salt tectonics in the United States Gulf Coast region.

Slump   The downslope movement of a mass of unconsolidated sediment as a coherent unit. Typically involves major dislocation along a few preferred planes of weakness.

Soft sand   Carbonate mud formed into soft pellets. In the depositional environment, the soft particles react to wave and current action as though they were sand grains. In the lab, they are washed through the sand sieve like any other mud.

Sole mark   The marking recorded on the bottom of a sandstone bed where the sand has filled in depressions and tool marks in the underlying shale bed. They are best preserved by the sandstone because the shale tends to flake into small pieces on outcrop, whereas the sandstone commonly breaks out as large blocks.

Solstice   The points in the ecliptic at which the perpendicular to the sun makes the largest angle with the plane of the equator. As seen from the northern hemisphere, the summer solstice takes place when the sun is "as far north as it comes" and the winter solstice occurs when the sun is "as far south as it goes."

Stromatolites   Planar to headlike laminated structures constructed by algal entrapment of sediment and usually found in the high intertidal and low supratidal zones. Most commonly preserved in carbonate sediments.

Subaerial   Formed or existing at or near a sediment surface significantly above sea level.

Subduction   Passage of material down the subduction zone, as opposed to obduction.

Submergence   The downward movement of a point within the sediment relative to sea level. The opposite of emergence.

Substrate   The sediment or rock upon which benthonic organisms reside. The distribution of rock, sand, and soft-mud substrates is a major factor controlling local distribution of marine benthonic organisms.

Subtidal   The shallow-marine environment lying below normal low tide.

Supratidal   The coastal environment lying above normal high tide.

Swale   A marshy depression in otherwise generally level land.

Talus   High-angle debris deposits at the base of a pronounced physiographic feature.

Tectonic   Pertaining to structural deformation of the earth's lithosphere.

Terrigenous clastic sediment   Sediment derived from the land by the weathering of preexisting rocks and transported by wind or water to the sedimentary environment in which we now find it. It contrasts with carbonates, evaporites, and some clays, all of which may form *in situ* by chemical or biochemical precipitation from aqueous solution.

Thallasia   A shallow-marine grass common in areas of Recent carbonate sedimentation.

Thrust sheet   Coherent units of sedimentary rocks tectonically emplaced into a sedimentary sequence.

Till   Poorly sorted sediment deposited under or immediately adjacent to a continental glacier.

Tillite   A sedimentary rock taken to have formed as till.

Time line   A line in a stratigraphic cross section so constructed that all points represent the same instant in geologic time. A time line is the vertical tracing of a time surface, but this latter word is seldom used.

Time transgressive   Said of a lithologic unit that is older in one area than it is in another area. For example, a prograding beach deposit is youngest toward the basin center and oldest toward the land. The lithologic unit crosses (that is, transgresses) time lines.

Tool mark   The structure produced on a bedding-plane surface where transported objects or turbulent currrents leave a mark indicating the direction of transport.

Toplap   Seismic startigraphy term indicating termination of strata against an overlying surface which is taken to represent nondeposition. Progradation of coastal clastics commonly results in toplap against a very thin layer of marsh deposits. An erosional surface separating younger strata from older rocks; as distinct from a hiatus, where time-rock units are missing because of nondeposition.

Transform fault   A strike-slip fault bounded at each end by an area of crustal spreading that tends to be more or less perpendicular to the strike-slip fault.

Transgression   A nonspecific term meaning that the shoreline moved away from the center of the basin. It may be caused by tectonic or eustatic submergence.

Turbidite   A graded bed of sand and mud that presumably was deposited by a turbidity current.

Turbidity current   A density current that owes its density to suspended mineral matter.

Upwelling   The rising of cold, nutrient-rich subsurface water brought about by interaction of prevailing winds with a nearby land mass. Especially important in cratonic stratigraphy because the low-oxygen content of upwelled waters may produce widespread black shale environments in relatively shallow seas.

Vadose   That portion of the hydrologic system where the pore space of a rock is occupied in part by fresh water and in part by air or soil gas; as opposed to the phreatic zone.

Varves   Distinctive thin-bedded sedimentary couplets presumably resulting from annual accumulation in relatively quiet water. Typically, sediment-rich layers accumulate from summer meltwater runoff and organic-rich layers accumulate while the water body is frozen over in winter.

Viscosity   The extent to which a fluid resists flow. As in, water has relatively low viscosity, whereas tar has relatively high viscosity.

Volcanic island arc   A spectacular topographic feature commonly developed on the up plate at a subduction zone.

Index